METHODS IN MOLECULAR BIOLOGY

Series Editor
John M. Walker
School of Life and Medical Sciences
University of Hertfordshire
Hatfield, Hertfordshire, AL10 9AB, UK

For further volumes:
http://www.springer.com/series/7651

Zebrafish

Methods and Protocols

Second Edition

Edited by

Koichi Kawakami

Division of Molecular and Developmental Biology, National Institute of Genetics, Mishima, Japan

E. Elizabeth Patton

Institute of Genetics and Molecular Medicine, MRC Human Genetics Unit & The University of Edinburgh Cancer Research UK Centre, University of Edinburgh, Edinburgh, UK

Michael Orger

Champalimaud Neuroscience Programme, Champalimaud Centre for the Unknown, Lisbon, Portugal

Humana Press

imaging of infection and host-pathogen interactions are described, leading to powerful understanding of how pathogens invade and survive in the host. Chapter 15 describes cardiac injury and regeneration and highlights the power of zebrafish as a model system to study regeneration. Finally, Chap. 16 describes techniques for the metabolism and transport of lipids, underscoring the value of zebrafish as a system to study metabolic disease.

Part III comprises methods for neuroscience. In particular, it covers techniques for studying the structure and function of neural circuits and their role in generating behavior, an area that has recently seen rapid growth within the zebrafish community. The first three chapters cover methods that allow the targeted labeling, manipulation, and interrogation of specific neurons in the small brain of the zebrafish. Chapter 17 describes the use of electroporation to deliver DNA constructs and other reagents to specific neurons in larval and adult zebrafish. Chapter 18 shows how laser microsurgery can be used to study regeneration in sensory axons, and Chap. 19 describes methods for making in vivo patch clamp recordings. Zebrafish are increasingly used to study complex social behavior, and Chap. 20 provides methods for quantifying aggressive encounters. There has been great progress recently in the use of optical approaches to record activity from the whole larval brain. Chapter 21 describes methods for relating neuronal calcium signals to behavior. Chapter 22 explains how to build a light-sheet microscope for fast volumetric imaging, and Chap. 23 gives a protocol for imaging from freely swimming fish. Methods to manipulate activity are essential to establish causal relationships between neuronal firing and behavior. Chapters 24 and 25 describe ways to specifically perturb or isolate genetically identified neuronal populations via optogenetics, ablations, and the expression of toxin-insensitive channels.

We hope this method book will be of help to all zebrafish researchers, especially scientists newly entering this ever-growing field.

We wish you all the best in your work with this truly exciting vertebrate animal system that continues to push forward the boundaries of science.

Mishima, Japan — Koichi Kawakami
Edinburgh, UK — E. Elizabeth Patton
Lisbon, Portugal — Michael Orger

The original version of the book frontmatter was revised. The Erratum to the book frontmatter is available at DOI 10.1007/978-1-4939-3771-4_26

Contents

Contributors

MISHA B. AHRENS • *Janelia Research Campus, Howard Hughes Medical Institute, Ashburn, VA, USA*

JAMES F. AMATRUDA • *Department of Pediatrics, UT Southwestern Medical Center, Dallas, TX, USA; Department of Molecular Biology, UT Southwestern Medical Center, Dallas, TX, USA; Department of Internal Medicine, UT Southwestern Medical Center, Dallas, TX, USA*

ARISTIDES B. ARRENBERG • *Institute of Neurobiology, Centre for Integrative Neuroscience, University of Tübingen, Tübingen, Germany; Developmental Biology, Faculty of Biology, Institute Biology I, Freiburg, Germany; Albert-Ludwigs-Universität Freiburg, Freiburg, Germany*

THOMAS O. AUER • *Neuronal Circuit Development Group, Unité de Génétique et Biologie du Développement, U934/ UMR3215, Pole de Biologie du Développement et Cancer, Institut Curie—Centre de Recherche, Paris, France; CNRS UMR 3215, Paris, France; INSERM U934, Paris, France; Centre for Organismal Studies Heidelberg, University of Heidelberg, Heidelberg, Germany; Center for Integrative Genomics, University of Lausanne, Lausanne, Switzerland*

DARIUS BALCIUNAS • *Department of Biology, Temple University, Philadelphia, PA, USA*

FILIPPO DEL BENE • *Neuronal Circuit Development Group, Unité de Génétique et Biologie du Développement, U934/ UMR3215, Pole de Biologie du Développement et Cancer, Institut Curie—Centre de Recherche, Paris, France; CNRS UMR 3215, Paris, France; INSERM U934, Paris, France*

DAVIS V. BENNETT • *Janelia Research Campus, Howard Hughes Medical Institute, Ashburn, VA, USA; University of Chicago, Chicago, IL, USA*

JASON N. BERMAN • *Department of Microbiology and Immunology, Dalhousie University, Halifax, NS, Canada; Department of Pathology, Dalhousie University, Halifax, NS, Canada; Department of Pediatrics, IWK Health Centre/Dalhousie University , Halifax, NS, Canada*

ISAAC H. BIANCO • *Department of Neuroscience, Physiology & Pharmacology, University College London, London, UK*

PATRICK R. BLACKBURN • *Mayo Clinic, Rochester, MN, USA*

COLLEEN A. BRADY • *Cardiovascular Research Center, Massachusetts General Hospital, Harvard Medical School, Charlestown, MA, USA; Department of Systems Biology, Harvard Medical School, Boston, MA, USA; Broad Institute, Cambridge, MA, USA*

HAROLD A. BURGESS • *Division of Developmental Biology, Eunice Kennedy Shriver National Institute of Child Health and Human Development, Bethesda, MD, USA; NIH, Bethesda, MD, USA*

CRAIG J. CEOL • *Program in Molecular Medicine and Department of Molecular, Cell and Cancer Biology, University of Massachusetts Medical School, Worcester, MA, USA*

LANPENG CHEN • *Institute of Biology, Leiden University, Leiden, The Netherlands*

YI CHEN • *Mayo Clinic, Rochester, MN, USA*

MAURICIO CORTES • *Beth Israel Deaconess Medical Center, Harvard Medical School, Boston, MA, USA*
AILANI DEEPAK • *Division of Molecular and Developmental Biology, National Institute of Genetics, Mishima, Japan; Department of Genetics, Sokendai (The Graduate University for Advanced Studies), Mishima, Japan*
GRAHAM DELLAIRE • *Department of Pathology, Dalhousie University, Halifax, NS, Canada; Department of Biochemistry & Molecular Biology, Dalhousie University, Halifax, NS, Canada*
JIU-LIN DU • *Institute of Neuroscience, State Key Laboratory of Neuroscience, Center for Excellence in Brain Science and Intelligence Technology, Shanghai Institutes for Biological Sciences, Chinese Academy of Sciences, Shanghai, China; University of Chinese Academy of Sciences, Shanghai, China; School of Life Science and Technology, ShanghaiTech University, Shanghai, China*
STEPHEN C. EKKER • *Mayo Clinic, Rochester, MN, USA*
WIETSKE VAN DER ENT • *Institute of Biology, Leiden University, Leiden, The Netherlands*
VIRGINIE ESAIN • *Beth Israel Deaconess Medical Center, Harvard Medical School, Boston, MA, USA*
STEVEN A. FARBER • *Department of Embryology, Carnegie Institution, Baltimore, MD, USA*
RAINER W. FRIEDRICH • *Friedrich Miescher Institute for Biomedical Research, Basel, Switzerland; University of Basel, Basel, Switzerland*
ARWIN GROENEWOUD • *Institute of Biology, Leiden University, Leiden, The Netherlands*
YAVOR HADZHIEV • *Institute of Cancer and Genomic Sciences, College of Medical and Dental Sciences, University of Birmingham, Birmingham, UK*
SHUNING HE • *Institute of Biology, Leiden University, Leiden, The Netherlands*
TERUMI HORIUCHI • *Department of Medical Genome Sciences, Graduate School of Frontier Sciences, The University of Tokyo, Chiba, Japan*
ERIC J. HORSTICK • *Division of Developmental Biology, Eunice Kennedy Shriver National Institute of Child Health and Human Development, Bethesda, MD, USA*
MARI ITOH • *Division of Molecular and Developmental Biology, National Institute of Genetics, Mishima, Japan*
DIANA C. JORDAN • *Division of Developmental Biology, Eunice Kennedy Shriver National Institute of Child Health and Human Development, Bethesda, MD, USA*
ATSUO KAWAHARA • *Laboratory for Developmental Biology, Center for Medical Education and Sciences, Graduate School of Medical Science, University of Yamanashi, Yamanashi, Japan*
KOICHI KAWAKAMI • *Division of Molecular and Developmental Biology, National Institute of Genetics, Mishima, Japan; Department of Genetics, Sokendai (The Graduate University for Advanced Studies), Mishima, Japan*
GENEVIEVE C. KENDALL • *Department of Pediatrics, UT Southwestern Medical Center, Dallas, TX, USA; Department of Molecular Biology, UT Southwestern Medical Center, Dallas, TX, USA*
PRADEEP LAL • *Division of Molecular and Developmental Biology, National Institute of Genetics, Mishima, Japan*
JAMES A. LISTER • *Department of Human and Molecular Genetics, Virginia Commonwealth University, School of Medicine Sanger Hall, Richmond, VA, UK*
HERNÁN LÓPEZ-SCHIER • *Sensory Biology & Organogenesis, Helmholtz Zentrum München, Munich, Germany*

Alvin C.H. Ma • *Department of Medicine, LKS Faculty of Medicine, University of Hong Kong, Pokfulam, Hong Kong*
Molly A. Matty • *Department of Molecular Genetics and Microbiology, Center for Host-Microbial Interactions, Duke University Medical Center, Durham, NC, USA; University Program in Genetics and Genomics, Duke University, Durham, NC, USA*
Annemarie H. Meijer • *Institute of Biology, Leiden University, Leiden, The Netherlands*
Irene Miguel-Escalada • *Institute of Cancer and Genomic Sciences, College of Medical and Dental Sciences, University of Birmingham, Birmingham, UK*
Ferenc Müller • *Institute of Cancer and Genomic Sciences, College of Medical and Dental Sciences, University of Birmingham, Birmingham, UK*
Akira Muto • *Division of Molecular and Developmental Biology, Department of Genetics, National Institute of Genetics, Sokendai (The Graduate University for Advanced Studies), Mishima, Japan*
Trista E. North • *Beth Israel Deaconess Medical Center, Harvard Medical School, Boston, MA, USA; Harvard Stem Cell Institute, Cambridge, MA, USA*
Stefan H. Oehlers • *Department of Molecular Genetics and Microbiology, Center for Host-Microbial Interactions, Duke University Medical Center, Durham, NC, USA*
Rui F. Oliveira • *ISPA—Instituto Universitário, Lisbon, Portugal; Instituto Gulbenkian de Ciência, Oeiras, Portugal; Champalimaud Neuroscience Program, Champalimaud Centre for the Unknown, Lisbon, Portugal*
Michael B. Orger • *Champalimaud Neuroscience Program, Champalimaud Centre for the Unknown, Lisbon, Portugal*
Satoshi Ota • *Laboratory for Developmental Biology, Center for Medical Education and Sciences, Graduate School of Medical Science, University of Yamanashi, Yamanashi, Japan*
E. Elizabeth Patton • *Institute of Genetics and Molecular Medicine, MRC Human Genetics Unit & The University of Edinburgh Cancer Research UK Centre, University of Edinburgh, Edinburgh, UK*
Randall T. Peterson • *Cardiovascular Research Center, Massachusetts General Hospital, Harvard Medical School, Charlestown, MA, USA; Department of Systems Biology, Harvard Medical School, Boston, MA, USA; Broad Institute, Cambridge, MA, USA*
Ruben Portugues • *Max Planck Institute of Neurobiology, Sensorimotor Control Research Group, Martinsried, Germany*
Kenneth D. Poss • *Department of Cell Biology, Duke University Medical Center, Durham, NC, USA*
Vinothkumar Rajan • *Department of Microbiology and Immunology, Dalhousie University, Halifax, NS, Canada*
Andrew J. Rennekamp • *Cardiovascular Research Center, Massachusetts General Hospital, Harvard Medical School, Charlestown, MA, USA; Department of Systems Biology, Harvard Medical School, Boston, MA, USA; Broad Institute, Cambridge, MA, USA*
Masahide Seki • *Department of Medical Genome Sciences, Graduate School of Frontier Sciences, The University of Tokyo, Chiba, Japan*
Minori Shinya • *Department of Biology, Keio University, Kanagawa, Japan*
B. Ewa Snaar-Jagalska • *Institute of Biology, Leiden University, Leiden, The Netherlands*
Herman P. Spaink • *Institute of Biology, Leiden University, Leiden, The Netherlands*
Yutaka Suzuki • *Department of Medical Genome Sciences, Graduate School of Frontier Sciences, The University of Tokyo, Chiba, Japan*

KATHRYN M. TABOR • *Division of Developmental Biology, Eunice Kennedy Shriver National Institute of Child Health and Human Development, Bethesda, MD, USA*
HIDEYUKI TANABE • *Division of Molecular and Developmental Biology, National Institute of Genetics, Mishima, Japan*
MAGDA C. TELES • *ISPA - Instituto Universitário, Lisbon, Portugal; Instituto Gulbenkian de Ciência, Oeiras, Portugal; Champalimaud Neuroscience Program, Champalimaud Centre for the Unknown, Lisbon, Portugal*
DAVID M. TOBIN • *Department of Molecular Genetics and Microbiology, Center for Host-Microbial Interactions, Duke University Medical Center, Durham, NC, USA; Department of Immunology, Duke University Medical Center, Durham, NC, USA*
CLAUDIA TULOTTA • *Institute of Biology, Leiden University, Leiden, The Netherlands*
JINHU WANG • *Department of Cell Biology, Duke University Medical Center, Durham, NC, USA*
SONIA WOJCIECHOWSKA • *MRC Human Genetics Unit and Edinburgh Cancer Research UK Centre, The Institute of Genetics and Molecular Medicine, Western General Hospital, University of Edinburgh, Edinburgh, UK*
AN XIAO • *Key Laboratory of Cell Proliferation and Differentiation of the Ministry of Education, College of Life Sciences, Peking University, Beijing, People's Republic of China; Department of Internal Medicine, Eastern Virginia Medical School, Norfolk, VA, USA*
YAN XIAO • *Sensory Biology & Organogenesis, Helmholtz Zentrum München, Munich, Germany*
ERIN M. ZEITUNI • *Department of Embryology, Carnegie Institution, Baltimore, MD, USA*
ZHIQIANG ZENG • *MRC Human Genetics Unit and Edinburgh Cancer Research UK Centre, The Institute of Genetics and Molecular Medicine, Western General Hospital, University of Edinburgh, Edinburgh, UK*
BO ZHANG • *Key Laboratory of Cell Proliferation and Differentiation of the Ministry of Education, College of Life Sciences, Peking University, Beijing, People's Republic of China*
RONG-WEI ZHANG • *Institute of Neuroscience, State Key Laboratory of Neuroscience, Center for Excellence in Brain Science and Intelligence Technology, Shanghai Institutes for Biological Sciences, Chinese Academy of Sciences, Shanghai, China*
MING ZOU • *Friedrich Miescher Institute for Biomedical Research, Basel, Switzerland; University of Basel, Basel, Switzerland*

Part I

Genetics and Genomics

Chapter 1

Chemical Screening in Zebrafish

Colleen A. Brady, Andrew J. Rennekamp, and Randall T. Peterson

Abstract

Phenotypic small molecule screens in zebrafish have gained popularity as an unbiased approach to probe biological processes. In this chapter we outline basic methods for performing chemical screens with larval zebrafish including breeding large numbers of embryos, plating larval fish into multi-well dishes, and adding small molecules to these wells. We also highlight important considerations when designing and interpreting the results of a phenotypic screen and possible follow-up approaches, including popular methods used to identify the mechanism of action of a chemical compound.

Key words Small molecules, Phenotypic screening, Multi-well plates

1 Overview

Small molecule screening allows for unbiased or targeted discovery of chemical modulators of biological pathways. In recent years, zebrafish have emerged as an excellent model system for chemical screening. Compared to cell-based screening, zebrafish screens can address more complex biological questions. Several features of the fish make them amenable to this type of experimentation, particularly their high fecundity and small size. These characteristics allow researchers to plate zebrafish in 96-well plates and screen through small molecule libraries for compounds that can elicit phenotypic changes.

Since the first zebrafish small molecule screen in 2000, over 60 additional screens have been published [1]. These screens have used diverse phenotypic readouts to identify interesting molecules that have shed light on a wide range of biological questions. For example, screens have identified compounds that can lower glucose [2], inhibit cardiomyopathy after chemotherapy treatment [3], alter developmental pathways [4], reduce leukemia burden [5, 6], change behaviors [7, 8], and influence stem cell numbers [9]. Some of these molecules have even entered into clinical trials, such as dimethyl prostaglandin E2 for expansion of hematopoietic

Koichi Kawakami et al. (eds.), *Zebrafish: Methods and Protocols*, Methods in Molecular Biology, vol. 1451,
DOI 10.1007/978-1-4939-3771-4_1, © Springer Science+Business Media New York 2016

stem cells before transplantation (reviewed in [10]). These experiments highlight the value and versatility of zebrafish screening.

Although screens vary in their phenotypic readout, certain steps are shared across experiments. Here, we will describe several of these common steps, including breeding zebrafish, pipetting the fish embryos or larvae into 96-well plates, and adding small molecules to the wells. Because the readout of each screen varies widely, we refer you to primary literature for examples of different types of screens. Our lab and others have also contributed several review articles that serve as useful resources when planning and executing a chemical screen [1, 11–14].

2 Materials

2.1 Zebrafish Mating

1. Zebrafish of breeding age from desired genetic background (the number of fish required will depend on the experiment).
2. Large aquaculture tanks (e.g., 10-L Aquatic Habitat "Ahab" tanks, Pentair Aquatic Eco-Systems).
3. Roll of plastic mesh (1/16 in., Pentair Aquatic Eco-Systems).
4. Pair of scissors.
5. Stapler with staples.
6. Binder clips (medium, 1¼ in. with 5/8 in. capacity, Staples).
7. Nylon mesh tea strainer (3 in., Harold Import Company).
8. Disposable polystyrene petri dishes (e.g., 100 mm × 15 mm, sterile, VWR).
9. Buffered E3 embryo medium (5 mM NaCl, 0.17 mM KCl, 10 mM HEPES, 0.33 mM $MgSO_4$, 0.33 mM $CaCl_2$; pH solution to 7.8; optionally add 0.00002 % methylene blue as an antifungal agent).
10. Manual P1000 pipette with disposable tips.
11. Disposable polyethylene transfer pipettes (e.g., 3.2 mL bulb volume, Fisherbrand).
12. 28 °C incubator (with light cycle if desired).

2.2 Raising Larval Zebrafish

1. Disposable polyethylene transfer pipettes (e.g., 3.2 mL bulb volume, Fisherbrand).
2. Buffered E3 water (5 mM NaCl, 0.17 mM KCl, 10 mM HEPES, 0.33 mM $MgSO_4$, 0.33 mM $CaCl_2$; pH solution to 7.8; optionally add 0.00002 % methylene blue).
3. Polypropylene wash bottle with cap and nozzle (e.g., 32 oz, 1 L capacity, Fisherbrand).
4. 1-phenyl-2-thiourea (PTU) 1000× chemical stock (75 mM).

2.3 Plating Larvae

1. Ice bucket with ice.
2. Polypropylene 50-mL conical centrifuge tubes (e.g., "Falcon" tubes, Corning).
3. Manual P1000 pipette with disposable tips.
4. Polystyrene 96-well plates (exact model depends on experimental output).

2.4 Small Molecule Treatment

1. 12- or 8-channel P10 or P20 pipette.
2. 12- or 8-channel P200 pipette.
3. 96-well plates of small molecules from library of interest. For most commonly used chemical libraries, *see* ref. 1. Chemical stocks and working dilutions of compounds (typically dissolved in DMSO) can be stored in polypropylene plates at −80 °C (e.g., 96-well conical bottom, Nunc).
4. Centrifuge capable of spinning 96-well plates.
5. Aluminum foil lids for 96-well plates (e.g., "Seal & Sample" lids, Beckman Coulter).

3 Methods

3.1 Large-Scale Mating of Zebrafish

There are several ways to obtain the high numbers of zebrafish embryos required for a chemical screen. When examining developmental stage-specific phenotypes, it is important to obtain synchronized fish through timed mating. For this method, one can set up fish in individual mating tanks where males and females are separated by plastic dividers and remove all of the dividers at the same time the following morning. However, setting up fish in this way requires a significant amount of time if very large numbers of embryos are required. To reduce this effort, Adatto and colleagues invented a specialized large-scale breeding system [15]. In trials of this system, the authors placed 100 males and 80 females in the tank in the evening and then removed the divider to combine all of the fish the following morning. They reported collecting an average of 8400 embryos (~105 eggs per female fish) after 10 min of breeding. This method still requires separation of males and females, which can be time-consuming when breeding large numbers of fish. While careful synchronization of embryos can be important when examining developmental phenotypes or using very early embryos, we have found that precise staging is not required for many experiments, especially when using larvae. Our lab prefers a simple and affordable method to breed large numbers of fish, as described below.

1. Select the strain of fish that will be used for the screen. Depending on desired phenotypic endpoint, selection may be

limited to fish with a unique genetic background necessary for the function of the screening assay (i.e., a mutant or transgenic line). But, select the most fecund strain available (usually the most outbred strain) to obtain optimal embryo yields.

2. Prepare breeding apparatus by fitting a 10-L tank with a mesh insert. Construct the mesh insert by cutting and stapling or sewing 1/16-in. plastic mesh to form a basket that fits inside the tank (*see* Fig. 1a for pattern). Use medium-sized binder clips to hold the basket in place (Fig. 1b).
3. Optional: Though not required, we recommend feeding the fish an extra meal 1 h before setting up the breeding tank. The extra meal will increase fecundity. The hour wait will give the fish enough time to consume the meal and excrete waste. Setting up the fish right after eating will increase waste products in the water, which will reduce breeding efficiency.
4. The night before embryo collection, mix male and female fish in a 10-L tank fitted with a mesh insert. In our experience, the

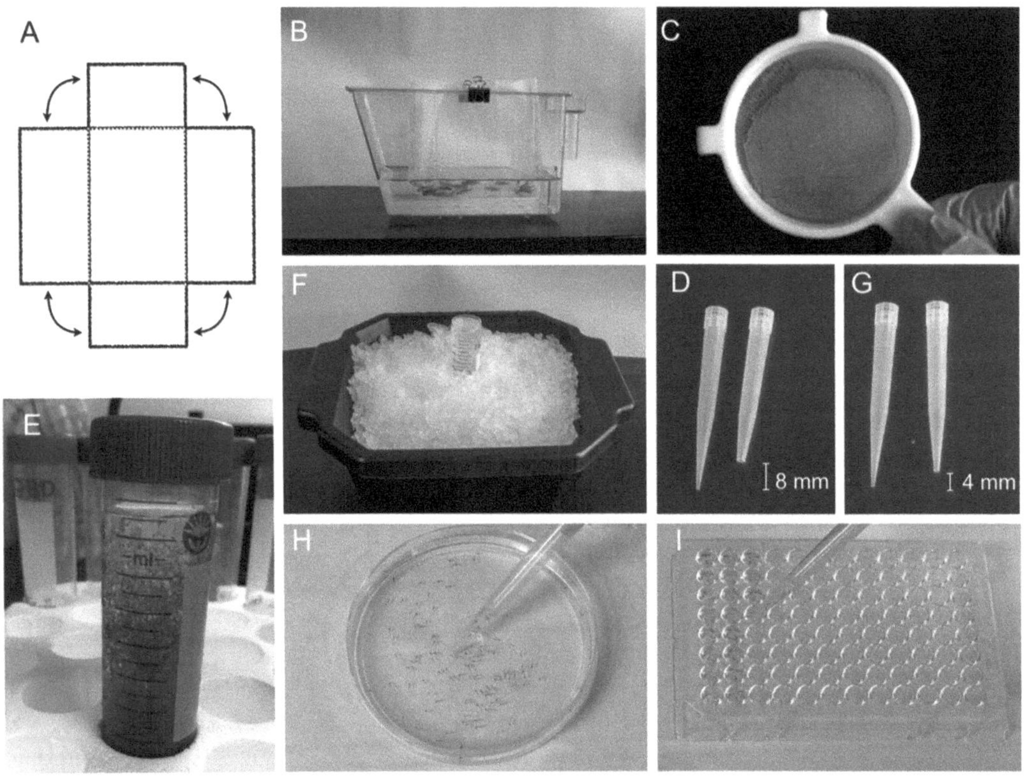

Fig. 1 (**a**) Pattern used to make mesh basket for breeding tanks. *Arrows* indicate that the edges should be joined and stapled. (**b**) Image of 10-L tank with a mesh insert held by binder clips filled with fish for mating. (**c**) Mesh tea strainer filled with embryos collected from several breeding tanks. (**d**) P1000 pipette tip cut for transfer of embryos into 10-cm petri dishes. (**e**) Larvae collected in a 50-mL tube. (**f**) Larvae put on ice to anesthetize them for cleaning and plating. (**g**) P1000 pipette tip cut for transfer of larvae into a 96-well dish. (**h**) Collection of embryos for plating using pipette tip in g. (**i**) Transfer of embryos into a 96-well dish for chemical treatment

optimal ratio of females to males is about 2:1, although 1:1 also works well. For ease of set up, we house the male and female fish together in 10-L tanks and simply transfer all of the fish from the aquaculture tank to a 10-L breeding tank by net. Ensure that the water level is fairly shallow, only about 2 in. or less above the mesh floor (Fig. 1b).

5. In the morning after the lights turn on for the day, allow fish to breed. If desired, synchronization of the embryos can be achieved in this system by setting up the fish in a full tank of water the night before and then transferring the fish with the mesh insert to a tank with fresh shallow water immediately after the lights turn on.
6. After the fish have laid eggs, remove adult fish from the tank by lifting out the mesh insert and simply pouring them back into their home aquaculture tank. Typically, a healthy tank of 15 female and 15 male zebrafish will produce 1000–2000 embryos per week using this method.
7. Ideally, collect the embryos within 3 h of the start of mating. To do so, pour the water from the breeding tank through a mesh tea strainer (Fig. 1c). Embryos left uncollected for more than 3 h may begin to die, and the health of the adults may also decline as the holding tank accumulates waste.
8. Using a wash bottle with the tip cut to increase flow, rinse embryos thoroughly with E3 medium or fish room water while in the strainer to remove all debris.
9. Transfer all of the embryos to a petri dish by inverting the tea strainer over the dish and rinsing with a wash bottle of E3 medium to remove all embryos from the mesh of the strainer.
10. Optional: Mix all embryos from similar clutches. Group mating will ensure diverse embryo populations, and thorough population mixing before plating can further minimize clutch/tank variations.
11. Cut about 8 mm off the end of a P1000 pipette tip to create a larger opening (Fig. 1d) and then use it to transfer about 200 embryos into 100-mm petri dishes filled with E3 medium. Transferring 800 μL of concentrated embryos per dish will yield approximately 200 embryos/dish.
12. Remove any dead embryos from the dish using a transfer pipette. Dead embryos are easily identified by their opaque or cloudy appearance, which can be readily visualized by eye against a dark background, such as a black bench top.

3.2 Raising Larval Fish

Zebrafish embryos become larvae at 3 days post fertilization. We find that larvae are ideal for many chemical screens—the fish have hatched from their chorions and have developed most organ systems. Ideally, we perform chemical screens in the earlier larval phases while the fish obtain sustenance from their yolks rather than from external food. Feeding larvae would be disadvantageous when such great numbers of fish are required, as the fish would need to be transferred to tanks, fed 3 times a day, and the tank water changed periodically to maintain cleanliness.

1. Incubate the dishes containing embryos at 28 °C. An incubator with a physiological light/dark cycle provides the optimal environment to ensure normal development.
2. Check the petri dishes daily for dead embryos and remove them using a transfer pipette. This step will help maintain the health of the rest of the embryos in the dish.
3. Optional: Add methylene blue to the medium to limit fungal growth, which may adversely affect larval health. Use of methylene blue is unadvisable in situations where a fluorescent readout is required for phenotyping because methylene blue can increase the autofluorescence of fish tissues.
4. Optional: If desired, zebrafish pigment can be eliminated by chemical treatment with 1-phenyl-2-thiourea (PTU). At 90% epiboly stage, add 1000× PTU stock (75 mM) to E3 medium for a final concentration of 75 μM. Add fresh PTU each time the E3 medium is changed.
5. Optional: Change the fish water as needed to maintain fish health. We have found that fish typically need fresh E3 medium on day 3 or 4 to remove chorion debris from the dishes. To change the medium, carefully pour off most of the old E3 medium without pouring out the embryos, and then add fresh medium using a wash bottle.

3.3 Plating Larval Fish in 96-Well Dishes

Each well of a 96-well plate can house several zebrafish larvae; the most common format for past zebrafish chemical screens has been three fish per well [1]. We manually array the fish, which is a time-consuming step, taking about 1 h per plate for someone inexperienced in this procedure but about 10–15 min for someone with experience. Manually pipetting the fish has some advantages; it allows us to reliably count the larvae and visually assess their health and developmental morphology before plating. Alternatively, a few automated mechanical options for loading zebrafish into multiwell plates are either on the market or in production [1, 16]. The type of plate used for the experiment will depend on the desired output; there are several types of commercially available 96-well plates, including those with thin plastic bottoms that are optimized for imaging, those with flat bottoms, and plates with round bottoms. Custom plates have also been used for chemical screening, such as plates with prisms in each well to allow for lateral fish viewing and alignment [17].

We have found that plates containing mesh inserts in each well (100 μm mesh filter plates, EMD Millipore) work well for rapidly washing the wells either to remove compound after a specified incubation period or for fixing and staining screening wells for whole mount in situ hybridization (for a detailed in situ protocol, *see* ref. 14). These mesh-bottom plates come in two pieces—a mesh insert that fits into a bottom plate with 96 wells. This setup allows for treatment of each well with a separate compound, but the mesh insert can be placed into a larger container, such as a pipette tip box lid, to wash the wells. These plates can be washed and reused a number of times, which significantly cuts down their expense. Below we outline the basic steps for pipetting zebrafish into 96-well plates.

1. Combine larvae from petri dishes by pouring them into a 50-mL conical tube (Fig. 1e).
2. Put this 50-mL conical tube on ice and allow the larvae to cool (Fig. 1f). This will cause them to fall to the lower part of the tube. Typically, the fish will stop moving and drop after 5–10 min, and we have found that they can tolerate the ice for 20–30 min, depending on the volume of medium in the tube.
3. Pour off all but 10 mL of liquid in the tube. During this step, most debris will decant off, such as remaining chorions.
4. Next, either transfer E3 medium and larval fish to a fresh petri dish or continue to add new dishes of fish to the conical tube. The number of plates combined per tube will depend on the speed at which the fish can be plated. Do not leave larvae on ice for longer than 30 min.
5. Place the petri dish on a white surface, such as a sheet of white paper, to aid in visualization of the fish.
6. Cut 4 mm off the opening of a P1000 pipette tip to allow the larval fish to easily enter (Fig. 1g). Then, use the pipette to collect the desired number of larvae in a specified volume of E3 medium. Be careful to pipette up the fish head or tail first to avoid damaging them. Avoid pipetting fish with an abnormal morphology or fish that have may have died. Consider the size of the well, length of time of the assay, and concentration of the chemical stocks when determining the volume to pipette in each well. Typical volumes for a 96-well plate experiment in our lab range from 200 to 400 μL.
7. Transfer the larvae well by well into the plate, using care when expelling the fish (Fig. 1h, i).
8. Allow the plate of fish to warm to 28 °C or desired assay temperature. Note: this step may occur before or after chemical treatment.

3.4 Selecting a Chemical Library

There are several important considerations when selecting a chemical library to screen. First decide how many compounds you would like to screen. This will depend on the throughput of your screening assay(s) and the cost you are willing to spend on the screen. Additionally, consider how many hits you expect from the screen and what time and resources you plan to devote to the following up of each hit.

Next, decide what type of chemical library you would like to screen. The theoretical chemical space is vast; even a very large library will only cover a very small fraction of it. No matter what library you choose, your selection will inherently bias your results. When possible, use the selection bias to your advantage. For example, if you want to discover a novel drug, choose a library that contains mostly novel compounds with unknown pharmacology. Or, if you are trying to learn something about the biology of the phenotype you are assaying, use a library composed of drugs with well-annotated pharmacology. If you have a behavioral endpoint, use a library of known neuroactive compounds.

Finally, choose a library that will be accessible to you in the future. To perform follow-up experiments, you will need to repurchase or resynthesize your lead compounds. Before performing your screen, it is advisable to determine the future availability of the small molecules in the collection and the cost of resupply. In our experience, libraries composed of molecules obtained from natural sources (usually plants, e.g., the Spectrum collection) often yield higher hit rates. However, these naturally occurring compounds often have complex three-dimensional structures, making them difficult to synthesize and therefore generally more expensive to resupply. In contrast, libraries made up of synthetic compounds (e.g., the ChemBridge collection) often contain simpler two-dimensional molecules that are less biologically active as a group, but far cheaper to resupply.

3.5 Treating Larvae with Small Molecules

Once the plates have been loaded with zebrafish larvae, chemicals can be added to the fish water. Many chemicals dissolved in the water easily penetrate the larval fish, and the larvae can tolerate moderate levels of dimethyl sulfoxide (DMSO) (typically up to 1%) without exhibiting toxic effects. The final concentration of DMSO will vary depending on the assay. It is important that each plate contains vehicle (solvent only) control wells. This will control for any possible effects of the DMSO, provide a way for assessing the quality of the data, and account for possible plate-to-plate variations. It is also helpful to include a positive control well(s) on each plate, if available.

Libraries should be stored at −80 °C in 96-well format to allow for well transfer with an 8- or 12-channel pipette. We use 96-well polypropylene plates with conical bottoms so that small compound volumes can be centrifuged to the center of the wells for easier pipetting. Thaw the plate containing the small molecule library in the dark at room temperature, in a desiccator when possible, wipe off extra moisture from the outside of the plate, and then spin it in a centrifuge to remove droplets from the lid before removing the plate seal. The library plate may need to be diluted to form a "working" plate, depending on the initial concentration of the molecules and the desired final concentration. We typically keep our concentrated stocks at 10 mM and make working plates at 0.5–1 mM.

Using a multichannel P10 pipette, transfer the desired volume of chemical compounds to the wells containing zebrafish. Typical zebrafish screens use a final concentration of 1–10 μM of the compounds [1] and a final concentration of DMSO solvent less than 1 %. For example, zebrafish plated in 200 μL of medium per well dosed with 1 μL of compound per well from a chemical stock plate at 1 mM in DMSO would achieve a final concentration of 5 μM of compound and 0.5 % DMSO. An ideal screening dose results in minimal toxicity (toxicity rate of <1 %) and provides a desired hit rate (we typically aim for a hit rate of 0.5–1 %).

1. Mix the wells with a multichannel P200 pipette by pipetting up and down a portion of the water in each well while carefully avoiding the larvae. If the larvae are swimming actively, this step may not be necessary, as the fish will sufficiently mix the solution on their own.
2. Incubate the zebrafish with the compound at 28 °C. Timing of this step is highly dependent on the biological process of interest.
3. Optional: If the incubation time exceeds 1 or 2 h, the plates may need to be covered and/or fresh medium with chemical may need to be added to prevent evaporation. Some plates have special mini wells around the outer rim, and adding water to these wells reduces evaporation (e.g., SCREENSTAR plates, Greiner Bio-One).
4. Cover the plates with foil or keep them in a dark incubator, as some of the compounds may be light sensitive.
5. Reseal the chemical library plate with an unused aluminum foil lid and refreeze the plate at −80 °C. Keep track of the number of thaws, as multiple freeze-thaw cycles may affect compound stability.

3.6 Phenotypic Readout of Treated Larval Fish

After chemical treatment, each well is scored for the desired phenotype. Published zebrafish screens can provide a good starting point for designing a phenotypic assay, and we recommend investing the time to optimize this assay before screening a full library of compounds. Methods that have been used previously include imaging transgenic lines [17, 18], examining fish morphology [19, 20], in situ hybridization for mRNA expression [9, 21], behavior recording [7, 8], luciferase screening [22], and biochemical outputs [2]. This diversity of chemical screens highlights the versatility of the zebrafish model.

It is important that the assay for phenotyping provides robust, quantitative results. Repetitions of control fish should yield similar results. Ideally results from control wells will be consistent from plate to plate and day to day. However, if there is a great deal of variation, you may be able to reduce the noise and identify hits by normalizing the data from the test wells to the control wells from the same plate.

If possible, use positive control wells in addition to negative wells. Positive controls should be easily discernable from negative controls, ideally falling three standard deviations away from the negative control mean. Ideally, the threshold used to identify hits will also be three standard deviations from the negative control mean. A three standard deviation cutoff will theoretically provide a false-positive rate of <0.3%. A false-positive rate greater than this may be tolerable, but it will result in the expenditure of additional time and resources during the follow-up phase. A common way to deal with a higher false-positive rate is to perform the screen using duplicate wells per compound. This is costly when performing a large-scale screen because each duplicate will also substantially increase the time, workload, and overall cost of the screen.

3.7 Data Assessment and Secondary Screening

Once chemical plates have been screened and potential candidate molecules identified, repetition and secondary screening should be used to confirm these candidates. All hit compounds of interest should be repurchased or resynthesized and retested in the original screening assay. This is important not only to weed out false positives but also to confirm the identity of the compound in the library stock plate. The compound should be tested at a range of doses encompassing the initial screening dose to demonstrate the dose dependence of the effect. We find that several two- or threefold dilutions often work well. For example, if the screening dose was 10 μM, we will typically perform a dose curve starting at 80 μM with twofold serial dilutions down to 0.625 μM.

Depending on the nature of the primary screen, a secondary screen may be required to further confirm the effect of a compound or to rule out alternative mechanisms. For example, a screen for a loss of function phenotype may require a secondary toxicity screen to confirm that the loss of function effect of the compound is specific to the biology of interest rather than a consequence of general toxicity, which may be uninteresting. For this reason, we prefer to conduct screens for compounds that produce new and distinctive phenotypes, rather than for compounds that cause common or nonspecific phenotypes.

Another example of when a secondary screen may be important is if the primary screen employs a florescent readout and several of the hit compounds are also fluorescent. A nonfluorescent secondary screen will help to identify false positives. Secondary screens are often more labor intensive than the primary screen, but need only be carried out on hits from the primary screen. They are useful in narrowing down the primary hit list to the most interesting leads, especially when the initial list is too broad.

3.8 Follow-Up Studies: Translation and Mechanism of Action

Small molecules discovered in zebrafish screens often exhibit conserved activity in analogous mammalian systems. A common follow-up experiment for compounds identified in zebrafish screens is testing in a secondary mammalian assay. The ideal mammalian assay is one that can be performed in vivo and resembles the original zebrafish phenotypic assay as closely as possible. If an in vivo assay is unavailable or impractical, then a mammalian cell-based experiment or in vitro assay using mammalian-derived substrates may be appropriate. In vitro experiments can be very useful when demonstrating applicability to human biology if they can be carried out on human targets and in human cells. In vitro experiments, however, can also be misleading if, for example, your small molecule of interest turns out to be a prodrug requiring in vivo metabolism or a systems-modulating compound requiring an in vivo circuit.

Many journals appear to prefer papers that provide details about the functional target (singular) of a hit compound identified in a zebrafish chemical screen. Identifying the target of a small molecule can be a rewarding process and provide valuable insight into the biology behind the phenotype being examined. Unfortunately, it can also be a somewhat misguided effort at times because it assumes that every biologically active small molecule has a simple, easily identifiable, mechanism of action involving a single relevant target. This kind of oversimplification leads to overstatements of the importance of single targets; drugs are rarely monoselective, and the mechanism of action for a small molecule often involves partial activity at multiple targets. Nevertheless, with these caveats in mind, we will provide a brief overview of the ways in which researchers have identified mechanisms of action for compounds discovered in zebrafish screens. For a more in-depth review of this topic, we point you to [23].

About half of all zebrafish chemical screens reported to date have used libraries of compounds with small molecules that have annotated mechanism(s) of action, and another 15% provided no mechanistic follow-up [1]. For the remaining 35%, five different but complementary approaches have been used to identify functional targets.

1. The most common approach has been the candidate-based approach where researchers made educated guesses about the targets of their novel compounds based on prior knowledge of the biology underlying their phenotype of interest. This approach

can be limited by insufficient prior knowledge, but is often a logical place to start. It may not provide new biological insight, but it can be a useful way to identify mechanisms of action or to identify structurally novel classes of compounds with mechanisms similar to known drugs.

2. Structure-based approaches have been used to computationally predict targets by comparing the chemical structures of newly discovered small molecules with databases of known chemical ligands. This approach may not identify new structurally interesting compounds, and its success is usually limited to cases involving well-understood targets. On the other hand, structure-based approaches can provide clues to help elucidate basic biology underlying the phenotype of interest.
3. Phenocopy approaches have provided clues about the mechanisms of action of novel small molecules pulled out of chemical libraries by zebrafish screens. These come in two flavors—pharmacologic and genetic phenocopy—and are most useful when then screen produces a wide range of phenotypic outcomes. For pharmacologic phenocopy, the phenotypes of novel compounds are compared to the phenotypes of compounds with known mechanisms of action. For genetic phenocopy, the phenotypes of novel compounds are compared to the phenotypes of genetically manipulated zebrafish. In either case, complex phenotypes can be subdivided into smaller features to create "barcode" signatures of each compound or mutant screened. These barcodes can then be clustered using a bioinformatics approach [24]. This approach is limited by the quality of preexisting phenotypic data and is less useful for simple screens that have binary outcomes.
4. Targets can also be identified using affinity purification assays. In these biochemical approaches, the small molecule of interest is used as bait to bind and isolate its protein target(s). To prepare the bait molecule, the compound of interest is covalently linked to a tag that serves as a high-affinity partner for an immobilized ligand. For this approach to work, the tag must be added in such a way that it does not disrupt the biological activity of the compound. Structure activity relationship (SAR) studies will be required to determine where the tag can be placed, and mass spectrometry will be needed to analyze the pulled-down proteins.
5. Clues about mechanism of action have also come from localization of activity approaches where researchers have honed in on where in the zebrafish the compound of interest is producing the effect (anatomical localization), at what time the compound is acting (chronological localization), and at which gene product (genetic localization). For examples of each of these, we refer you to Rennekamp and Peterson (2015).

Each of these approaches can provide clues concerning the mechanism(s) of action of a compound, but it will still be necessary to provide evidence of the functional relevance of each target identified. One may be able to demonstrate that the target is necessary for the chemical's effect by knocking out the candidate target using a reverse genetics approach, such as CRISPR/Cas9-mediated mutagenesis [25, 26]. If the chemical no longer produces an effect in the mutant, you will have good evidence that the target is necessary for the effects of the compound. If the chemical phenotype is replicated by the mutation, you will have good evidence that the chemical effect on that target is sufficient to explain the phenotypic effects of the compound. If mutation is deleterious, you may not be able to draw any conclusions.

Another possibility may be to counteract the effect of the identified compound with a second compound that has a well-established effect on the same target or within the same pathway. For example, a known receptor antagonist may counteract the effect of a novel receptor agonist. If you can indeed counteract the effect of your novel compound, you have good evidence the target is necessary for the observed phenotype. If the effect of your compound can be replicated with a drug that has good specificity for that target, you will have good evidence that perturbation of that chemical target is sufficient for the phenotypic effect.

References

1. Rennekamp AJ, Peterson RT (2015) 15 years of zebrafish chemical screening. Curr Opin Chem Biol 24C:58–70. doi:10.1016/j.cbpa.2014.10.025
2. Nath AK, Ryu JH, Jin YN, Roberts LD, Dejam A, Gerszten RE, Peterson RT (2015) PTPMT1 inhibition lowers glucose through succinate dehydrogenase phosphorylation. Cell Rep. doi:10.1016/j.celrep.2015.01.010
3. Liu Y, Asnani A, Zou L, Bentley VL, Yu M, Wang Y, Dellaire G, Sarkar KS, Dai M, Chen HH, Sosnovik DE, Shin JT, Haber DA, Berman JN, Chao W, Peterson RT (2014) Visnagin protects against doxorubicin-induced cardiomyopathy through modulation of mitochondrial malate dehydrogenase. Sci Transl Med 6(266):266ra170. doi:10.1126/scitranslmed.3010189
4. Yu PB, Hong CC, Sachidanandan C, Babitt JL, Deng DY, Hoyng SA, Lin HY, Bloch KD, Peterson RT (2008) Dorsomorphin inhibits BMP signals required for embryogenesis and iron metabolism. Nat Chem Biol 4(1):33–41
5. Ridges S, Heaton WL, Joshi D, Choi H, Eiring A, Batchelor L, Choudhry P, Manos EJ, Sofla H, Sanati A, Welborn S, Agarwal A, Spangrude GJ, Miles RR, Cox JE, Frazer JK, Deininger M, Balan K, Sigman M, Muschen M, Perova T, Johnson R, Montpellier B, Guidos CJ, Jones DA, Trede NS (2012) Zebrafish screen identifies novel compound with selective toxicity against leukemia. Blood 119(24):5621–5631. doi:10.1182/blood-2011-12-398818
6. Gutierrez A, Pan L, Groen RW, Baleydier F, Kentsis A, Marineau J, Grebliunaite R, Kozakewich E, Reed C, Pflumio F, Poglio S, Uzan B, Clemons P, VerPlank L, An F, Burbank J, Norton S, Tolliday N, Steen H, Weng AP, Yuan H, Bradner JE, Mitsiades C, Look AT, Aster JC (2014) Phenothiazines induce PP2A-mediated apoptosis in T cell acute lymphoblastic leukemia. J Clin Invest 124(2):644–655. doi:10.1172/jci65093
7. Rihel J, Prober DA, Arvanites A, Lam K, Zimmerman S, Jang S, Haggarty SJ, Kokel D, Rubin LL, Peterson RT, Schier AF (2010) Zebrafish behavioral profiling links drugs to biological targets and rest/wake regulation. Science 327(5963):348–351. doi:10.1126/science.1183090
8. Kokel D, Bryan J, Laggner C, White R, Cheung CY, Mateus R, Healey D, Kim S, Werdich AA, Haggarty SJ, Macrae CA, Shoichet B, Peterson RT (2010) Rapid

behavior-based identification of neuroactive small molecules in the zebrafish. Nat Chem Biol 6(3):231–237
9. North TE, Goessling W, Walkley CR, Lengerke C, Kopani KR, Lord AM, Weber GJ, Bowman TV, Jang IH, Grosser T, Fitzgerald GA, Daley GQ, Orkin SH, Zon LI (2007) Prostaglandin E2 regulates vertebrate haematopoietic stem cell homeostasis. Nature 447(7147): 1007–1011
10. Hagedorn EJ, Durand EM, Fast EM, Zon LI (2014) Getting more for your marrow: boosting hematopoietic stem cell numbers with PGE2. Exp Cell Res 329(2):220–226. doi:10.1016/j.yexcr.2014.07.030
11. Zon LI, Peterson RT (2005) In vivo drug discovery in the zebrafish. Nat Rev Drug Discov 4(1):35–44. doi:10.1038/nrd1606
12. Taylor KL, Grant NJ, Temperley ND, Patton EE (2010) Small molecule screening in zebrafish: an in vivo approach to identifying new chemical tools and drug leads. Cell Commun Signal 8:11. doi:10.1186/1478-811X-8-11
13. Peterson RT, Fishman MC (2011) Designing zebrafish chemical screens. Methods Cell Biol 105:525–541. doi:10.1016/B978-0-12-381320-6.00023-0
14. Jing L, Durand EM, Ezzio C, Pagliuca SM, Zon LI (2012) In situ hybridization assay-based small molecule screening in zebrafish. Curr Protoc Chem Biol 4(2):110236. doi:10.1002/9780470559277.ch110236
15. Adatto I, Lawrence C, Thompson M, Zon LI (2011) A new system for the rapid collection of large numbers of developmentally staged zebrafish embryos. PLoS One 6(6):e21715. doi:10.1371/journal.pone.0021715
16. Bruni G, Lakhani P, Kokel D (2014) Discovering novel neuroactive drugs through high-throughput behavior-based chemical screening in the zebrafish. Front Pharmacol 5:153. doi:10.3389/fphar.2014.00153
17. Rovira M, Huang W, Yusuff S, Shim JS, Ferrante AA, Liu JO, Parsons MJ (2011) Chemical screen identifies FDA-approved drugs and target pathways that induce precocious pancreatic endocrine differentiation. Proc Natl Acad Sci U S A 108(48):19264–19269. doi:10.1073/pnas.1113081108
18. Molina G, Vogt A, Bakan A, Dai W, Queiroz de Oliveira P, Znosko W, Smithgall TE, Bahar I, Lazo JS, Day BW, Tsang M (2009) Zebrafish chemical screening reveals an inhibitor of Dusp6 that expands cardiac cell lineages. Nat Chem Biol 5(9):680–687. doi:10.1038/nchembio.190
19. Peal DS, Mills RW, Lynch SN, Mosley JM, Lim E, Ellinor PT, January CT, Peterson RT, Milan DJ (2011) Novel chemical suppressors of long QT syndrome identified by an in vivo functional screen. Circulation 123(1):23–30. doi:10.1161/circulationaha.110.003731
20. Peterson RT, Link BA, Dowling JE, Schreiber SL (2000) Small molecule developmental screens reveal the logic and timing of vertebrate development. Proc Natl Acad Sci U S A 97(24):12965–12969
21. Yeh JR, Munson KM, Chao YL, Peterson QP, Macrae CA, Peterson RT (2008) AML1-ETO reprograms hematopoietic cell fate by downregulating scl expression. Development 135(2):401–410. doi:10.1242/dev.008904
22. Gut P, Baeza-Raja B, Andersson O, Hasenkamp L, Hsiao J, Hesselson D, Akassoglou K, Verdin E, Hirschey MD, Stainier DY (2013) Whole-organism screening for gluconeogenesis identifies activators of fasting metabolism. Nat Chem Biol 9(2):97–104. doi:10.1038/nchembio.1136
23. Rennekamp AJ, Peterson RT (2013) From phenotype to mechanism after zebrafish small molecule screens. Drug Discov Today Dis Models 10(1):e51–e55. doi:10.1016/j.ddmod.2012.02.002
24. Kokel D, Rennekamp AJ, Shah AH, Liebel U, Peterson RT (2012) Behavioral barcoding in the cloud: embracing data-intensive digital phenotyping in neuropharmacology. Trends Biotechnol 30(8):421–425. doi:10.1016/j.tibtech.2012.05.001
25. Hwang WY, Fu Y, Reyon D, Maeder ML, Tsai SQ, Sander JD, Peterson RT, Yeh JR, Joung JK (2013) Efficient genome editing in zebrafish using a CRISPR-Cas system. Nat Biotechnol 31(3):227–229. doi:10.1038/nbt.2501
26. Hwang WY, Fu Y, Reyon D, Maeder ML, Kaini P, Sander JD, Joung JK, Peterson RT, Yeh JR (2013) Heritable and precise zebrafish genome editing using a CRISPR-Cas system. PLoS One 8(7):e68708. doi:10.1371/journal.pone.0068708

Chapter 2

TALEN-Mediated Mutagenesis and Genome Editing

Alvin C.H. Ma, Yi Chen, Patrick R. Blackburn, and Stephen C. Ekker

Abstract

Transcription activator-like effectors (TALEs) are important genomic tools with customizable DNA-binding motifs for locus-specific modifications. In particular, TALE nucleases or TALENs have been successfully used in the zebrafish model system to introduce targeted mutations via repair of double-stranded breaks (DSBs) either through nonhomologous end joining (NHEJ) or by homology-directed repair (HDR) and homology-independent repair in the presence of a donor template. Compared with other customizable nucleases, TALENs offer high binding specificity and fewer sequence constraints in targeting the genome, with comparable mutagenic activity. Here, we describe a detailed in silico design tool for zebrafish genome editing for TALENs and CRISPR/Cas9 custom restriction enzymes using Mojo Hand 2.0 software.

Key words TALEN, Customized nucleases, Zebrafish, Genome editing, Golden Gate, FusX

1 Introduction

TALEs are naturally occurring transcription factors isolated from plant pathogen *Xanthomonas* [1, 2]. Each TALE has a DNA-recognizing TALE domain made up of a tract of almost identical repetitive units (33–35 amino acid residues) and a partial (or half) repeat unit at the end. Within each unit, the two repeat-variable di-residues (RVDs) are solely responsible for the binding specificity of the unit toward a DNA nucleotide in a highly predictable fashion [3, 4]. Commonly used RVDs include NI and NN for adenine; HD for cytosine; NK, NN, and NH for guanine; and NG for thymine [3–6]. Because of the 1:1 RVD to nucleotide modularity of the TALE domain, it can be engineered to target almost any DNA sequence in the genome and can be fused with different functional domains including nuclease, transcription activator/repressor, and methyltransferases. TALEs represent important genomic tools for locus-specific modifications [7–14]. In particular, TALENs have been extensively used for targeted mutations in vitro and in different model organisms [15–22].

Koichi Kawakami et al. (eds.), *Zebrafish: Methods and Protocols*, Methods in Molecular Biology, vol. 1451,
DOI 10.1007/978-1-4939-3771-4_2, © Springer Science+Business Media New York 2016

Diverse methodologies have been developed to assemble the modular TALE domain, with the Golden Gate TALEN assembling method (Golden Gate TALEN Kit 2.0) being widely used because of its flexibility, low start-up cost, and requirement of minimal, common molecular cloning reagents [23]. We previously reported the first use of GoldyTALEN in targeted zebrafish genome editing through both NEHJ and HDR [8]. We also described a simple and highly active GoldyTALEN design with only 15 RVDs (or 14.5 TALE repeats) [22]. To further facilitate TALEN-mediated high-throughput genome editing, we subsequently developed a modified Golden Gate TALEN assembling FusX system (Ma et al., manuscript in preparation). The new system increased assembling efficiency, but shortened assembling time without affecting mutagenic activity and compatibility.

With the rapid development of novel genome engineering tools such as TALENs and CRISPR/Cas9 systems [24], new software tools are needed to aid biologists in designing and constructing high-efficiency reagents that can be used to make tailored changes within any model system of interest. Through a better understanding of the cell's endogenous DNA repair mechanisms, we can improve reagent design and targeting to achieve predictable outcomes. Microhomology-mediated end joining (MMEJ) appears to be a dominant repair pathway for TALEN, and RNA-guided engineered nuclease (RGEN) induced double-stranded breaks and has been used to generate predictable out-of-frame deletions and to incorporate donor DNA sequences in a highly efficient manner [25].

We previously presented the web-based Mojo Hand designer tool [26]. In the latest version 2.0, algorithm adheres to the same general steps that the original algorithm follows with the integration of new features including .bed file creation, microhomology, and out-of-frame scoring. Another major consideration was the incorporation of user-generated next-generation sequencing data in reagent design to deal with the tremendous inter- and intrastrain genetic variation during zebrafish genome targeting. In the current version, high-depth RNAseq datasets were integrated to simplify design and reduce time and cost through the avoidance of regions rich in single nucleotide polymorphisms (SNPs). Here, we describe a detail protocol of targeted zebrafish genome editing through NHEJ and HDR, respectively, using TALENs or CRISPR/Cas9 using the open access Mojo Hand 2.0 software.

2 Materials

2.1 Zebrafish Embryo Genotyping and RFLP Assay

1. Genomic DNA extraction buffer: 10 mM Tris–HCl, pH 8.3, 50 mM KCl.
2. 10% Tween-20.

3. 10% NP-40.
4. Proteinase K solution (recombinant, PCR grade, 14–22 mg/mL in 10 mM Tris–HCl, pH 7.5, Roche Life Science).
5. PCR reaction mix (*see* **Note 1**).
6. Restriction enzyme.
7. Agarose.
8. TAE buffer (1×): 40 mM Tris–HCl, 20 mM acetic acid, 1 mM EDTA, pH 8.4.
9. Standard gel electrophoresis system.

2.2 TALEN Assembling

1. FusX collection (pFusX1–4 and pFusX_B2) (Addgene *in progress*).
2. Last half-repeat components pLR-NI, pLR-HD, pLR-NN, and pLR-NG (Addgene #31006, #30984, #31017, #30995).
3. RCIscript-GoldyTALEN backbone (Addgene, cat# 38142).
4. T4 DNA ligase (2,000,000 U/mL, New England Biolabs).
5. BsmBI (New England Biolabs) (optional, *see* **Note 2**).
6. Esp3I (Thermo Scientific).
7. Standard thermocycler.
8. Competent *E. coli* cell.
9. LB agar plate with ampicillin (100 μg/mL).
10. LB medium with ampicillin (100 μg/mL).
11. 20 mg/mL X-Gal (5-bromo-4-chloro-3-indolyl-β-D-galactoside).
12. 0.1 M IPTG (isopropylthio-β-galactoside).
13. Colony PCR screening primers: TAL_F1 (ttggcgtcggcaaacagtgg) and TAL_R2 (ggcgacgaggtggtcgttgg) [23].
14. Sequencing primers: TAL_F1, TAL_R2, RVD-MM-F (ctcacacccgatcaggtc), and RVD-MM-R (gacctgatcgggtgtgag) (*see* **Note 3**) [24].

2.3 In Vitro Transcription

1. SacI (New England Biolabs).
2. 3 M sodium acetate, pH 5.0.
3. 70% ethanol.
4. Ambion mMESSAGE mMACHINE® T3 Transcription Kit (Life Technologies).
5. Deionized water.
6. Lithium chloride precipitate solution: 7.5 M LiCl, 50 mM EDTA, pH 8.0.

3 Methods

3.1 Designing TALEN with Mojo Hand 2.0 (Fig. 1; See Note 4)

1. Select genomic region for TALEN targeting (Fig. 2; *see* **Note 5**).
2. Sequence input into Mojo Hand 2.0 (http://talendesign.org/).
3. Identification of binding sites with the following parameters (*see* **Note 6**):

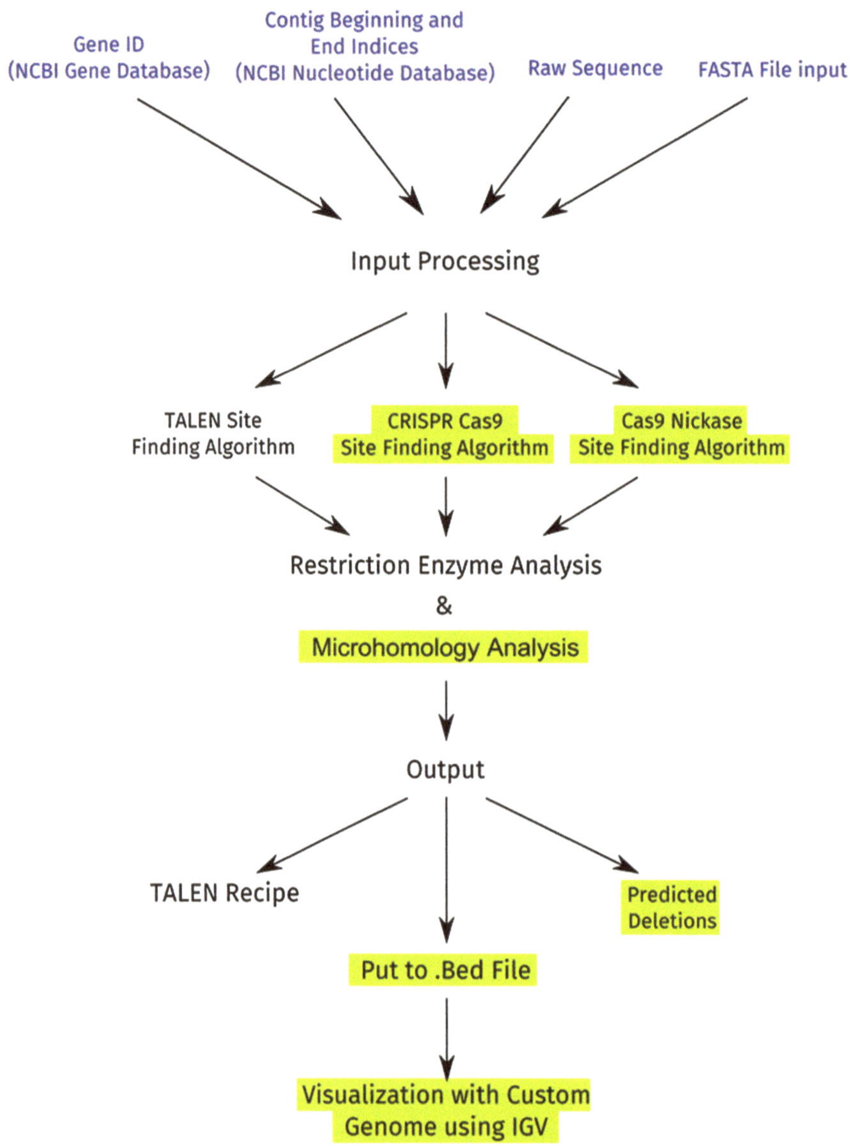

Fig. 1 Flowchart of Mojo Hand 2.0. Input formats are in *blue*. Features new to Mojo Hand 2.0 are highlighted in *yellow*. Output is a report containing potential binding sequences, RVDs for TALENS, oligos for CRISPR/Cas9 nucleases or nickases, microhomology score, and out-of-frame scoring. The output can be further processed to create a .bed file, which can be loaded into other tools such as IGV, generate customized recipes for each TALEN, or analyzed for predicted deletions

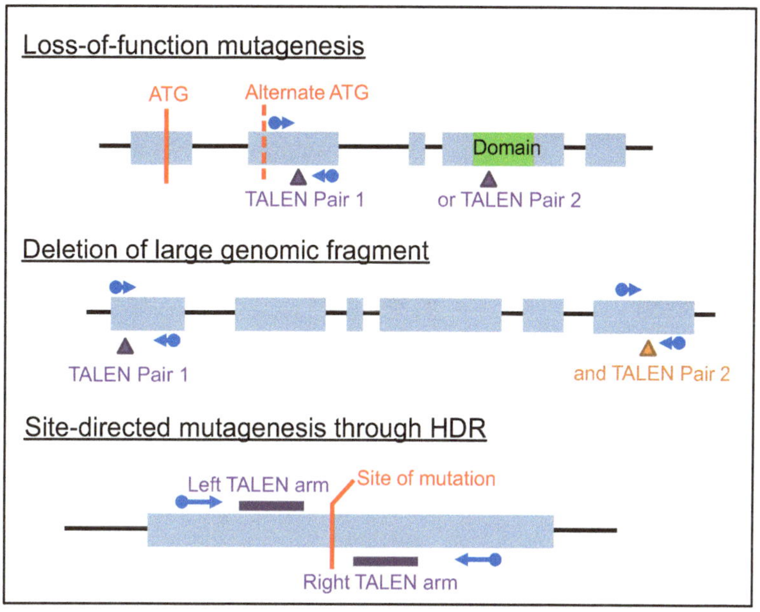

Fig. 2 Typical genomic region for TALEN targeting in different types of mutagenesis. Either TALEN pair 1 or 2 can be used in case of loss-of-function mutagenesis, and TALEN pairs 1 and 2 are used together for deletion of large genomic fragment. *Blue arrow* indicated primer pairs for RFLP or PCR screening of mutagenesis

(a) Length of TAL-binding domain 15 RVDs

(b) Spacer length between 14 and 18 bp

(c) Unique restriction site within the spacer for RFLP assay of NHEJ-mediated mutagenesis (optional for large genomic fragment deletion using two pairs of TAELN; *see* **Note 7**)

(d) T nucleotide upstream of both TAL-binding domains

4. Restriction enzyme analysis.
5. Mojo Hand output.
6. Select TALEN design with desired microhomology score above or out-of-frame score if predictable deletion through MMEJ is desirable (*see* **Note 8**).
7. Generate BED file to be used in conjunction with Integrated Genomics Viewer (IGV) (*see* **Note 9**).

3.2 Genotyping Targeted Genomic Locus

1. Design primers to amplify the targeted locus (*see* **Note 10**).
2. Extract genomic DNA from zebrafish embryos of the targeted fish line (*see* **Note 11**):

(a) To prepare 1 mL working extraction buffer, freshly add 30 μL 10% Tween-20, 30 μL 10% NP-40, and 10 μL proteinase K to 950 μL genomic DNA extraction buffer.

(b) Transfer embryos to centrifuge tube and remove excess embryo water.

   (c) Add working extraction buffer (50 μL per embryo).
   (d) Incubate at 55 °C with shaking ≥ 4 h.
   (e) Incubate at 98 °C for 10 min to inactivate proteinase K.
   (f) Store genomic DNA at −20 °C until PCR.
   (g) Typically, 5 μL of genomic DNA solution is used in 25 μL PCR.
3. PCR amplify the target locus.
4. Test RFLP assay:
   (a) PCR with RFLP assay primers (*see* **Note 10**, Fig. 2).
   (b) Digest 10 μL PCR product with appropriate restriction enzyme.
   (c) Resolve digested product on 1.5 % agarose gel.
   (d) PCR product should be completely digested into two correctly sized bands.
5. Confirm sequence of the targeted locus by Sanger sequencing, identify any polymorphic region affecting TALEN-binding sites, and redesign TALEN if necessary.

3.3 Design Short Single-Stranded Donor Oligo

1. Design donor oligo with the following parameters:
   (a) Around 50 base pairs in length
   (b) Mutated nucleotide(s) in the middle part of the oligo
   (c) Unique restriction site added in the middle of the oligo by introducing silent mutations to allow easy screening of donor incorporation with RFLP assay

3.4 TALEN Assembling with FusX System (3 Days)

3.4.1 Day 1

1. Break down the 15-RVD TALE domain from 5′ to 3′ into six building blocks from different libraries of the FusX kit according to the formula 3 (pFusX-1) + 3 (pFusX-2) + 3(pFusX-3) + 3 (pFusX-4) + 2 (pFus_B2) + 1 (pLR) (Fig. 3).

   For example,

   a TALEN arm with the following targeting sequence: 5′-ATTGACTTCAGAGAG-3′.

   Corresponding RVD sequences: NI NG NG NN NI HD NG NG HD NI NN NI NN NI NN.

   List of building blocks required for each TAL:

Library	RVD sequence
pFusX-1	NI NG NG
pFusX-2	NN NI HD
pFusX-3	NG NG HD
pFusX-4	NI NN NI
pFus_B2	NN NI
pLR	NN

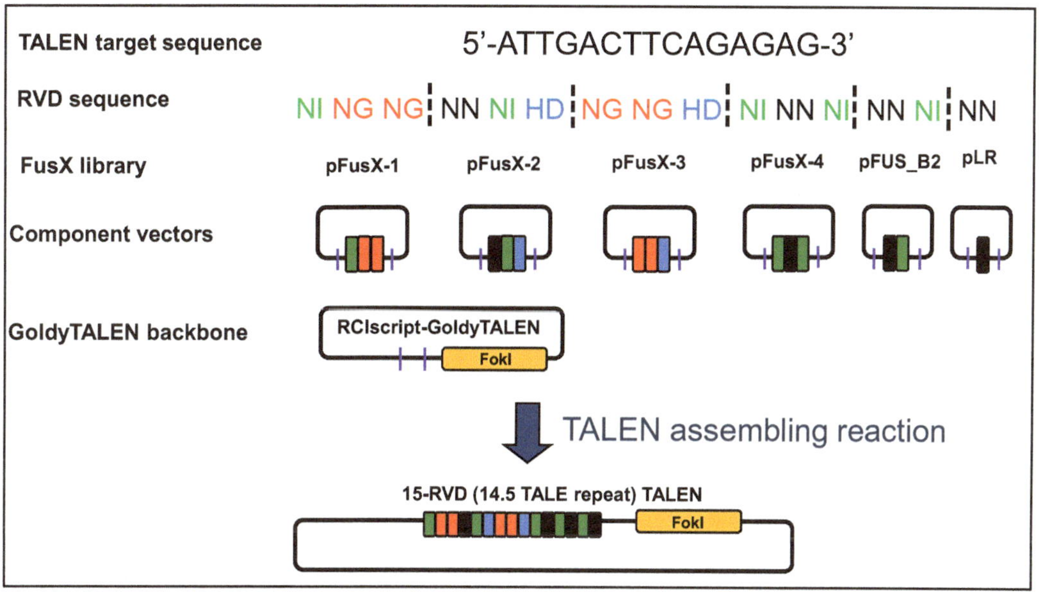

Fig. 3 Picking corresponding component vectors from FusX libraries to assembly 15-RVD GoldyTALEN

2. Mix 25–50 ng of each vector in a PCR tube with 50 ng RCIscript-GoldyTALEN backbone (*see* **Note 12**).
3. (Optional) Add to each reaction 1 μL 10× NEBuffer 3.1 and 0.5 μL BsmBI, and make up to 10 μL with deionized water (*see* **Note 2**).
4. (Optional) Incubate at 55 °C for 30 min (*see* **Note 2**).
5. Add to each reaction 1.5 μL 10× T4 DNA ligase reaction buffer, 0.5 μL T4 DNA ligase, and 0.5 μL Esp3I, and make up to 15 μL with deionized water.
6. Run the following program in thermocycler:
   (a) 37 °C, 5 min, and 16 °C, 10 min → ten cycles
   (b) 37 °C, 15 min
   (c) 80 °C, 5 min
   (d) 4 °C forever
7. Transform 3–5 μl of the reaction product, and plate ~1/5 of the recovered transformants on LB agar plate with ampicillin, 40 μL X-Gal (20 mg/mL), and 40 μL 0.1 M IPTG.
8. Incubate LB agar plate at 37 °C overnight.

3.4.2 Day 2

1. Pick 2–4 white colonies for colony PCR with primers TAL_F1 and TAL_R2.
2. PCR with the following program (*see* **Note 13**):
   (a) 95 °C, 10 min
   (b) 95 °C, 30 s; 55 °C, 30 s; and 72 °C, 3 min → 30 cycles

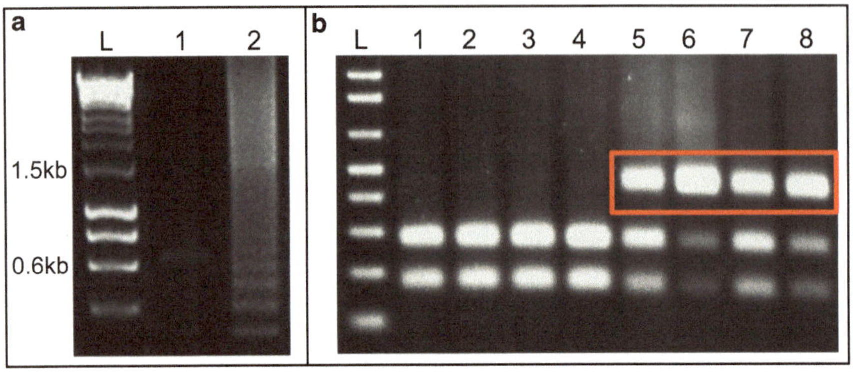

Fig. 4 (**a**) Typical colony PCR result after TALEN assembly. *Lane 1* is a negative clone with empty GoldyTALEN backbone showing a ~0.65 kb band, and *lane 2* is a positive TALEN clone showing the laddering effect with a band at ~1.5 kb. (**b**) Typical RFLP assay result of single embryos. *Lanes 1–4* are uninjected control with completely digested PCR product, and *lanes 5–8* are embryos injected with TALEN showing undigested products (*red box*). *L* ladder

   (c) 72 °C, 5 min
   (d) 4 °C forever
3. Resolve PCR product in 1% agarose gel and identify positive clones (Fig. 4a).
4. Culture positive colonies overnight 37 °C in LB with ampicillin.

3.4.3 Day 3

1. Mini-prep overnight cultures of selected positive clones.
2. Verify assembled TALEN by Sanger sequencing with TAL_F1 and TAL_R2 (*see* **Note 3**).

3.5 Synthesizing TALEN-Encoding mRNA and Microinjection into One-Cell Zebrafish Embryos

1. Linearize TALEN-encoding plasmid with SacI.
2. Purify linearized plasmid by ethanol precipitation and quantify purified plasmids.
3. Set up in vitro transcription reaction with Ambion mMESSAGE mMACHINE® T3 Transcription Kit according to manufacturer's instruction (*see* **Note 14**).
4. Purify and quantify transcribed mRNA:
   (a) Add 50 μL LiCl precipitation solution to each transcription reaction.
   (b) Precipitate at −20 °C ≥ 1 h.
   (c) Centrifuge at 4 °C, 12,000 × g, for 15 min.
   (d) Remove supernatant and wash with 70% ethanol.
   (e) Centrifuge at 4 °C, 12,000 × g, for 5 min.
   (f) Remove supernatant and air-dry pellet.
   (g) Resuspend pellet in 50 μL deionized water and quantify mRNA.

5. Make working stock for microinjection by mixing and diluting both mRNA encoding the TALEN pair (final concentration ~20 ng/μl of each TALEN mRNA, 20 pg × 2; *see* **Note 15**).
6. Microinject (20–100 pg each TAELN arm; *see* **Note 16**) into the yolk of one-cell embryos.

3.6 Examine Somatic TALEN Activity by RFLP Assay or PCR to Detect a Large Deletion

1. Extract genomic DNA from control (uninjected) and TALEN-injected embryos (*see* **Note 17**) as described in Subheading 3.2.
2. PCR amplify the target locus.
3. Digest 10 μL PCR product with appropriate restriction enzyme, and resolve digested product on 1.5 % agarose gel (Fig. 4b).
4. To detect a large deletion generated by two TALEN pairs, extract genomic DNA from control (uninjected) and TALEN-injected embryos (*see* **Note 18**) as described in Subheading 3.2.
5. PCR amplifies the target locus with appropriate primers (*see* **Note 18**, Fig. 2), and resolve PCR product on agarose gel.

3.7 Screening of Germline Transmission for Stable Mutants

For loss-of-function mutagenesis using a single TALEN pair, germline transmission efficiency correlated with TALEN mutagenic activity. Usually founder fish will be identified within screening of ten injected fishes when working with a moderately active TALEN (~60 % mutagenic activity in RFLP assay). In large deletion with two TALEN pairs, efficiency is typically two- to fivefold lower, also depending on the activity of TALEN pairs. In the case of site-directed mutagenesis through HDR, efficiency will be ~100-fold lower, and a much larger number of injected fish will have to be screened.

1. Raise potential batches of injected embryos (siblings showing expected somatic mutation).
2. Genotype juvenile fishes (around 4–6 weeks old) by tail fin biopsy (*see* **Note 19**).
3. Extract genomic DNA from fin tissue following Subheading 3.2, and screen with RFLP or PCR assay for maintenance of induced as described in Subheading 3.6.
4. Raise juveniles with stable somatic mutations to sexually mature and outcross with wild type to obtain F1 embryos.
5. Extract genomic DNA from individual F1 embryos following Subheading 3.2, and genotype with RFLP or PCR assay.
6. Raise potential batches of F1 embryos (siblings showing heterozygous mutation).
7. Genotype juvenile F1 as described in **steps 2** and **3**.
8. Confirm mutation carried in F1 by Sanger sequencing (*see* **Note 20**).

4 Notes

1. Any PCR reagents could be used and ready-to-use PCR master mix will be efficient in high-throughput screening.
2. BsmBI and Esp3I are isoschizomers that have different optimum reaction temperature (55 °C and 37 °C, respectively). While it is not recommended to use in cycling reactions with T4 DNA ligase, optional predigestion with BsmBI at 55 °C will significantly enhance the efficiency of TALEN assembly, reducing the number of blue colonies.
3. For TALENs with 15 RVDs, Sanger sequencing with TAL_F1 and TAL_R2 will typically cover all 14.5 repeat units. In case units are unread in sequencing, RVD-MM-F and RVD-MM-R primers, with sequences specific to RVD-8, can be used.
4. Mojo Hand is available as a web service at www.talendesign.org. The site allows access to the program without the trouble of installation and with the ease of a familiar interface. Point-of-use help is available for each field. The source code and spreadsheet are also available for noncommercial use with applicable license.
5. For loss-of-function mutations, TALENs should be designed against early conserved exons after the start codon (and alternate start codon) or important functional domain(s) such that small indels will be introduced through NHEJ and resulted in frame-shifting/premature termination. For deletion of a large genomic fragment with two pairs of TALENs, simply design two pairs of TALENs flanking the genomic fragment to be deleted. For site-directed mutagenesis through HDR, TALENs should target the site to be mutated.
6. Templates for each system can be changed to user specifications. Notation for templates has been slightly changed from "." representing a non-preferential base to "N" representing any base. The default template for TALENs remains TsN*e, which constrains TAL-binding sites to an initial 5′ T bp.
7. For deletion of a large genomic fragment with two pairs of TALENs, unique restriction site in spacer for RFLP assay is not necessary since deletion can be simply detected by PCR (*see* **Note 5**). However, inclusion of restriction site in the design of both TALEN pairs is recommended such that the activity of each TALEN pair can be confirmed with RFLP assay before co-injection.
8. Microhomology-mediated end joining (MMEJ) is a Ku- and ligase IV-independent DNA repair mechanism that utilizes regions of microhomology adjacent to the site of DSB.

Because in-frame deletions can sometimes lower the efficiency of loss-of-function mutagenesis, we integrated an algorithm developed by Bae et al. [25] into Mojo Hand that calculates a microhomology score and an out-of-frame score for each binding site. The microhomology score is an aggregate of each pattern score associated with each microhomology between two and eight bases long, and the pattern score is calculated based on the length of the microhomology and deletion. Higher microhomology scores correspond with binding sites with stronger microhomologies. Out-of-frame score is the percentage of microhomology score from frameshifting microhomologies for each binding site. Predicted deletions give a list of all homologies within a binding site, with their sequences, deletion lengths, and pattern scores, and whether or not they cause frameshifts. Higher pattern scores correlate with a higher chance of any particular deletion occurring due to microhomology-mediated end joining. This prediction does not take into account deletions that occur due to NHEJ.

9. Integrated Genomics Viewer (IGV) is a tool that allows users to visualize their own genomic datasets and load tracks and other features in a variety of formats. We utilized the BED file format to store user designs for site-specific nucleases, which can then be loaded as a searchable feature within the track line of IGV. This allows users to visualize potential TALEN candidates in tandem with their own in-house next-generation sequencing datasets in an efficient and intuitive manner. BLAT search maps each potential binding site across the genome, which allows users to visualize and avoid designs that are not unique. In addition this function can be used to avoid designs that bind within polymorphic stretches of the genome that may negatively impact cutting efficiency. BED files are created by using the BLAT tool [27] to map binding sites and restriction enzymes to a genome specified by the user. Current genomes supported by Mojo Hand include *D. rerio* and *C. elegans* due to current hosting limitations. A detailed specification of BED file format is available at http://genome.ucsc.edu/FAQ/FAQformat.html#format1.

10. Although there is no restriction on primer design for initial genotyping purposes, primer pair can be designed such that they could also be used for RFLP assay. Typically, primers with amplicon size around 300–500 base pairs work well for RFLP assay. Avoid having the unique restriction site for RFLP assay in the middle of the amplicon, which, otherwise, would give two similar-sized digestion products difficult to be resolved in electrophoresis.

11. To identify potential polymorphic region, genomic DNA can be extracted different batches of non-sibling embryos.
12. Assembling reaction works well even if component vectors varied in amounts within range. Equal volume of each vector could be mixed to simplify reaction setup even if their concentrations are different.
13. PCR cycle can be further optimized based on the PCR reagent used.
14. An initial 10 μL half in vitro transcription reaction resuspended in 25 μL final volume will typically yield mRNA at concentration around 500–1000 ng/μL, which is more than enough in most applications.
15. Working mRNA solution should be stored in small aliquots and avoid repeated freeze-thaw.
16. It is recommended to conduct dose–response trials within the range from 20 to 100 pg per TALEN arm such that the optimum dose can be chosen which resulted in survival of around 50 % of normally developed embryos.
17. Genomic DNA could be extracted from single embryo to examine mutagenic activity in individual embryo or from a group of five or ten embryos to assay the average mutagenic activity of the TALEN.
18. For screening large genomic deletion, forward primer used to genotype TALEN pair 1 and reverse primer used TALEN pair 2 can be used together to screen for a large deletion resulting in a smaller-sized PCR product compared with the larger or absent PCR product in control. Reverse primer from pair 1 and forward primer from pair 2 can also be used together to screen for very rare "flipping" events where the targeted genomic fragment was excised but inversely inserted back into the genomic lesion. Since the PCR screening is only qualitative and does not reflect mutagenic activity, genomic DNA can be extracted from a single embryo instead of a group of embryos.
19. This round of fin biopsy is optional. However, prescreening for stable somatic mutation can significantly increase the percentage of founder in the pool. Therefore, it is recommended in examples of large fragment deletion and site-directed mutagenesis, where germline transmission efficiency is considerably lower.
20. F1 carrying desirable mutation will be selected. For example, small indels resulted in frameshifting or premature stop in case of loss-of-function mutagenesis and precisely incorporated donor sequence in site-directed mutagenesis.

Acknowledgments

We thank all of the great user comments for refining this genome engineering design system. This project is supported by the State of Minnesota (University of Minnesota/Mayo Clinic Gene Targeting Partnership grant H001274506-3 to SCE), the National Institutes of Health grants GM63904 and P30DK084567 to SCE, the Mayo Foundation, the Health and Medical Research Fund (HMRF02132326) to ACM, and the HKU Seed Funding for Basic Research (201401159004, 201411159098) to ACM.

References

1. Bai J, Choi SH, Ponciano G et al (2000) Xanthomonas oryzae pv. oryzae avirulence genes contribute differently and specifically to pathogen aggressiveness. Mol Plant Microbe Interact 13:1322–1329
2. Yang B, White FF (2004) Diverse members of the AvrBs3/PthA family of type III effectors are major virulence determinants in bacterial blight disease of rice. Mol Plant Microbe Interact 17:1192–1200
3. Boch J, Scholze H, Schornack S et al (2009) Breaking the code of DNA binding specificity of TAL-type III effectors. Science 326:1509–1512
4. Moscou MJ, Bogdanove AJ (2009) A simple cipher governs DNA recognition by TAL effectors. Science 326:1501
5. Christian ML, Demorest ZL, Starker CG et al (2012) Targeting G with TAL effectors: a comparison of activities of TALENs constructed with NN and NK repeat variable di-residues. PLoS One 7:e45383
6. Streubel J, Blu¨cher C, Landgraf A et al (2012) TAL effector RVD specificities and efficiencies. Nat Biotechnol 30:593–595
7. Miller JC, Tan S, Qiao G et al (2011) A TALE nuclease architecture for efficient genome editing. Nat Biotechnol 29:143–148
8. Bedell VM, Wang Y, Campbell JM et al (2012) In vivo genome editing using a high-efficiency TALEN system. Nature 491:114–118
9. Zhang F, Cong L, Lodato S et al (2011) Efficient construction of sequence-specific TAL effectors for modulating mammalian transcription. Nat Biotechnol 29:149–153
10. Cong L, Zhou R, Kuo YC et al (2012) Comprehensive interrogation of natural TALE DNA-binding modules and transcriptional repressor domains. Nat Commun 3:968
11. Crocker J, Stern DL (2013) TALE-mediated modulation of transcriptional enhancers in vivo. Nat Methods 10:762–767
12. Maeder ML, Angstman JF, Richardson ME et al (2013) Targeted DNA demethylation and activation of endogenous genes using programmable TALE-TET1 fusion proteins. Nat Biotechnol 31:1137–1142
13. Mendenhall EM, Williamson KE, Reyon D et al (2013) Locus-specific editing of histone modifications at endogenous enhancers. Nat Biotechnol 31:1133–1136
14. Thanisch K, Schneider K, Morbitzer R et al (2014) Targeting and tracing of specific DNA sequences with dTALEs in living cells. Nucleic Acids Res 42:e38
15. Hockemeyer D, Wang H, Kiani S et al (2011) Genetic engineering of human pluripotent cells using TALE nucleases. Nat Biotechnol 29:731–733
16. Tesson L, Usal C, Ménoret S et al (2011) Knockout rats generated by embryo microinjection of TALENs. Nat Biotechnol 29:695–696
17. Carlson DF, Tan W, Lillico SG et al (2012) Efficient TALEN-mediated gene knockout in livestock. Proc Natl Acad Sci U S A 109:17382–17387
18. Liu J, Li C, Yu Z et al (2012) Efficient and specific modifications of the Drosophila genome by means of an easy TALEN strategy. J Genet Genomics 39:209–215
19. Zhang Y, Zhang F, Li X et al (2013) Transcription activator-like effector nucleases enable efficient plant genome engineering. Plant Physiol 161:20–27
20. Ma S, Zhang S, Wang F et al (2012) Highly efficient and specific genome editing in silkworm using custom TALENs. PLoS One 7:e45035

21. Sung YH, Baek IJ, Kim DH et al (2013) Knockout mice created by TALEN-mediated gene targeting. Nat Biotechnol 31:23–24
22. Ma AC, Lee HB, Clark KJ et al (2013) High efficiency in vivo genome engineering with a simplified 15-RVD GoldyTALEN design. PLoS One 8:e65259
23. Cermak T, Doyle EL, Christian M et al (2011) Efficient design and assembly of custom TALEN and other TAL effector-based constructs for DNA targeting. Nucleic Acids Res 39:e82
24. Peng Y, Clark KJ, Campbell JM, Panetta MR, Guo Y, Ekker SC (2014) Making designer mutants in model organisms. Development 141:4042–4054
25. Bae S, Kweon J, Kim SK et al (2014) Microhomology-based choice of Cas9 nuclease target sites. Nat Methods 11:705–706
26. Neff KL, Argue DP, Ma AC et al (2013) Mojo hand, a TALEN design tool for genome editing applications. BMC Bioinformatics 14:1
27. Kent WJ (2002) BLAT - the BLAST-like alignment tool. Genome Res 12:656–664

Chapter 3

Homology-Independent Integration of Plasmid DNA into the Zebrafish Genome

Thomas O. Auer and Filippo Del Bene

Abstract

Targeting nucleases like zinc-finger nucleases (ZFNs), transcription activator-like effector nucleases (TALENs), and the clustered regularly interspaced short palindromic repeats/CRISPR-associated (CRISPR/Cas) system have revolutionized genome-editing possibilities in many model organisms. They allow the generation of loss-of-function alleles by the introduction of double-strand breaks at defined sites within genes, but also more sophisticated genome-editing approaches have become possible. These include the integration of donor plasmid DNA into the genome by homology-independent repair mechanisms after CRISPR/Cas9-mediated cleavage. Here we present a protocol outlining the most important steps to target a genomic site and to integrate a donor plasmid at this defined locus.

Key words Zebrafish, CRISPR/Cas, Targeted transgene integration, Genome modification, Genome editing, Genome engineering, Site-specific nuclease, Homology-independent repair

1 Introduction

Zebrafish (*Danio rerio*) is becoming a more and more widely used and increasingly powerful model system for many fields of modern biomedical research including vertebrate development, small molecule screening for drug discovery, disease modeling, and functional neuroscience. Its transparency and extrauterine development allow sophisticated in vivo imaging approaches, and its relative short generation time and high number of progeny are only some of the reasons that constitute the success of this model organism. Genetic studies provide insights into vertebrate gene function and help to increase our understanding of basic mechanisms of normal development and human disease.

In the early stages of zebrafish research, genetic methods were limited to the analysis of mutant phenotypes originating from forward genetic mutagenesis screens using gamma irradiation, viral infections, and chemical mutagens [1–6]. These screens lead to the

Koichi Kawakami et al. (eds.), *Zebrafish: Methods and Protocols*, Methods in Molecular Biology, vol. 1451,
DOI 10.1007/978-1-4939-3771-4_3, © Springer Science+Business Media New York 2016

identification of a broad spectrum of various mutant phenotypes [7, 8], and the subsequent positional cloning of the causal genomic loci delivered precious insights into vertebrate gene function.

1.1 Gene-Trap Screens

Following the introduction of the transposase Tol2 as efficient transgenesis tool in zebrafish [9], a series of Tol2-mediated gene-trap screens were performed combining mutagenesis with a visual readout of target gene expression [10–12]. A typical gene-trap vector thereby contains an n-terminal splice acceptor site followed by a reporter gene—flanked by Tol2 sites for efficient transgenesis. After random insertion of the gene-trap vector into the zebrafish genome, intronic insertions lead to the disruption of the endogenous splicing product but instead produce of a functional reporter in the endogenous expression domain. As no promoter nor start codon of the reporter gene is present in the gene-trap vector, exogenic insertions will result neither in transcription nor translation of the reporter.

While in earlier screens often the fluorescent protein eGFP was used as reporter, later versions built on the Gal4/UAS transactivation system allowing the combination of various UAS-transgenic lines with the Gal4 gene-trap allele [13, 14]. Depending on the exact location of the insertion (within an early or later intron of the target gene), the function of the target gene can be impaired or disrupted what allows the analysis of gene function. Since recently, more sophisticated gene-trap screens are in progress enabling the creation of conditional loss-of-function alleles by using the Cre/loxP or the FLp/FRT system [15–17], or other transgenesis methods are used to circumvent insertional biases of the Tol2 transposon system [18, 19].

1.2 Enhancer-Trap Screens

In parallel to gene-trap screens where the transgenesis vector interferes with gene function upon insertion, enhancer-trap approaches were developed that recapitulate the expression pattern of endogenous genes without disrupting their function [20–23]. While both rely on the efficient integration of the donor vector into the genome, enhancer-trap vectors do not require integration within intronic sequences. Upstream of a reporter open-reading frame with its own start codon, a minimal promoter is placed with a low basal expression level. After integration into a genomic locus, the surrounding *cis*-regulatory landscape at the target site influences the transcriptional activity of the basal promoter and can lead to activation of reporter gene expression. Functional readouts of the reporter are produced when integrated up- and downstream of open-reading frames, within intronic sequences, or even when integrated into exons. The latter though might also interfere with gene function.

Using enhancer-trap approaches, a vast repertoire of reporter lines labeling various cell types and anatomical structures have been created which are available to the zebrafish community [24, 25].

The main disadvantage of the two aforementioned trapping approaches though is the random nature of transgene integration. To label a specific cell type on demand remains an obstacle—with luck screening of a large repertoire of created lines might result in the identification of a suitable pattern. Furthermore, it is not possible to target a specific gene of interest, and gene-trap insertions for a great part of the zebrafish genome are still missing. While the design of bacterial artificial chromosome (BAC) reporters, containing large stretches of the zebrafish genome, delivered a solution to the first problem [26–28], targeted genome editing remained still a hindrance until recently.

1.3 Genome-Targeting Methods

This luckily changed with the development and successful application of zinc-finger nucleases (ZFNs) [29, 30] and was further facilitated by the use of transcription activator-like effector nucleases (TALENs) [31, 32] for genome editing in zebrafish. Both systems direct the nuclease activity of the proteolytic domain of a FokI endonuclease to a genomic target site and thereby trigger the introduction of double-strand breaks (DSBs) at a defined genomic locus. This DSB is thereafter detected and repaired by the error-prone nonhomologous end joining (NHEJ) DNA repair pathway resulting often in the insertion or deletion (indels) of base pairs at the target site. Depending on the kind of indels, frameshift and premature stop codon causing alleles, sequence modifications resulting in single amino acid substitutions, or hypomorphic alleles caused by deletions can be selected.

Various studies confirmed the transmission of alleles created with ZFNs or TALENs to the next generation confirming their applicability to generate stable loss-of-function alleles of any gene of interest in a relatively short period of time [31, 33–35].

With the introduction of the CRISPR/Cas (clustered regularly interspaced palindromic repeats/CRISPR-associated) system to zebrafish [36], this development was even accelerated. Not relying on protein-DNA interactions but using Watson-Crick base pairing between DNA and RNA as targeting principle, the design and preparation of genome targeting reagents can now be done in a couple of days. While the Cas enzyme delivers the necessary nuclease activity for DNA cleavage, a single-guide RNA (sgRNA) directs the nuclease to specific genomic locations. The sgRNA consists of a 5′ sequence complementary to the desired target site and a secondary structure at its 3′ end for interaction with the Cas nuclease. To customize the sgRNA and specifically target a genomic locus of interest, it is sufficient to replace around 20 bp responsible for target site binding with a sequence of choice. The transcribed, customized sgRNA binds to its complementary sequence in the genome and recruits the Cas nuclease that will consequently introduce a DSB at the genomic target site. A three-nucleotide spacer sequence called protospacer adjacent motif (PAM) located just outside the region of DNA-RNA hybridization in the target locus is required, and its composition varies based on the CRISPR/Cas system used. The most widely used Cas9 is efficiently working with NGG as PAM sequence and a guide RNA length of 20 nucleotides. Consequently, the typical target site for this system displays a N_{21}-GG consensus, and the nuclease cleaves 3 bp upstream of the PAM sequence. Microinjection of RNA encoding the *sgRNA* and *Cas9* mRNA leads to formation of a functional sgRNA-Cas9 complex, and cleavage at the complementary genomic target site is promoted. An especially valuable feature of the CRISPR/Cas9 system is the possibility to co-inject multiple sgRNAs and thus edit multiple target sites simultaneously [37].

As off-target cleavage or binding of this system has shown to happen rather frequently in various experimental systems [38–41], careful bioinformatical analysis of potential off-target sites and in vivo evaluation is recommended. By using zebrafish codon-optimized Cas9 versions and recombinant Cas9 protein [42, 43] or by adding of a second nuclear localization signal to the *Cas9* open-reading frame [37, 44], several optimizations have been performed even increasing the efficiency of this genome targeting system.

1.4 Genome Engineering

In addition to the generation of loss-of-function alleles, more sophisticated genome engineering methods offer the chance for detailed analysis of gene function [45]. In zebrafish this was achieved by homologous recombination (HR)-based [46–48] and homology-independent targeting approaches using TALENs and the CRISPR/Cas9 system [49, 50]. Both strategies rely on the efficient cleavage of an endogenous genomic target site by targeted nucleases. In the homologous recombination-based approach, a linearized double-stranded DNA vector with homology flanks to the genomic target site on both sides of a reporter gene cassette is co-injected with the required genome engineering reagents [46–48]. The homology flanks show perfect homology to the endogenous locus, and HR leads to the integration of the reporter gene at the target locus. In the homology-independent approach, the endogenous target site is integrated in a so-called "bait" sequence upstream of a functional reporter cassette in the donor plasmid (Fig. 1a). After injection into one-cell stage embryos, the nuclease activity not only leads to cleavage of the endogenous target site but also linearizes the co-injected donor vector (Fig. 1b). Homology-independent repair mechanisms thereafter lead to the integration of the donor plasmid into the endogenous locus. Indel mutations at the 5′ and 3′ junctions of the genomic target site and the donor

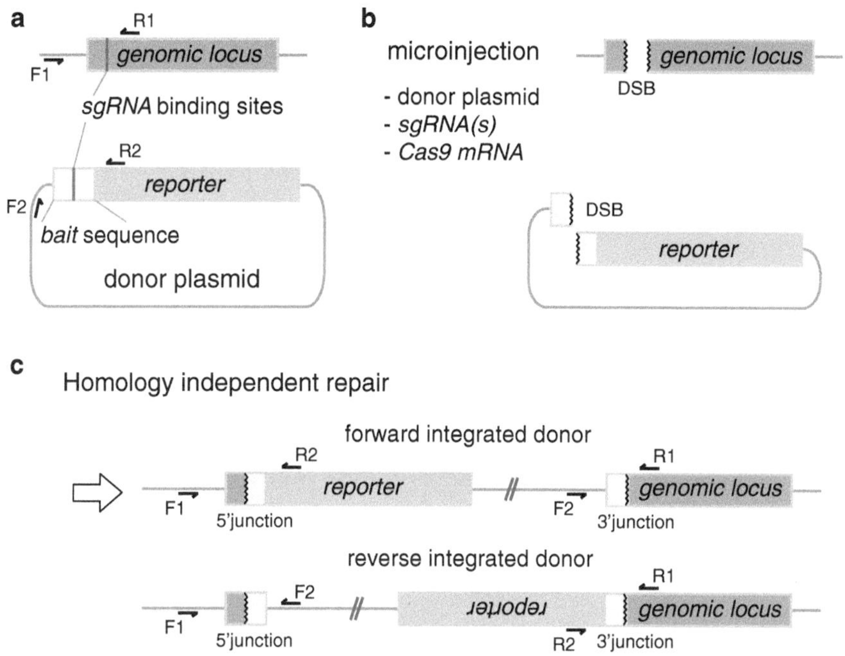

Fig. 1 Overview of the basic components and steps for integration of plasmid DNA at a defined genomic locus. (**a**) A specific single -guide RNA (sgRNA) is designed to induce double-strand breaks (DSBs) in the zebrafish genome at a desired target locus. This sgRNA-binding site is also placed upstream of a reporter cassette within the donor plasmid in the so-called *bait* sequence. The bait can be as small as one sgRNA-binding site (around 20 bp) or contain multiple sgRNA-binding sites. *Red bar*=sgRNA-binding site. *F1*=locus-specific forward primer. *R1*=locus-specific reverse primer. *F2*=donor plasmid-specific forward primer. *R2*=donor plasmid-specific reverse primer. (**b**) Upon microinjection of the donor plasmid together with the bait and genomic locus-specific *sgRNA* and *Cas9* mRNA into one-cell-stage embryos, DSBs are induced in the plasmid and at the genomic target site. *DSB* = double-strand break. (**c**) After concurrent cleavage of the donor plasmid and the genomic locus, homology-independent repair mechanisms promote the integration of the donor plasmid at the target locus. This integration is orientation independent and forward and reverse integration of the donor plasmid can happen. At the 5′ and 3′ junctions, insertions or deletions of additional base pairs can be observed. *F1*=locus-specific forward primer. *R1*=locus-specific reverse primer. *F2*=donor plasmid-specific forward primer. *R2*=donor plasmid-specific reverse primer

vector occur due to the imperfect nature of the NHEJ pathway. Besides, forward as well as reverse integration of the donor plasmid can happen (Fig. 1c).

Based on this homology-independent strategy, two different setups for locus-specific targeting have been developed (Fig. 2). In a gene-trap-like approach pioneered in our lab, we integrated a donor vector carrying a *Gal4* reporter cassette into the exonic sequence of *eGFP* transgenes or endogenous target genes [49, 51]. The successful activation of reporter gene expression thereby depends on the forward and in-frame integration of the donor vector at the target locus. At the same time, the function of the target gene is disrupted; the expression pattern of the reporter reflects the endogenous expression of the target gene (Fig. 2a).

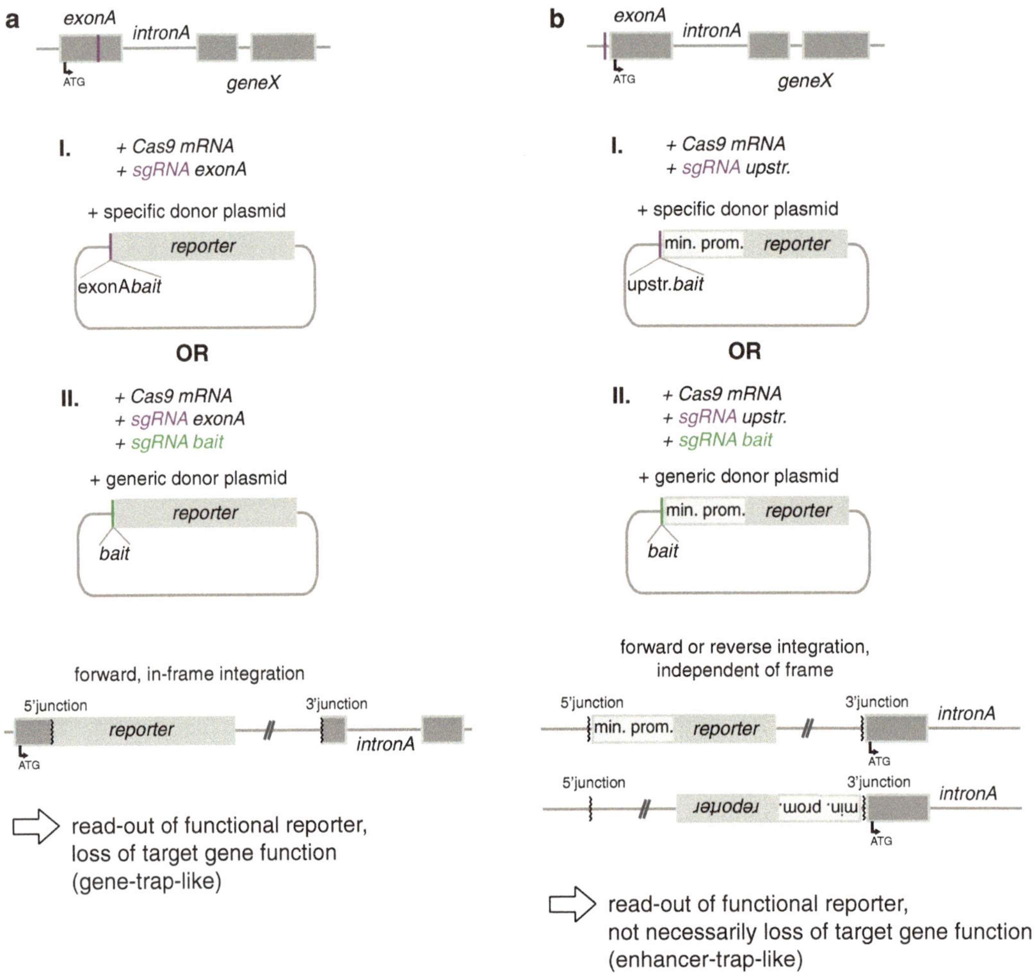

Fig. 2 Two possible strategies to generate reporter alleles by homology-independent repair. (**a**) Gene-trap-like targeting approach. The sgRNA-binding site is located within an early exon and 3′ to the start codon of the target gene. The donor plasmid is designed in a way that the activation of the reporter cassette downstream of the bait sequence depends on forward, in-frame integration of the donor plasmid at the target locus. The reporter gene does not contain an own start codon and translation depends on the endogenous start codon. The target gene function is disrupted using this approach. *Purple* and *green bar* = sgRNA-binding sites. *See* ref. 49 for details. (**b**) Enhancer-trap-like targeting approach. The sgRNA-binding site is located upstream of the transcription start site (200–600 bp). The donor plasmid contains an upstream bait sequence followed by a minimal promoter (min. prom.) and a reporter open-reading frame. Upon integration, the donor plasmid generates a functional readout independent of orientation and frame of integration. The function of the target gene is not necessarily disrupted. *See* ref. 50 for details. For both, the gene-trap- and the enhancer-trap-like approach, two different strategies have been established to introduce DSBs in the genome and the donor plasmid (compare (**a**) and (**b**)): (*I*) A single *sgRNA* is injected with a specific donor plasmid containing a locus-specific bait sequence. The same *sgRNA* directs Cas9 nuclease activity to the endogenous genomic locus and the donor plasmid. (*II*) Alternatively, a locus-specific *sgRNA* induces DSBs in the genome. A second *sgRNA* binding to the bait sequence of a generic donor vector is injected together with the corresponding generic donor plasmid and promotes its linearization. This strategy does not require the re-design of the donor plasmid as already established generic donor plasmids with their corresponding *sgRNAs* can be employed. *Purple* and *green bar* = sgRNA-binding sites

A follow-up study by Kimura et al. [50] expanded the use of homology-independent targeting in an enhancer-trap-like approach. While the bait sequence remains identical, the reporter cassette also carries a minimal promoter rendering expression of the *Gal4* reporter independent of frame and direction of insertion. By directing the location of insertion upstream of the transcription start site of a defined target gene, integration results in *Gal4* expression driven by the *cis*-regulatory input to the minimal promoter. The function of the target gene thereby most probably is not disrupted, but interference with its expression might be observed (Fig. 2b). As off-target integration of the donor vector is more probable to result in a functional readout of the donor vector in the second approach, enhancer-trap lines at different genomic locations might be generated as side product of injections [50].

For both strategies a donor vector can be designed that carries the sgRNA-binding site of the endogenous target gene in its bait sequence. As a consequence only one *sgRNA* has to be injected together with *Cas9* mRNA and the donor vector. Alternatively, a generic donor plasmid can be employed that carries an *eGFP*-specific sgRNA-binding site in its bait sequence. This requires the co-injection of an endogenous *sgRNA*, an *eGFP*-specific *sgRNA* together with *Cas9* mRNA, and the donor vector as circular DNA. While the endogenous *sgRNA* directs Cas9 activity to the genomic target site, the *eGFP*-specific *sgRNA* leads to linearization of the donor vector (Fig. 2).

1.5 Outline of the Procedure

The protocol we present here delivers a detailed description of how to insert a donor vector by homology-independent repair into a specific genomic target locus. As a first step, a locus-specific sgRNA has to be designed. After in vitro transcription of the *sgRNA* and *Cas9* mRNA, its efficiency in introducing site-specific DSBs is estimated. Based on the result, either a locus-specific donor vector is designed or a generic donor vector is chosen for insertion. With all three reagents—*sgRNA*, *Cas9* mRNA, and donor vector—ready, injections into UAS-transgenic embryos are performed. Next to fluorescent screening for successful integration events, PCR is used to confirm integration events. After raising potential founder fish to adulthood, germline-transmitted integration alleles are identified by outcrossing, fluorescent screening, and PCR analysis.

2 Materials

2.1 Zebrafish Strains Used in This Protocol

1. Zebrafish (*Danio rerio*) AB wild-type strain.
2. Zebrafish *Tg(UAS:RFP, cry:eGFP)* [13] (available from F.DB upon request) or any other *Tg(UAS:fluorescent protein)*-transgenic line suited to screen for integration of Gal4 donor plasmids.

2.2 Plasmids Used in This Protocol

1. *eGFP*bait containing donor plasmids with Gal4 reporter cassettes [49, 50].
2. sgRNA expression plasmid [36] (Addgene, DR274, Plasmid #42250) for cloning of locus-specific sgRNAs.
3. sgRNA expression plasmid [49] for expression of an *eGFP*-specific sgRNA.
4. Cas9 expression plasmid [36] (Addgene, JDS246, Plasmid #43861).

2.3 Kits Used in This Protocol

1. Gel extraction kit.
2. PCR purification kit.
3. QIAGEN Plasmid Midi kit.
4. mMessage mMachine T7 Ultra kit (Life Technologies).
5. MEGAscript T7 kit (Life Technologies).
6. Rneasy Mini Kit (Qiagen).
7. TOPO TA Cloning Kit, Dual Promoter (Life Technologies).
8. Miniprep kit.

2.4 Buffers and Other Reagents

1. 1× Danieau solution: 58 mM NaCl, 0.7 mM KCl, 0.4 mM $MgSO_4$, 0.6 mM $Ca(NO_3)_2$, and 5 mM HEPES; adjust pH to 7.6 using NaOH; filter sterilized. Store at 25 °C for up to 1 year.
2. Embryo water: 5 mM NaCl, 0.17 mM KCl, 0.33 mM $CaCl_2$, and 0.33 mM $MgSO_4$. Store at 25 °C for up to 2 months.
3. Ampicillin stock solution: Dissolve 1 g of ampicillin in 5 ml of H_2O and 5 ml of ethanol absolute. Stock can be stored at −20 °C for several years.
4. IPTG stock solution: Dissolve 1 g of IPTG in 42 ml of H_2O (final concentration: 100 mM). Prepare 1 ml aliquots which can be stored at −20 °C for up to 1 year.
5. X-Gal stock solution: Dissolve 1 g of X-Gal in 25 ml of 100% DMSO (final concentration: 40 mg/ml). Prepare 1 ml aliquots which can be stored at −20 °C, protected from light, for up to 1 year.
6. LB-kanamycin agar plate: Mix 40 g of LB agar and 1 g of agar in 1 l of H_2O. Autoclave. Add 1 ml of 50 mg/ml kanamycin stock solution and pour into 40 10-cm petri dishes. Store at 4 °C for up to 2 months.
7. LB-ampicillin agar plate: Mix 40 g of LB agar and 1 g of agar in 1 l of H_2O. Autoclave. Add 2 ml of 100 mg/ml ampicillin stock solution and pour into 40 10-cm petri dishes. Store at 4 °C for up to 2 months.
8. LB-ampicillin medium: Mix 25 g of LB broth into 1 l of H_2O. Autoclave. Add 1 ml of 100 mg/ml ampicillin stock

solution to 1 l of autoclaved LB broth. Store at 4 °C for up to 2 months.

9. LB-kanamycin medium: Mix 25 g of LB broth into 1 l of H_2O. Autoclave. Add 1 ml of 50 mg/ml kanamycin stock solution to 1 l of autoclaved LB broth. Store at 4 °C for up to 2 months.
10. TE buffer: 10 mM Tris–HCl pH 8 and 1 mM EDTA pH 8. Store at 25 °C for up to 1 year.
11. Genomic lysis buffer: 100 mM Tris–HCl pH 9.2, 100 mM NaCl, 50 mM EDTA pH 8, 0.5% SDS (wt/vol), and 200 mM sucrose. Store at 25 °C for up to 1 year.
12. DNA polymerase.
13. Nucleotide set.
14. Ethidium bromide.
15. Agarose.
16. Subcloning efficiency DH5a competent cells.
17. TOP10 chemically competent cells.
18. Ethanol absolute 99.9%.
19. Isopropanol.
20. Restriction enzymes: *DraI, BsaI, BbsI*, and *EcoRI.*
21. T7 endonuclease I.
22. UltraPure Phenol:Chloroform:Isoamyl Alcohol (25:24:1) (Life Technologies).
23. Proteinase K, recombinant.
24. Petri dishes 94 mm.
25. Borosilicate glass with filament (Sutter Instrument).

2.5 Equipment

1. Microwave oven.
2. Aquatic facility with 3-l and breeding tanks.
3. Thermocycler.
4. Incubator at 37 and 28.5 °C.
5. Biological shaker at 37 °C.
6. Benchtop centrifuge.
7. Nanovue (Dutscher) or any other suited spectrophotometer.
8. DNA electrophoresis system.
9. Fluorescent stereomicroscope equipped with suitable filters.
10. Thermomixer.
11. Electrophoresis generator.
12. Microinjection apparatus as previously described [52].

3 Methods

3.1 Design of Locus-Specific sgRNAs (See Note 1)

1. Binding sites range from 17 to 21 bases in length.
2. Include a 5′ NGG sequence for T7-mediated transcription.
3. Select for binding sites within an early exon or upstream of the transcriptional start site of the gene of interest depending on the targeting strategy (*see* **Notes 2–4**, Fig. 2, and Ref [49–51] for two published targeting strategies).
4. Order the oligonucleotides containing the locus-specific sgRNA sequence (*see* **Note 5**).
5. Clone the two oligonucleotides after annealing into the BsaI-digested sgRNA expression plasmid [36] for T7-mediated transcription (*see* **Note 6**).
6. Transform bacteria, using standard laboratory procedure, with the sgRNA expression plasmid containing the locus-specific sgRNA sequence.
7. Using a single colony, grow 3 ml of the bacteria containing the sgRNA expression vector, and purify the plasmid using a spin miniprep kit.

3.2 Preparation of Plasmid DNAs for In Vitro Transcription

1. Linearize about 5 μg of Cas9 and sgRNA expression vector DNAs with BbsI or DraI, respectively, at 37 °C overnight (*see* **Notes** 7 and **8**).
2. Run 1 μl of the digestion mixture on a 1 % agarose gel to check for complete digestion. If incomplete, redo the digestion, and check beforehand the expiration date and functionality of the employed restriction enzymes.
3. In case of complete digestion, purify the linearized DNAs using a PCR purification kit, and elute in a final volume of 15 μl. Measure the concentration of the eluted DNA with a spectrophotometer. To generate RNA for *sgRNAs*, continue with Subheading 3.3. To generate *Cas9* mRNA, continue with Subheading 3.4. Both syntheses can be performed in parallel.

3.3 Transcription of sgRNAs

1. Transcribe the *sgRNA* using the T7 MEGAscript kit (Life Technologies) as described in the manufacturer's instructions, and incubate the reaction mixture at 37 °C for 4 h.
2. Add 1 μl TURBO DNase (provided in the kit) to the reaction mixture and incubate at 37 °C for 15 min.
3. To recover the *sgRNA*, purify the reaction mix using the RNeasy Mini kit according to the manufacturer's RNA cleanup instructions. Elute two consecutive times with 30 μl each step. Measure the RNA concentration using a spectrophotometer. The concentration should be around 600 ng/μl (*see* **Notes 9** and **10**).

3.4 Transcription of Cas9 mRNA

1. Transcribe *Cas9* mRNA using the mMessage mMachine T7 Ultra kit. Use at least 2 μg linearized template as input in a total volume of 20 μl, and incubate the reaction mixture at 37 °C for 2 h (*see* **Note 11**).
2. Add 1 μl of TURBO DNase (provided in the kit) to the reaction mixture and incubate at 37 °C for 15 min.
3. Perform a poly(A) tailing reaction as described in the manufacturer's instructions, and incubate the reaction mix at 37 °C for 45 min.
4. Perform a lithium chloride precipitation and consecutive ethanol washes as described in the kit.
5. Resuspend the pellet in 60 μl of nuclease-free water.
6. Incubate the resuspended RNA sample at 65 °C for 10 min to dissolve the RNA. Measure the RNA concentration using a spectrophotometer. Typically the concentration should be around 1 μg/μl (*see* **Notes 12–14**).

3.5 Efficiency Estimation of sgRNA

1. Design primers flanking each endogenous sgRNA-binding site and amplifying around 500 bp of sequence (*see* **Note 1**; *see* Fig. 1a for primer locations, F1 and R1).
2. Set up mating couples of a wild-type zebrafish strain (we use AB) for spawning the evening before injection as described previously [53].
3. On the day of injection, prepare the microinjection setup as described previously [52]. Prepare an injection mix containing 150 ng/μl *Cas9* mRNA and around 7 ng/μl of *sgRNA* (*see* **Notes 15** and **16**).
4. Centrifuge the injection mixture at 14,000×*g* for 1 min at 4 °C. Load the mixture into the injection needle [52].
5. Inject about 1 nl of the solution into the cytoplasm of at least 100 one-cell-stage embryos. Retain about 50 embryos as an un-injected control group [26] (*see* **Note 17**).
6. Place injected and un-injected embryos at 28.5 °C. Remove dead embryos in the afternoon of injection and in the morning of the next day. If necessary transfer the embryos into a new dish, and change the embryo water to prevent bacterial growth and to maintain embryo health.
7. Select and pool 10–20 injected and 10–20 un-injected control embryos at 48 h post fertilization.
8. Isolate genomic DNA from the two pools of embryos as described previously [26, 54].
9. Perform PCR on the target genomic locus in both samples (*see* **Note 18**).

10. Run the PCR products on a 1 % agarose gel to check for the size of the PCR products.
11. Cut and purify the PCR products of the amplified locus using a gel extraction kit, and elute in a final volume of 15 μl.
12. Subclone the PCR products using the Topo® TA Cloning® kit, dual promoter, following the manufacturer's instructions (*see* **Note 19**).
13. Transform the entire TOPO reaction using chemically competent TOP10 cells according to the manufacturer's instructions, and plate them on an IPTG/X-Gal containing LB-kanamycin plate. Incubate the plate at 37 °C overnight.
14. On the next day, pick at least 20 white colonies from the transformation plate of the injected sample and two white colonies from the control plate. Inoculate each single colony into 2 ml of LB-kanamycin medium, and incubate at 37 °C overnight with agitation.
15. Isolate plasmid DNA from the overnight culture using a miniprep kit and perform an EcoRI test digest.
16. Run the digested miniprep plasmid DNAs on a 1 % agarose gel and check for the presence of the PCR product.
17. Analyze at least ten plasmids containing the PCR product of the injected sample and the two controls by Sanger sequencing using the Sp6 or T7 sequencing primer.
18. Align the sequencing reads of the controls to the injected sample. Calculate the somatic mutation rate of the *Cas9* mRNA/*sgRNA* at the target locus using the following formula (*see* **Note 20**):

Somatic mutation rate
= # clones with indel mutations / # clones with PCR insert

3.6 Design of the Donor Vector and Preparation of Donor Plasmid DNA (See Note 21)

1. To generate a locus-specific bait sequence, order primers that amplify the endogenous target site, and introduce flanking EcoRI restriction sites (*see* **Note 22**).
2. Perform a PCR on genomic DNA, and subclone the PCR amplicon after EcoRI digest into the *eGFPbait-E2A-KalTA4-pA*-containing, EcoRI-digested donor vector using standard protocols (*see* **Notes 23** and **24**).
3. Verify the sequence of the donor plasmid by Sanger sequencing.
4. Grow 100 ml of an overnight culture from a freshly streaked individual bacterial colony.
5. Prepare donor plasmid DNA using the QIAGEN Plasmid Midi kit, and resuspend the donor vector DNA in 100 μl nuclease-free water (*see* **Notes 25** and **26**).

3.7 Creation of Transgenic Zebrafish Embryos with Locus-Specific Integration

1. Set up mating couples of the *Tg(UAS:RFP, cry:eGFP)* [13]-transgenic line or any other *Tg(UAS:fluorescent protein)*-transgenic line for spawning the evening before injection as described previously [53] (*see* **Note 27**).
2. On the day of injection, prepare the microinjection setup as described previously [52] and an injection mixture containing 150 ng/μl *Cas9* mRNA, 7 ng/μl genomic locus- and bait-specific *sgRNA*, and 7 ng/μl donor plasmid DNA (*see* **Notes 15** and **28**).
3. Centrifuge the injection mixture at 14,000 × *g* for 1 min at 4 °C. Load the mixture into the injection needle [52].
4. Inject about 1 nl of the solution into the cytoplasm of at least 100 one-cell-stage embryos. Retain about 50 embryos as an un-injected control group [26] (*see* **Notes 16** and **17**).
5. Place injected and un-injected embryos at 28.5 °C. Remove dead embryos in the afternoon of injection and in the morning of the next day. If necessary transfer the embryos into a new dish, and change the embryo water to prevent bacterial growth and to maintain embryo health.

3.8 Fluorescent Screening and Confirmation of Locus-Specific Integration by PCR

1. Monitor the injected embryos daily under an epifluorescence binocular microscope equipped with a suitable filter set for RFP detection (or suitable to detect the fluorophore present in the UAS-transgenic line used for injection). RFP expression is expected to be observed in a mosaic manner in the expression domain of the endogenous target gene. The onset of RFP expression will be delayed compared to the onset of the endogenous target gene expression as the *UAS:RFP* transgene has to be transactivated (*see* **Notes 29–31**).
2. Screen the injected embryos for consecutive days, and select RFP-positive embryos for raising as described in ref. 26.
3. Take 3–5 RFP-positive embryos and 3–5 un-injected control embryos, and extract genomic DNA (*see* **Note 32**).
4. Design donor vector-specific primers that flank the bait sequence within the donor, and amplify around 500 bp (*see* **Note 1**, compare Fig. 1c for primer locations F2 and R2).
5. Perform three separate PCR reactions on the extracted genomic DNA of each embryo (un-injected controls and injected, RFP-positive selected) with three different primer pairs. The first pair, the two locus-specific primers F1 and R1, serves as positive control for successful DNA extraction. The second pair, the locus-specific forward primer F1 and the donor vector-specific reverse primer R2, amplifies the 5′ junction sites of the donor vector insertion site. The third pair, the locus-specific reverse primer R1 and the donor vector-specific forward primer F2, amplifies the 3′ junction of the donor vec-

tor insertion site (*see* Fig. 1c for primer localizations). The last two PCR reactions will only amplify a product in the case of locus-specific, forward integration of the donor vector. PCR is performed following standard laboratory protocols.

6. Run the PCR products on a 1% agarose gel. Cut out the bands migrating at the expected amplicon sizes, and extract the PCR product from the gel. No PCR amplicons should be detected for the last two primer combinations in the un-injected control samples.
7. Perform a TOPO cloning and bacterial transformation as described in Subheading 3.5 (*see* **Note 19**).
8. Pick five white colonies, isolate plasmid DNA, and analyze five subcloned PCR products of each junction site by Sanger sequencing.
9. Align the sequencing reads to the theoretically expected sequence after successful and precise integration of the donor plasmid into the target locus. This depends on the bait and sgRNA sequence used. Typically insertions and deletions of base pairs can be observed at the junction sites.
10. If integration events of the donor plasmid are confirmed, raise up to 100 injected embryos positively selected for donor vector integration by fluorescence to adulthood. We observed a positive correlation between degree of RFP-expressing cells and germline transmission rate—the more RFP-positive cells detected in F_0 embryos, the higher the chance for stable germline transmission of the integration event (*see* **Note 33**).

3.9 Screening for Germline Transmission in Injected Embryos by Fluorescence and PCR

1. When the injected fish reach sexual maturity (after 2–3 months of age), set up mating pairs between F_0 and *Tg(UAS:RFP, cry:eGFP)*-transgenic zebrafish.
2. Collect each clutch of F_1 embryos on the next day in a petri dish. Put the potential F_0 founder fish into a separate tank. Unambiguously label the potential founder fish and the corresponding clutch of embryos.
3. Grow the F_1 embryos at 28.5 °C, and monitor them for RFP expression in the expression domain of the endogenous target gene. If RFP-positive embryos are observed, keep the founder, and raise the embryos showing locus-specific integration of the donor vector.
4. Confirm the sequence at the 5′ junction site of fluorescent positive F_1 embryos by DNA extraction of single embryos and PCR and sequence analysis as described in Subheading 3.7 (*see* **Notes 34–36**).
5. Identify at least two different insertion alleles per integration site, and maintain transgenic fish lines with stable integration of the donor plasmid.

4 Notes

1. We recommend to amplify the genomic target locus and to verify its sequence before starting the procedure. Depending on the genetic background of the zebrafish strain used, polymorphisms at the sgRNA-binding site might occur that can reduce the sgRNA-binding efficiency. We recommend designing primers with free online software such as Primer3 (http://frodo.wi.mit.edu/primer3/). Ideally, primers should have a melting temperature of ~60 °C and a length of 25 bp and amplify about 1 kbp of genomic sequence.
2. To design gene-specific sgRNAs, a variety of web tools has been made available that helps to identify potential sgRNA-binding sites within the zebrafish genome. Possible sequence constraints are already implemented [36, 43, 55–57].
3. Many tools already include the NGG sequence required for T7-mediated transcription in their output sequences. Nevertheless, if desired, also other transcription methods can be chosen.
4. We recommend the ZiFit website for sgRNA design (http://zifit.partners.org/ZiFiT/ChoiceMenu.aspx), and we will describe in this protocol the procedure for T7-mediated transcription.
5. We recommend testing at least three different sgRNAs per target locus. The efficiency to induce insertions/deletions at the target locus should be at least 40%.
6. *See* ref. 36 for details.
7. We recommend to transcribe also a previously established *eGFP*-specific sgRNA [49]. It can either serve as a positive control for successful transcription and injection when injected into an *eGFP*-transgenic line together with *Cas9* mRNA. Alternatively it can be used in combination with generic donor plasmids that have been tested and integrated successfully in previous studies [49–51].
8. We strongly recommend not to use ready-to-load buffers for the linearization of the vector DNA as they might interfere with the subsequent RNA synthesis.
9. Be sure to use RNase-free microfuge tubes to avoid RNase contamination.
10. *sgRNAs* can be stored at −80 °C for several weeks; aliquot to avoid multiple freeze-thaw cycles.
11. In this protocol we use a *Cas9* version previously described in ref. 36. Various other versions of *Cas9* have been successfully used in zebrafish and should be compatible with the described method. These include zebrafish codon-optimized versions

[37, 44] that also carry additional nuclear localization signals for efficient translocation into the nucleus. Furthermore two studies show that recombinant Cas9 protein can be injected directly with sgRNAs into the one-cell-stage embryo [37, 43].

12. Be sure to use RNase-free microfuge tubes to avoid RNase contamination.
13. Cas9 mRNA can be stored at −80 °C for several weeks; aliquot to avoid multiple freeze-thaw cycles.
14. To check the integrity of the purified *sgRNA* (Subheading 3.3) and *Cas9* mRNA, take 2 μl aliquots, mix them with formaldehyde loading dye (provided in the mMessage mMachine T7 Ultra kit), and run them on a 1 % agarose gel. The transcribed RNA should appear as a distinct band without smearing.
15. All components for injection should be kept on ice, and the injection mixture should be prepared freshly each time.
16. 1 % phenol red solution can be added to the injection mixture to monitor the location and amount of the injected droplet.
17. Injections should be performed as soon as possible after fertilization into the cytoplasm of one-cell-stage embryos.
18. One possible way to check for introduced mutations at the target locus is a T7E1 endonuclease digest. For this, split each PCR reaction in two 25 μl aliquots, and perform the digest as described in ref. 51 with one aliquot of injected and un-injected PCR sample.
19. Use fresh PCR products for TOPO cloning as freezing and thawing or storage of the PCR product leads to reduced efficiencies. TOPO reactions require an A-overhang—make sure that the polymerase used for the PCR generates these overhangs.
20. To perform efficient integration of donor plasmids, we recommend using sgRNAs that show an efficiency of introducing indel mutations at the target locus between 60 and 100 %. If the sgRNAs targeting the endogenous locus are less efficient, we recommend to co-inject two *sgRNAs*—one targeting the endogenous locus and a second sgRNA specific for the bait sequence of a generic donor plasmid used previously [49–51].
21. The donor plasmid should contain an n-terminal bait sequence followed by a reporter cassette. Depending on the location of the sgRNA-binding site in the genome and the reporter cassette chosen, various designs are possible as discussed in the introduction. Here we describe the design of a donor vector targeting an exonic sequence with an exon-specific bait sequence.
22. For primer design *see* **Note 1**.

23. The donor vector used here contains an *eGFP*-specific bait sequence followed by an *E2A* sequence upstream of a *Gal4* ORF. The viral 2A sequence [58] allows the expression of two ORFs from the same mRNA. It prevents interference of the eGFP peptide (produced n-terminally after successful donor integration) with the activity of the Gal4 protein. 5′ of the *Gal4* ORF follows a poly-A signal for efficient translation. The plasmid backbone originates from the pCRII-TOPO vector (Invitrogen); for a plasmid map, *see* ref. 51.
24. Instead of PCR amplification of a bait sequence, it can also be annealed from two complementary oligonucleotides containing the sgRNA-binding sites flanked by the two EcoRI restriction sites.
25. For all donor plasmids, the DNA should be absolutely RNase-free and prepared freshly. Decreased integration frequencies have been observed when using DNA that was stored for more than 3 weeks or that has undergone multiple freeze-thaw cycles. An additional phenol-chloroform purification step (using standard protocols as described in ref. 59) can be performed to ensure a high purity of the donor vector.
26. The concentration of the midiprep DNA should be about 1.5 μg/μl.
27. We recommend the use of a line with strong RFP expression upon UAS activation to increase detection sensitivity.
28. If the locus-specific sgRNAs show a low efficiency, we recommend co-injection of the *eGFP*-specific *sgRNA* with the *eGFP*-bait containing generic donor vector.
29. To screen for successful integration events, the expression domain of the endogenous target gene has to be known. If it is unknown or cannot be found in any database like www.zfin.org, we recommend to perform in situ hybridization as described in ref. 60.
30. If another fluorescent protein than RFP is used as reporter, screen for the corresponding fluorophore. Be aware that *UAS:eGFP*-transgenic lines are not compatible with *eGFP*bait containing donor plasmids as the *eGFP*-specific *sgRNA* will also introduce indel mutations in the *UAS:eGFP* transgene what might obscure the readout of integration events.
31. We generally observe in a low amount (about 3%) of injected embryos ectopic expression of the *UAS* transgene in muscle and skin cells. We suggest not to select these embryos for further analysis as it most probably reflects artifactual expression from the donor plasmid.
32. Genomic DNA extraction on single embryos can be performed as described in ref. 26, 54.

33. If integration of the donor vector at the genomic target locus is confirmed by PCR in a subset of RFP-positive embryos, we assume that the remaining RFP-positive embryos will also show locus-specific integration events.
34. In the described gene-trap-like strategy, out-of-frame and reverse insertions of the donor plasmid will not result in reporter gene expression. To screen for these events, take a clutch of F_1 embryos showing no RFP expression, isolate genomic DNA from the pool of embryos, and check for donor plasmid integration by PCR using suitable primer combinations.
35. In some cases we detected concatemerization of the donor plasmid prior to integration at the target locus. This can be analyzed by PCR or Southern blotting analysis. For more details *see* ref. 49. As a consequence and depending on the nature of the concatemer, the PCR at the 3′ junction might not deliver an amplicon. Nevertheless, an in-frame forward integration of the donor plasmid at the 5′ junction might still result in an integration allele recapitulating the expression of the endogenous target gene.
36. Southern blotting might be employed to confirm the integration of the donor plasmid at the target locus. For more details *see* ref. 49, 51, 61.

Acknowledgments

A special thanks to J. P. Concordet, K. Duroure, and A. De Cian for helping with the development and initial establishment of the homology-independent targeting strategy. We would like to thank J. Wittbrodt for scientific discussion and support and members of the Del Bene lab for general discussion and comments. The Del Bene lab "Neural Circuits Development" is part of the Laboratoire d'Excellence (LabEx) entitled DEEP (ANR-11-LABX-0044) and the Ecole des Neurosciences de Paris. T.O.A. was supported by a Boehringer Ingelheim Fonds Ph.D. fellowship. This work has been supported by ATIP/AVENIR program starting grant (FDB), ERC-StG #311159 (FDB), CNRS, INSERM, and Institut Curie.

References

1. Walker C, Streisinger G (1983) Induction of mutations by gamma-rays in pregonial germ cells of zebrafish embryos. Genetics 103: 125–136
2. Chakrabarti S, Streisinger G, Singer F, Walker C (1983) Frequency of gamma-ray induced specific locus and recessive lethal mutations in mature germ cells of the zebrafish, BRACHYDANIO RERIO. Genetics 103: 109–123
3. Lin S, Gaiano N, Culp P, Burns JC, Friedmann T, Yee JK, Hopkins N (1994) Integration and

germ-line transmission of a pseudotyped retroviral vector in zebrafish. Science 265:666–669

4. Mullins MC, Hammerschmidt M, Haffter P, Nusslein-Volhard C (1994) Large-scale mutagenesis in the zebrafish: in search of genes controlling development in a vertebrate. Curr Biol 4:189–202
5. Solnica-Krezel L, Schier AF, Driever W (1994) Efficient recovery of ENU-induced mutations from the zebrafish germline. Genetics 136:1401–1420
6. Amsterdam A, Hopkins N (1999) Retrovirus-mediated insertional mutagenesis in zebrafish. Methods Cell Biol 60:87–98
7. Driever W, Solnica-Krezel L, Schier AF, Neuhauss SC, Malicki J, Stemple DL, Stainier DY, Zwartkruis F, Abdelilah S, Rangini Z et al (1996) A genetic screen for mutations affecting embryogenesis in zebrafish. Development 123:37–46
8. Haffter P, Granato M, Brand M, Mullins MC, Hammerschmidt M, Kane DA, Odenthal J, van Eeden FJ, Jiang YJ, Heisenberg CP et al (1996) The identification of genes with unique and essential functions in the development of the zebrafish, Danio rerio. Development 123:1–36
9. Kawakami K, Shima A, Kawakami N (2000) Identification of a functional transposase of the Tol2 element, an Ac-like element from the Japanese medaka fish, and its transposition in the zebrafish germ lineage. Proc Natl Acad Sci U S A 97:11403–11408
10. Kawakami K (2004) Transgenesis and gene trap methods in zebrafish by using the Tol2 transposable element. Methods Cell Biol 77:201–222
11. Kawakami K, Takeda H, Kawakami N, Kobayashi M, Matsuda N, Mishina M (2004) A transposon-mediated gene trap approach identifies developmentally regulated genes in zebrafish. Dev Cell 7:133–144
12. Balciunas D, Ekker SC (2005) Trapping fish genes with transposons. Zebrafish 1:335–341
13. Balciuniene J, Balciunas D (2013) Gene trapping using gal4 in zebrafish. J Vis Exp (79):e50113
14. Balciuniene J, Nagelberg D, Walsh KT, Camerota D, Georlette D, Biemar F, Bellipanni G, Balciunas D (2013) Efficient disruption of Zebrafish genes using a Gal4-containing gene trap. BMC Genomics 14:619
15. Maddison LA, Li M, Chen W (2014) Conditional gene-trap mutagenesis in zebrafish. Methods Mol Biol 1101:393–411
16. Maddison LA, Lu J, Chen W (2011) Generating conditional mutations in zebrafish using gene-trap mutagenesis. Methods Cell Biol 104:1–22
17. Ni TT, Lu J, Zhu M, Maddison LA, Boyd KL, Huskey L, Ju B, Hesselson D, Zhong TP, Page-McCaw PS et al (2012) Conditional control of gene function by an invertible gene trap in zebrafish. Proc Natl Acad Sci U S A 109:15389–15394
18. Song G, Li Q, Long Y, Gu Q, Hackett PB, Cui Z (2012) Effective gene trapping mediated by Sleeping Beauty transposon. PLoS One 7:e44123
19. Song G, Li Q, Long Y, Hackett PB, Cui Z (2012) Effective expression-independent gene trapping and mutagenesis mediated by Sleeping Beauty transposon. J Genet Genomics 39:503–520
20. Scott EK, Mason L, Arrenberg AB, Ziv L, Gosse NJ, Xiao T, Chi NC, Asakawa K, Kawakami K, Baier H (2007) Targeting neural circuitry in zebrafish using GAL4 enhancer trapping. Nat Methods 4:323–326
21. le Trinh A, Fraser SE (2013) Enhancer and gene traps for molecular imaging and genetic analysis in zebrafish. Develop Growth Differ 55:434–445
22. Balciunas D, Davidson AE, Sivasubbu S, Hermanson SB, Welle Z, Ekker SC (2004) Enhancer trapping in zebrafish using the Sleeping Beauty transposon. BMC Genomics 5:62
23. Parinov S, Kondrichin I, Korzh V, Emelyanov A (2004) Tol2 transposon-mediated enhancer trap to identify developmentally regulated zebrafish genes in vivo. Dev Dyn 231:449–459
24. Choo BG, Kondrichin I, Parinov S, Emelyanov A, Go W, Toh WC, Korzh V (2006) Zebrafish transgenic Enhancer TRAP line database (ZETRAP). BMC Dev Biol 6:5
25. Kondrychyn I, Teh C, Garcia-Lecea M, Guan Y, Kang A, Korzh V (2011) Zebrafish enhancer TRAP transgenic line database ZETRAP 2.0. Zebrafish 8:181–182
26. Suster ML, Abe G, Schouw A, Kawakami K (2011) Transposon-mediated BAC transgenesis in zebrafish. Nat Protoc 6:1998–2021
27. Bussmann J, Schulte-Merker S (2011) Rapid BAC selection for tol2-mediated transgenesis in zebrafish. Development 138:4327–4332
28. Suster ML, Sumiyama K, Kawakami K (2009) Transposon-mediated BAC transgenesis in zebrafish and mice. BMC Genomics 10:477
29. Meng X, Noyes MB, Zhu LJ, Lawson ND, Wolfe SA (2008) Targeted gene inactivation in zebrafish using engineered zinc-finger nucleases. Nat Biotechnol 26:695–701

30. Doyon Y, McCammon JM, Miller JC, Faraji F, Ngo C, Katibah GE, Amora R, Hocking TD, Zhang L, Rebar EJ et al (2008) Heritable targeted gene disruption in zebrafish using designed zinc-finger nucleases. Nat Biotechnol 26:702–708
31. Huang P, Xiao A, Zhou M, Zhu Z, Lin S, Zhang B (2011) Heritable gene targeting in zebrafish using customized TALENs. Nat Biotechnol 29:699–700
32. Sander JD, Cade L, Khayter C, Reyon D, Peterson RT, Joung JK, Yeh JR (2011) Targeted gene disruption in somatic zebrafish cells using engineered TALENs. Nat Biotechnol 29:697–698
33. Cade L, Reyon D, Hwang WY, Tsai SQ, Patel S, Khayter C, Joung JK, Sander JD, Peterson RT, Yeh JR (2012) Highly efficient generation of heritable zebrafish gene mutations using homo- and heterodimeric TALENs. Nucleic Acids Res 40:8001–8010
34. Dahlem TJ, Hoshijima K, Jurynec MJ, Gunther D, Starker CG, Locke AS, Weis AM, Voytas DF, Grunwald DJ (2012) Simple methods for generating and detecting locus-specific mutations induced with TALENs in the zebrafish genome. PLoS Genet 8:e1002861
35. Moore FE, Reyon D, Sander JD, Martinez SA, Blackburn JS, Khayter C, Ramirez CL, Joung JK, Langenau DM (2012) Improved somatic mutagenesis in zebrafish using transcription activator-like effector nucleases (TALENs). PLoS One 7:e37877
36. Hwang WY, Fu Y, Reyon D, Maeder ML, Tsai SQ, Sander JD, Peterson RT, Yeh J-RJ, Joung JK (2013) Efficient genome editing in zebrafish using a CRISPR-Cas system. Nat Biotechnol 31:227–229
37. Jao LE, Wente SR, Chen W (2013) Efficient multiplex biallelic zebrafish genome editing using a CRISPR nuclease system. Proc Natl Acad Sci U S A 110:13904–13909
38. Fu Y, Foden JA, Khayter C, Maeder ML, Reyon D, Joung JK, Sander JD (2013) High-frequency off-target mutagenesis induced by CRISPR-Cas nucleases in human cells. Nat Biotechnol 31:822–826
39. Hsu PD, Scott DA, Weinstein JA, Ran FA, Konermann S, Agarwala V, Li Y, Fine EJ, Wu X, Shalem O et al (2013) DNA targeting specificity of RNA-guided Cas9 nucleases. Nat Biotechnol 31:827–832
40. Mali P, Aach J, Stranges PB, Esvelt KM, Moosburner M, Kosuri S, Yang L, Church GM (2013) CAS9 transcriptional activators for target specificity screening and paired nickases for cooperative genome engineering. Nat Biotechnol 31:833–838
41. Pattanayak V, Lin S, Guilinger JP, Ma E, Doudna JA, Liu DR (2013) High-throughput profiling of off-target DNA cleavage reveals RNA-programmed Cas9 nuclease specificity. Nat Biotechnol 31:839–843
42. Sung YH, Kim JM, Kim HT, Lee J, Jeon J, Jin Y, Choi JH, Ban YH, Ha SJ, Kim CH et al (2014) Highly efficient gene knockout in mice and zebrafish with RNA-guided endonucleases. Genome Res 24(1):125–131
43. Gagnon JA, Valen E, Thyme SB, Huang P, Ahkmetova L, Pauli A, Montague TG, Zimmerman S, Richter C, Schier AF (2014) Efficient mutagenesis by Cas9 protein-mediated oligonucleotide insertion and large-scale assessment of single-guide RNAs. PLoS One 9:e98186
44. Liu D, Wang Z, Xiao A, Zhang Y, Li W, Zu Y, Yao S, Lin S, Zhang B (2014) Efficient gene targeting in zebrafish mediated by a zebrafish-codon-optimized cas9 and evaluation of off-targeting effect. J Genet Genomics 41:43–46
45. Auer TO, Del Bene F (2014) CRISPR/Cas9 and TALEN-mediated knock-in approaches in zebrafish. Methods 69(2):142–150
46. Shin J, Chen J, Solnica-Krezel L (2014) Efficient homologous recombination-mediated genome engineering in zebrafish using TALE nucleases. Development 141: 3807–3818
47. Zu Y, Tong X, Wang Z, Liu D, Pan R, Li Z, Hu Y, Luo Z, Huang P, Wu Q et al (2013) TALEN-mediated precise genome modification by homologous recombination in zebrafish. Nat Methods 10:329–331
48. Irion U, Krauss J, Nusslein-Volhard C (2014) Precise and efficient genome editing in zebrafish using the CRISPR/Cas9 system. Development 141:4827–4830
49. Auer TO, Duroure K, De Cian A, Concordet JP, Del Bene F (2014) Highly efficient CRISPR/Cas9-mediated knock-in in zebrafish by homology-independent DNA repair. Genome Res 24:142–153
50. Kimura Y, Hisano Y, Kawahara A, Higashijima S (2014) Efficient generation of knock-in transgenic zebrafish carrying reporter/driver genes by CRISPR/Cas9-mediated genome engineering. Sci Rep 4:6545
51. Auer TO, Duroure K, Concordet JP, Del Bene F (2014) CRISPR/Cas9-mediated conversion of eGFP- into Gal4-transgenic lines in zebrafish. Nat Protoc 9:2823–2840

52. Rembold M, Lahiri K, Foulkes NS, Wittbrodt J (2006) Transgenesis in fish: efficient selection of transgenic fish by co-injection with a fluorescent reporter construct. Nat Protoc 1:1133–1139
53. Westerfield M (ed) (2000) The zebrafish book. A guide for the laboratory use of zebrafish (Danio rerio), 4th edn. University of Orgeon Press, Oregon
54. Foley JE, Maeder ML, Pearlberg J, Joung JK, Peterson RT, Yeh JR (2009) Targeted mutagenesis in zebrafish using customized zinc-finger nucleases. Nat Protoc 4:1855–1867
55. Hwang WY, Fu Y, Reyon D, Maeder ML, Kaini P, Sander JD, Joung JK, Peterson RT, Yeh JR (2013) Heritable and precise zebrafish genome editing using a CRISPR-Cas system. PLoS One 8:e68708
56. Montague TG, Cruz JM, Gagnon JA, Church GM, Valen E (2014) CHOPCHOP: a CRISPR/Cas9 and TALEN web tool for genome editing. Nucleic Acids Res 42:W401–W407
57. Naito Y, Hino K, Bono H, Ui-Tei K (2014) CRISPRdirect: software for designing CRISPR/Cas guide RNA with reduced off-target sites. Bioinformatics 31(7):1120–1123
58. Szymczak AL, Workman CJ, Wang Y, Vignali KM, Dilioglou S, Vanin EF, Vignali DA (2004) Correction of multi-gene deficiency in vivo using a single 'self-cleaving' 2A peptide-based retroviral vector. Nat Biotechnol 22:589–594
59. Sambrook J, Russell DW (2006) Purification of nucleic acids by extraction with phenol:chloroform. CSH Protoc 2006
60. Di Donato V, Auer TO, Duroure K, Del Bene F (2013) Characterization of the calcium binding protein family in zebrafish. PLoS One 8:e53299
61. Abe G, Suster ML, Kawakami K (2011) Tol2-mediated transgenesis, gene trapping, enhancer trapping, and the Gal4-UAS system. Methods Cell Biol 104:23–49

Chapter 4

Detection of Multiple Genome Modifications Induced by the CRISPR/Cas9 System

Satoshi Ota and Atsuo Kawahara

Abstract

The recent remarkable innovation of an RNA-guided nuclease system, the clustered regularly interspaced short palindromic repeat (CRISPR)/CRISPR-associated (Cas) system, enables us the modification of specific genomic loci in various model animals including zebrafish. With this system, multiple guide RNAs simultaneously injected with the Cas9 nuclease into zebrafish embryos cause multiple genome modifications at different genomic loci with high efficiency; therefore, a simple method to detect individual mutations at distinct loci is desired. In this chapter, we describe a procedure for inducing multiple CRISPR/Cas9-mediated genome modifications in zebrafish and a convenient method to detect CRISPR/Cas9-induced insertion and/or deletion (indel) mutations using a heteroduplex mobility assay (HMA).

Key words Zebrafish, CRISPR/Cas9, Genome modifications, HMA, Multi-locus HMA

1 Introduction

Bacteria and archaea have evolved an adaptive immune system known as the clustered regularly interspaced short palindromic repeat (CRISPR)/CRISPR-associated (Cas) system that induces the degradation of foreign nucleic acids, such as viruses and plasmid DNA [1]. Recently, the type II CRISPR/Cas system was developed as a powerful tool for genome editing in mammalian cells [2, 3]. The CRISPR/Cas9 system consists of two components: a Cas9 nuclease and a guide RNA (gRNA) that contains a 20-base complimentary sequence for the targeted genomic locus [4]. The genomic target sequence must be followed by the protospacer adjacent motif (PAM), which is required for Cas9 nuclease cleavage activity [4] and varies among bacterial species [5, 6]. At present, *Streptococcus pyogenes* Cas9 is widely used for genome engineering, and its PAM sequence is NGG, where N is any nucleotide [6].

DNA double-strand breaks (DSBs) induced by the Cas9-gRNA complex can be repaired by nonhomologous end-joining (NHEJ) or homologous recombination (HR). NHEJ is an error-prone

Koichi Kawakami et al. (eds.), *Zebrafish: Methods and Protocols*, Methods in Molecular Biology, vol. 1451, DOI 10.1007/978-1-4939-3771-4_4, © Springer Science+Business Media New York 2016

repair system that often yields insertion and/or deletion (indel) mutations in targeted genomic loci, resulting in frameshift-mediated mutations in the coding regions of target genes [7]. When the DSBs are repaired in the presence of donor DNA containing homologous sequences, targeted knock-in mediated by HR can be achieved [7]. Because the gRNA is a small molecule and several gRNAs are easily injected along with the common Cas9 nuclease, the CRISPR/Cas system is suitable for multiplexed genome editing at different loci [8, 9]. In fact, we successfully induced multiple CRISPR/Cas9-mediated genome modifications at five distinct genomic targets [10].

After establishing target gene-disrupted animals (knockouts) using the CRISPR/Cas9 system, it is very important to determine the frequency and content of indel mutations induced at individual

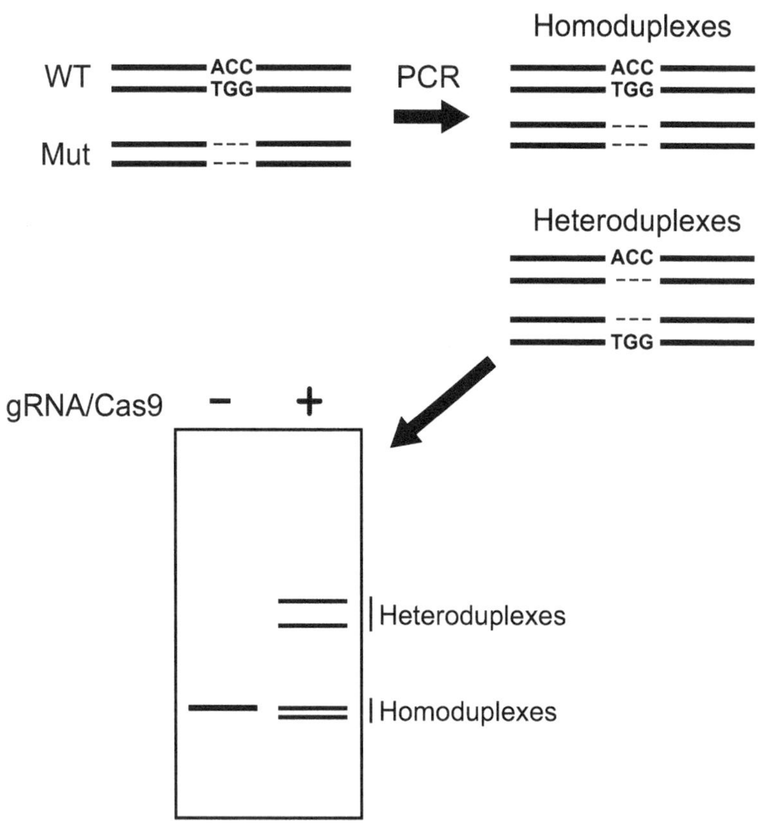

Fig. 1 A schematic representation of the heteroduplex mobility assay (HMA). When a target site is amplified by PCR from genomic DNA, including a wild-type allele and a mutant allele (3-bp deletion in the figure), the resultant PCR products contain both heteroduplexes and homoduplexes after denaturation and annealing. Heteroduplexes, which often contain an open single-stranded configuration at the mismatched region, migrate more slowly than homoduplexes during polyacrylamide gel electrophoresis

target sites. Several methods, such as the cell assay [11], T7 endonuclease 1 (T7E1) assay [12], and high-resolution melt analysis (HRMA) [13], are acceptable for the detection of indel mutations induced by genome editing. Both cell and T7E1 are nucleases that recognize and digest mismatch sequences in heteroduplexes. The digested fragment can be separated from the wild-type fragment by agarose gel electrophoresis [11, 12]. HRMA is a method that detects indel mutations by making use of the different melting temperatures between homoduplexes and heteroduplexes [13]. We recently found that the heteroduplex mobility assay (HMA) is a very simple and versatile detection method for indel mutations induced by genome editing technology [14]. When genomic DNA contains a wild-type allele and a mutant allele, both heteroduplexes and homoduplexes are generated during the denaturation and annealing steps that occur during PCR amplification using locus-specific primers. Heteroduplexes migrate more slowly than homoduplexes during polyacrylamide gel electrophoresis because of their open secondary structure (Fig. 1). Importantly, HMA is acceptable for three distinct processes in generating knockout animals: the evaluation of CRISPR/Cas9 cleaving activity in F0 embryos, the identification of potential F0 founders producing with mutant alleles, and the genotyping of mutants in the F1 generation [14]. Here, we describe the induction of multiple genome modifications mediated by CRISPR/Cas9 and the efficient identification of CRISPR/Cas9-induced indel mutations using HMA.

2 Materials

2.1 Construction of gRNA Expression Plasmids

1. Plasmid: a gRNA cloning vector, pDR274 [15] (Addgene plasmid number: 42250) (*see* **Note 1**).
2. Restriction enzyme: *Bsa*I-HF (NEB).
3. Ligation reaction: Ligation high ver.2 (TOYOBO).
4. Kit: QIAprep Spin Miniprep Kit (QIAGEN).
5. Primer: M13 reverse primer: 5′-CAGGAAACAGCTATGAC-3′.
6. Equipment: heat block set at 98 °C, water bath set at 65 °C.
7. Medium: LB medium (1 % polypeptone, 0.5 % yeast extract, 1 % NaCl).
8. Antibiotics: kanamycin.

2.2 Preparation of gRNA and Cas9 mRNA

1. Plasmid: pCS2+hSpCas9 [16] (Addgene plasmid number: 51815) (*see* **Note 1**).
2. Restriction enzymes: *Dra*I (NEB), *Not*I-HF (NEB).
3. Reagents: 20 mg/mL proteinase K, 10 % SDS, phenol/chloroform/isoamyl alcohol (25:24:1), 3 M sodium acetate, 5 M ammonium acetate, ethanol, 70 % ethanol.

4. Kits: MAXIscript T7 Transcription Kit (Life Technologies), mMESSAGE mMACHINE SP6 Transcription Kit (Life Technologies).
5. Column: PD SpinTrap™ G-25 (GE Healthcare).

2.3 Microinjection

1. Buffer: Injection buffer (40 mM HEPES [pH 7.4], 240 mM KCl, 0.5 % phenol red), E3 medium (5 mM NaCl, 0.17 mM KCl, 0.33 mM $CaCl_2$, 0.33 mM $MgSO_4$).
2. Tool: Glass capillaries with filament GD-1 (Narishige).
3. Equipment: Puller PC-10 (Narishige), Microinjector IM 300 (Narishige), Manipulator MMN-8 (Narishige).

2.4 Preparation of Genomic DNA

1. Buffer: Lysis buffer (10 mM Tris–HCl [pH 8.0], 1 mM EDTA, 0.3 % Tween-20, 0.3 % NP40, and 200 μg/mL proteinase K).
2. Equipment: Heat block or hybridization oven set at 55 °C.

2.5 Heteroduplex Mobility Assay (HMA)

1. Enzyme: *Ex Taq* DNA polymerase (TaKaRa) (*see* **Note 2**).
2. Buffer: Running buffer (25 mM Tris–HCl, 192 mM glycine).
3. Reagents: 10× PCR buffer (TaKaRa), 2.5 mM dNTP mix (TaKaRa), GeneRuler 50 bp DNA Ladder (Thermo), ethidium bromide solution (10 mg/mL), SuperSep DNA, 15 % (Wako).
4. Primers: Gene-specific primers should be designed to amplify 100–150 bp fragments of the targeted genomic locus by PCR.
5. Equipment: PowerPac universal power supply (BIO-RAD).

2.6 Identification of F0 Founders by Genotyping F1 Embryos

1. Plasmid: pGEM-T Easy Vector System (Promega).
2. Primer: M13 forward primer: 5′-GTAAAACGACGGCCAGT-3′.
3. Kit: MinElute PCR Purification Kit (QIAGEN).

3 Methods

3.1 Construction of gRNA Expression Plasmids

1. Identify the gRNA target sequence for each target locus. Seek a NGG sequence in the coding exon of the target gene. Use the 20 nucleotides upstream of the NGG as the gRNA target. Replace NN nucleotide of the 5′ end of the target sequence to GG nucleotides, even if the 5′ end of the target does not already include GG (*see* **Note 3**). Prepare oligonucleotides encoding the gRNA sequences: a sense oligonucleotide 5′-TAGGN_{18}-3′ and an antisense oligonucleotide 5′-AAACN_{18}-3′, for each target locus. These overhangs are compatible for subcloning into the *Bsa*I site of the pDR274 vector.
2. Mix 15 μL of sense oligonucleotides (1 μM) and 15 μL of antisense oligonucleotides (1 μM). Heat the mixture in a heart block at 98 °C for 3 min and then incubate in a water bath at 65 °C for 10 min.

3. Transfer the mixture to another container containing hot water and allow to cool to room temperature.
4. Mix 1 μL of the annealed oligonucleotides, 1 μL of *Bsa*I-digested pDR274 vector (10 ng/μL), and 2 μL of ligation high ver.2, and incubate at 16 °C for 30 min.
5. Transform 2 μL of the ligation mixture into competent *E. coli* (20 μL).
6. Grow bacterial colonies on LB medium containing kanamycin.
7. Purify the recombinant plasmid DNA from bacterial cultures using the QIAprep Spin Miniprep Kit according to the manufacturer's instruction.
8. Determine the sequence of the plasmid using the M13 reverse primer.

3.2 Preparation of gRNA

1. Linearize 5 μg of the plasmid DNA containing the gRNA target sequence by *Dra*I digestion.
2. Incubate the linearized DNA with proteinase K (200 μg/ml) and 0.5 % SDS at 50 °C for 30 min to remove residual RNase.
3. Extract the DNA using phenol/chloroform/isoamyl alcohol (25:24:1).
4. Transfer the supernatant to a 1.5-mL tube, add two volumes of ethanol and 1/10 volumes of 3 M sodium acetate, and centrifuge at 15,000 × *g*.
5. Rinse the pellet with 70 % ethanol.
6. Dissolve the pellet in RNase-free water.
7. Transcribe gRNA from the linearized DNA (1 μg) using the MAXIscript T7 kit according to the manufacturer's instruction.
8. To remove the template DNA, add 1 μL of TURBO DNase I and incubate at 37 °C for 15 min.
9. Extract the RNA with phenol/chloroform/isoamyl alcohol (25:24:1).
10. Remove unincorporated nucleotides using the PD SpinTrap™ G-25 column.
11. Precipitate the RNA with two volumes of ethanol and 1/10 volumes of 5 M ammonium acetate.
12. Dissolve the pellet in RNase-free water (*see* **Note 4**).

3.3 Preparation of Cas9 mRNA

1. Linearize 5 μg of pCS2+hSpCas9 by *Not*I digestion.
2. Incubate the linearized DNA with proteinase K (200 μg/mL) and 0.5 % SDS at 50 °C for 30 min to remove residual RNase.
3. Extract the DNA with phenol/chloroform/isoamyl alcohol (25:24:1).

4. Transfer the supernatant to a 1.5-mL tube, and add two volumes of ethanol and 1/10 volume of 3 M sodium acetate, and centrifuge at 15,000 ×*g*.
5. Transcribe mRNA from the linearized DNA (1 μg) using the mMESSAGE mMACHINE SP6 kit according to the manufacturer's instruction.
6. Follow the same subsequent steps as in Subheading 3.2.

3.4 Microinjection

1. Prepare the injection needle using a Puller PC-10.
2. Cut the tip of the needle with a razor.
3. Dilute the gRNA and Cas9 mRNA solutions with injection buffer (gRNA, 25 ng/μL each; Cas9 mRNA, 250 ng/μL) to prepare the injection solution (*see* **Note 5**).
4. Fill the needle with the injection solution.
5. Inject approximately 1 nL of the injection solution into the 1- to 2-cell stage zebrafish embryos.
6. Raise and incubate the injected embryos in E3 medium at 28.5 °C.

3.5 Preparation of Genomic DNA

1. Put five embryos (1–2 days post-fertilization, dpf) into a 1.5-mL tube, and remove as much liquid as possible.
2. Add lysis buffer (100 μl) and gently mix the tube (*see* **Note 6**).
3. Incubate at 55 °C for 2 h to overnight.
4. Incubate at 100 °C for 10 min to inactivate the proteinase K.
5. Centrifuge briefly in a microcentrifuge. Use 1 μL of genomic DNA lysate as the PCR template.
6. Store at 4 °C.

3.6 HMA

1. Amplify the DNA fragment (100–150 bp) containing the gRNA target region by PCR (10 μL reaction scale) using the locus-specific primers (*see* **Note 7**).
2. Add 2 μL of 6× loading dye to the PCR samples, and separate PCR products on a 15 % polyacrylamide gel at 30 mA for 60 min.
3. Remove the gel from the glass plate and soak it for approximately 10 min in ethidium bromide solution (10 μg/ml).
4. Take an image of the gel using a transilluminator (*see* **Note 8**) (Fig. 2).
5. Raise potential F0 founders to adult fish.

Fig. 2 (continued) (*white line*), suggesting the presence of various indel mutations. (**b**) CRISPR/Cas9-induced indel mutations at the *golden* (*gol*), *s1pr2*, and *spns2* loci. The PAM sequences are *underlined*, and the gRNA target sequences are indicated by *bold letters*. Insertions are in *lower case*, and deletion mutations are indicated by *dashes*. The numbers of deleted (−) and/or inserted (+) nucleotides are indicated at the *right* along with the number of mutants detected with each change. The mutation rate is indicated at the top of each set of sequences

a

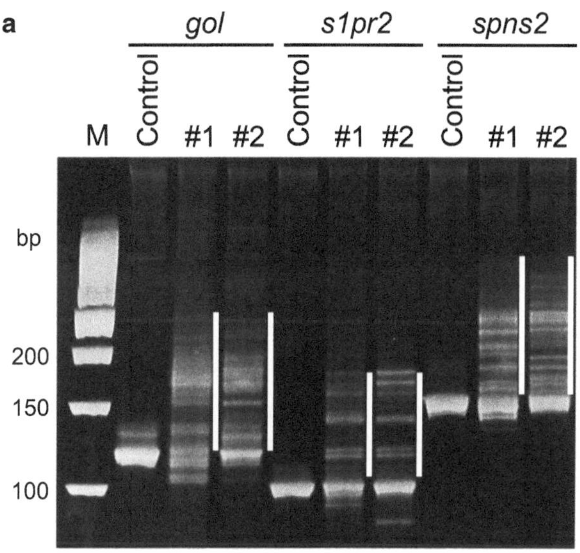

b

gol

mutation rate: 12/14 = 86%

```
CTCTAGTGGGCTGGAGTGACGACAGCAGCCTGCGGGTCCAGCGCCGTT WT
CTCTAGTGGGCTGGAGTGACGACAGCAG----CGGGTCCAGCGCCGTT -4 x2
CTCTAGTGGGCTGGAGTGACGACAGCAGggc------CCAGCGCCGTT -6(-9/+3)
CTCTAGTGGGCTGGAGTGACGAC-------TGCGGGTCCAGCGCCGTT -7
CTCTAGTGGGCTGGAGTGACGACAG---------GGTCCAGCGCCGTT -9 x2
CTCTAGTGGGCTGGAGTG--------------CGGGTCCAGCGCCGTT -14 x5
CTCTAGTGGGCTGGAGTGACGACAGCAGCgccgCGGGTCCAGCGCCGT +1(-3/+4)
```

s1pr2

mutation rate: 12/29 = 41%

```
TCAACAAGACCCTCATCCAGGTCCACTATCTCACTGCTAAGGAGATGA WT
TCAACAAGACCC---TCCAGGTCCACTATCTCACTGCTAAGGAGATGA -3 x3
TCAACAAGACCCT----CAGGTCCACTATCTCACTGCTAAGGAGATGA -4 x4
TCAACAAGACCCTCA-----GTCCACTATCTCACTGCTAAGGAGATGA -5 x1
TCAACAAGACCC------AGGTCCACTATCTCACTGCTAAGGAGATGA -6 x1
TCAACAAGACCCT----------CACTATCTCACTGCTAAGGAGATGA -10 x1
TCAACAAGACCC------------------TCACTGTTAAGGAGATGA -18 x1
TCAACAAGACCCTCATtCCAGGTCCACTATCTCACTGCTAAGGAGATG +1 x1
```

spns2

mutation rate: 12/12 = 100%

```
GCTCTCTCTCCGCAGCGGTCTGGGCTACATCCTGGGATCAATTGCTAA WT
GCTCTCTCTCCGCAca--TCTGGGCTACATCCTGGGATCAATTGCTAA -2(-4/+2)
GCTCTCTCTCCGCAG---TCTGGGCTACATCCTGGGATCAATTGCTAA -3 x3
GCTCTCTCTCctctgca------GGCTACATCCTGGGATCAATTGCTAA -6(-12/+6)
GCTCTCTCTCCGCA-------GGGCTACATCCTGGGATCAATTGCTAA -7 x4
GCTCTCTCTCCGCA-----------TACATCCTGGGATCAATTGCTAA -11
GCTCTCTCTCC---------------------TGGGATCAATTGCTAA -21
GCTCTCTCTCCGCAGtctgacggaCTGGGCTACATCCTGGGATCAAAA +5(-4/+9)
```

Fig. 2 Multiple genome modifications. (**a**) Multiple gRNAs (*gol*-gRNA, *s1pr2*-gRNA, and *spns2*-gRNA) and Cas9 mRNA were coinjected into zebrafish embryos, and HMA was performed for individual target sites using genomic DNA derived from the injected embryos. Injected embryos, but not control embryos, showed multiple heteroduplexes

3.7 Identification of F0 Founders Using F1 Embryos (See Note 9)

1. Cross the potential F0 founder with wild-type fish.
2. Transfer at least eight embryos (1–2 dpf) to individual PCR tubes and remove as much liquid as possible.
3. Prepare genomic DNA as described above in a 50-μL scale using a thermocycler.
4. Perform PCR using the locus-specific primers as described above (*see* **Note 10**).
5. Separate PCR products on a 15 % polyacrylamide gel (*see* **Note 11**).
6. Take an image of the gel using a transilluminator (Fig. 3).
7. Determine the potential mutant founders by judging the generation of heteroduplexes in the HMA.

3.8 Determination of the Mutation by Sequencing

1. Amplify the genomic target region from genomic DNA of the potential F0 founders by PCR (50-μL reaction scale) using the locus-specific primers.
2. Purify the PCR products using the MinElute PCR Purification Kit according to the manufacturer's instruction.

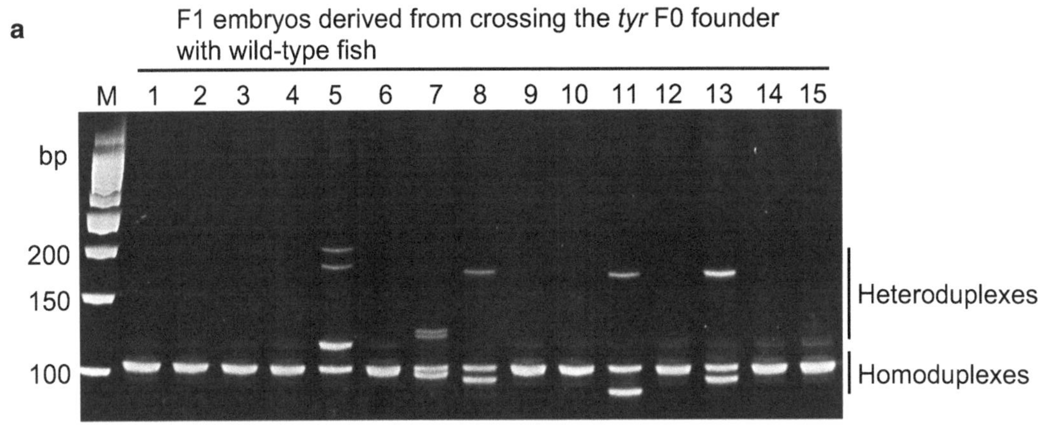

b

AAACGCTGC**TGTCCAGTCTGGCCCGGCGA**CGGCTCTGTGTGCGGCGTCCAGTCAGGTC WT

#5 AAACGCTGC**TGTCCAGTCTGGCCCG**tcgtgtaaaaacac**GCGA**CGGCTCTGTGTGCGG +14

#7 AAACGCTGC**TGTCCAGTCTGGCCCG**---**A**CGGCTCTGTGTGCGGCGTCCAGTCAGGTC -3

#8 AAACGCTGC**TGTCCAGTCTGGCC**------CGGCTCTGTGTGCGGCGTCCAGTCAGGTC -6

#11 AAACGCTGC**TGTCCAG**------------**A**CGGCTCTGTGTGCGGCGTCCAGTCAGGTC -12

#13 AAACGCTGC**TGTCCAGTCTGGCC**------CGGCTCTGTGTGCGGCGTCCAGTCAGGTC -6

Fig. 3 Identification of the *tyrosinase* (*tyr*) F0 founder using the heteroduplex mobility assay (HMA). (**a**) Genomic DNAs were prepared from individual F1 embryos derived from crossing the *tyr* F0 founder with wild-type fish. The results of HMA showed that five F1 embryos (numbers 5, 7, 8, 11, and 13) exhibited one or two heteroduplexes and two homoduplexes. (**b**) The nucleotide sequences of individual *tyr* F1 embryos containing indel mutations. The migration patterns of the homoduplexes and heteroduplexes in F1 embryos #8 and #13 were similar, and both contain the same deletion mutation. The PAM sequences are *underlined*, and the gRNA target sequences are indicated by *bold letters*. Insertions are in *lower case*, and deletion mutations are indicated by *dashes*. The numbers of deleted (–) and/or inserted (+) nucleotides are indicated at the *right*

3. Ligate the purified PCR products into the pGEM-T Easy vector.
4. Transform the ligation solution into competent *E. coli.*
5. Pick up bacterial colonies and amplify the cloned genomic target regions by PCR using the M13 forward and reverse primers.
6. Purify the PCR products using the MinElute PCR Purification Kit according to the manufacturer's instruction.
7. Sequence the purified PCR products and identify the F0 founders that can produce a mutant allele.
8. Raise F1 embryos derived from the F0 founders to adult fish.

4 Notes

1. Currently, several gRNA expression vectors and Cas9 plasmids can be obtained from Addgene. We used pDR274 [15] and pCS2+hSpCas9 [16], which is designed according to human codon usage.
2. Because the PCR products can be directly cloned into the pGEM-T Easy vector, we used the *Ex Taq* DNA polymerase in the process of HMA.
3. Typically, a sequence of the form 5′-GG-N_{18}-NGG-3′ is selected for a CRISPR/Cas9 target. However, recent studies have demonstrated that the 5′ sequence of the target is not always required for target recognition [16, 17]. Therefore, we generally replaced NN at the 5′ end of the target with GG for construction of the gRNA expression plasmid.
4. Do not allow the RNA pellet to dry completely, because it will be difficult to redissolve.
5. If you want to perform multiple genome modifications, multiple gRNAs are injected into the zebrafish embryos along with the Cas9 mRNA.
6. Alternatively, an alkaline lysis method may be used for preparation of genomic DNA. Put five embryos (1–2 dpf) into a 1.5-mL tube and remove as much liquid as possible. Add 100 μL of 50 mM NaOH. Heat at 98 °C for 10 min and add 12 μL of 1 M Tris–HCl (pH 8.0).
7. To select the locus-specific primers (approximately 25 bp oligos), we used the primer design tool "Primer-BLAST" (http://www.ncbi.nlm.nih.gov/tools/primer-blast/).
8. If an allele of the targeted locus contains indel mutations, heteroduplexes can be seen above the homoduplexes.
9. When multiple loci are simultaneously targeted by the CRISPR/Cas9 system, multi-locus HMA is more effective in identifying the F0 founders.

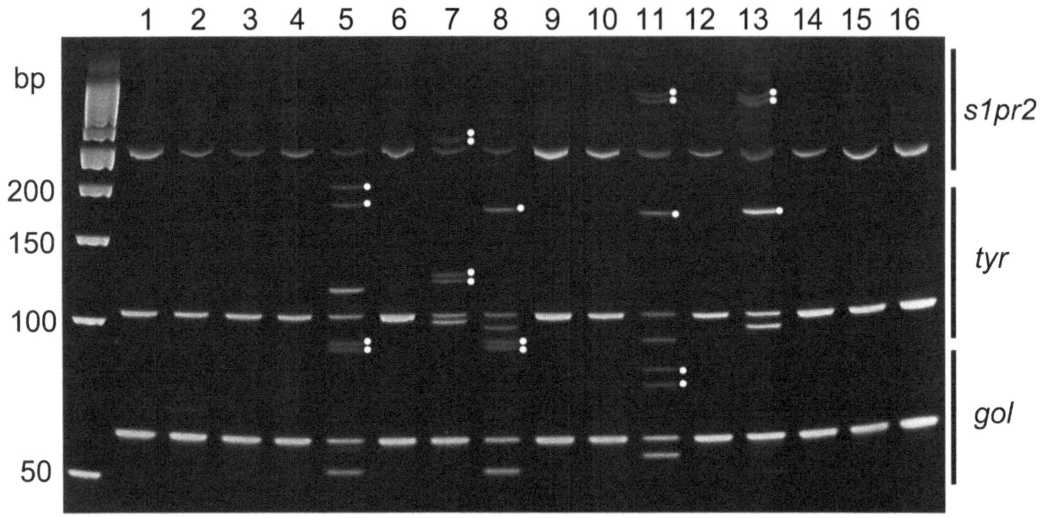

Fig. 4 Multi-locus HMA. Three gRNAs (*s1pr2*-gRNA, *tyr*-gRNA, *gol*-gRNA) and the Cas9 mRNA were coinjected into zebrafish embryos, and the embryos were grown up to adult fish. The potential F0 founders were crossed with wild-type fish, and genomic DNAs were prepared from the F1 embryos. The three target genomic loci were independently amplified by PCR using locus-specific primers that were designed to produce products of varying sizes (*s1pr2*; 253 bp, *tyrosinase*; 104 bp, *golden*; 60 bp). A mixture of the individual PCR amplicons (*s1pr2*, *tyrosinase*, and *golden*) was applied to a single lane and separated on a 15 % polyacrylamide gel. All three target loci had heteroduplexes (*circles*)

10. For multi-locus HMA, design locus-specific primers that produce different sizes of PCR products (e.g., *gol*, 60 bp; *tyr*, 104 bp; *s1pr2*, 253 bp in Fig. 4).
11. For multi-locus HMA, the PCR products are mixed at an appropriate rate (e.g., the ratio of the PCR products for the *gol*, *tyr*, and *s1pr2* targets is 6:3:1 in Fig. 4) and loaded on a gel.

Acknowledgment

This work was supported by the Japan Society for the Promotion of Science and the Program for Next-Generation World-Leading Researchers (NEXT Program).

References

1. Wiedenheft B, Sternberg SH, Doudna JA (2012) RNA-guided genetic silencing systems in bacteria and archaea. Nature 482: 331–338
2. Cong L, Ran FA, Cox D et al (2013) Multiplex genome engineering using CRISPR/Cas systems. Science 339:819–823
3. Mali P, Yang L, Esvelt KM et al (2013) RNA-guided human genome engineering via Cas9. Science 339:823–826
4. Jinek M, Chylinski K, Fonfara I et al (2012) A programmable dual-RNA-guided DNA endonuclease in adaptive bacterial immunity. Science 337:816–821

5. Deveau H, Barrangou R, Garneau JE et al (2008) Phage response to CRISPR-encoded resistance in Streptococcus thermophilus. J Bacteriol 190:1390–1400
6. Mojica FJ, Diez-Villaseñor C, García-Martínez J et al (2009) Short motif sequences determine the targets of the prokaryotic CRISPR defence system. Microbiology 155:733–740
7. Hisano Y, Ota S, Kawahara A (2014) Genome editing using artificial site-specific nucleases in zebrafish. Dev Growth Differ 56:26–33
8. Wang H, Yang H, Shivalila CS et al (2013) One-step generation of mice carrying mutations in multiple genes by CRISPR/Cas-mediated genome engineering. Cell 153: 910–918
9. Jao LE, Wente SR, Chen W (2013) Efficient multiplex biallelic zebrafish genome editing using a CRISPR nuclease system. Proc Natl Acad Sci U S A 110:13904–13909
10. Ota S, Hisano Y, Ikawa Y et al (2014) Multiple genome modifications by the CRISPR/Cas9 system in zebrafish. Genes Cells 19:555–564
11. Miller JC, Holmes MC, Wang J et al (2007) An improved zinc-finger nuclease architecture for highly specific genome editing. Nat Biotechnol 25:778–785
12. Kim HJ, Lee HJ, Kim H et al (2009) Targeted genome editing in human cells with zinc finger nucleases constructed via modular assembly. Genome Res 19:1279–1288
13. Dahlem TJ, Hoshijima K, Jurynec MJ et al (2012) Simple methods for generating and detecting locus-specific mutations induced with TALENs in the zebrafish genome. PLoS Genet 8:e1002861
14. Ota S, Hisano Y, Muraki M et al (2013) Efficient identification of TALEN-mediated genome modifications using heteroduplex mobility assays. Genes Cells 18:450–458
15. Hwang WY, Fu Y, Reyon D et al (2013) Efficient genome editing in zebrafish using a CRISPR-Cas system. Nat Biotechnol 31: 227–229
16. Ansai S, Kinoshita M (2014) Targeted mutagenesis using CRISPR/Cas system in medaka. Biol Open 3:362–371
17. Hwang WY, Fu Y, Reyon D et al (2013) Heritable and precise zebrafish genome editing using a CRISPR-Cas system. PLoS One 8:e68708

Chapter 5

Generation of Targeted Genomic Deletions Through CRISPR/Cas System in Zebrafish

An Xiao and Bo Zhang

Abstract

Using TALEN or CRISPR/Cas system to induce small indels into coding sequences has been implicated in broad applications for reverse genetic studies of many organisms including zebrafish. However, complete deletion of a large gene or noncoding gene(s) or removing a large genomic fragment spanning several genes or other chromosomal elements is preferred in various cases, as well as inducing chromosomal inversions. Here, we describe the detailed protocols for the generation of chromosomal deletion mutations mediated by Cas9 and a pair of gRNAs and the evaluation for the efficiencies in F_0 founder fish and of germline transmission.

Key words Zebrafish, TALEN, CRISPR/Cas, Gene targeting, Genome manipulation, Gene disruption, Chromosomal deletion

1 Introduction

As an ideal vertebrate model organism, zebrafish (*Danio rerio*) is valuable to study gene functions during embryonic development and organ regeneration, as well as in modeling human diseases. Transcription activator-like effector nucleases (TALENs) or clustered regularly interspaced short palindromic repeats (CRISPR)/CRISPR-associated (Cas) systems have been widely used to induce double-strand breaks (DSBs) in target genes and subsequent nonhomologous end-joining (NHEJ)-mediated repairing mutagenesis, leading to disruption of protein-coding genes, usually due to frameshift, in a variety of organisms including zebrafish [1–4]. NHEJ is error prone and tends to generate different types of small insertions and/or deletions (indels) around the DSB region, with the length ranging from a few to tens, or sometimes hundreds, of nucleotides.

However, the indel mutation strategy is not suitable or satisfactory for all purposes. For example, it is usually not sufficient to disrupt the function of noncoding genes, untranslated or regulatory

Koichi Kawakami et al. (eds.), *Zebrafish: Methods and Protocols*, Methods in Molecular Biology, vol. 1451,
DOI 10.1007/978-1-4939-3771-4_5,

regions in a genome; it is also tricky to use this approach to disrupt multiple adjacent genes or gene clusters simultaneously. Even for a single coding gene, indel mutations may not guarantee to disrupt the function of the target gene completely, especially for large genes, due to the existence of multiple transcripts or splice variants, or unexpected upstream/downstream alternative start codons. In all these cases, complete deletion of the whole gene(s) or sequence should be a reliable and better strategy to solve the above problems, which can be easily achieved by using two pairs of TALENs or two guide RNAs (gRNAs) with Cas9 targeting two sites flanking the region to be deleted. This strategy may also be used to precisely remove specific functional elements, e.g., a protein domain, from the coding sequence without affecting other parts of the gene, which is difficult to achieve for a single pair of TALENs or one CRISPR/Cas (or called Cas9/gRNA) system when the target sequence is more than several hundreds of nucleotides long. Furthermore, in order to disrupt the target gene completely, the first several exons are usually selected for targeting to generate indel mutations and screen for frameshift alleles which could lead to early premature stop codons. However, this restriction for the targeting region may prevent the selection for high-efficient target sites. In contrast, the targeted deletion strategy provides more choices for the target sites, which could extend the options for the target site to introns and intergenic sequences, in addition to exons [5]. In addition to generating large genomic deletions, one can also create genomic inversions by using two pairs of TALENs or paired gRNAs with Cas9. In this chapter, we provide the detailed protocols for using two gRNAs together with the Cas9 mRNA to generate large chromosomal deletions in the zebrafish genome.

2 Materials

2.1 Reagents and Solutions for Molecular Biology Experiments

1. zCas9 expression vector (pGH-T7-NLS-zCas9-NLS) [6].
2. gRNA template plasmid (e.g., pX459-V2.0, Addgene #62988 [7]).
3. mMessage mMachine T7 kit or mMessage mMachine T7 Ultra kit (Ambion, USA).
4. MEGAshortscript T7 Kit (Ambion, USA).
5. mirVana miRNA Isolation Kit (Ambion, USA).
6. XbaI restriction enzyme (New England Biolabs, USA).
7. PCR purification kit (Qiagen, USA).
8. Gel extraction kit (Invitrogen, USA).

9. TOPO TA-cloning vectors (Invitrogen, USA).
10. 50 mM NaOH.
11. 1 M Tris–HCl (pH 8.0).
12. Nuclease-free water or RNase-free water.
13. Hi-Taq DNA polymerase master mix.
14. Forward primer to amplify the DNA template for gRNA synthesis: 5′-TAATACGACTCACTATA**GXXXXXXXXXXXXXXXXXXX** GTTTTAGAGCTAGAAATAGC-3′. The nucleotides in bold letters (and framed) represent the 20-nt protospacer sequence and are variable according to different target sites.
15. Reverse primer to amplify the DNA template for gRNA synthesis: 5′-AAAAAAGCACCGACTCGGTGCCAC-3′. Universal for all gRNAs.
16. Real-time PCR primers and probes: For evaluation of the efficiency of chromosomal deletions. Variable according to different target regions.

2.2 Equipment, Reagents, and Consumable Materials for Zebrafish Husbandry and Microinjection

1. Wild-type zebrafish or other desired zebrafish strains.
2. E3 embryo buffer: 5 mmol/L NaCl, 0.17 mmol/L KCl, 0.33 mmol/L $CaCl_2$, 0.33 mmol/L $MgSO_4$.
3. 0.5 % Phenol red.
4. Glass capillaries: For making injection needles (e.g., O.D. 1.0 mm, I.D. 0.58 mm; Harvard Apparatus, USA).
5. Dumont #5 Tweezer (Inox, 11 cm) (Word Precision Instruments, Inc.) or other equivalents.
6. 1-μL disposable capillaries (R: 0.25 %, CV: 0.6 %) (CAMAG, Switzerland).
7. Microloader tips (Eppendorf, USA).
8. Microinjection molds or other equivalents to hold zebrafish embryos for microinjection.
9. Mating tanks.
10. Stereo microscope.
11. 28.5 °C incubator.
12. PN-30 Puller (Narishige, Japan) or other equivalents.
13. Nitrogen gas and tank.
14. MPPI-2 Pressure Injector (Applied Scientific Instrumentation, USA), or PLI-90Pico-Injector (Harvard Apparatus, USA) or other equivalents.

3 Methods

3.1 Design and Verification of Cas9/gRNA Target Sites

1. Each Cas9/gRNA target site (Cas9/gRNA recognition and binding site) consists of a 20-nt protospacer followed by a 3-nt PAM (protospacer adjacent motif) sequence (i.e., 5′-NGG-3′). Therefore, the general formula for a single target site is 5′-NNNNNNNNNNNNNNNNNNNN -NGG-3′, where N represents any nucleotide and the first 20 nucleotides (framed) represent the sequence of the protospacer. In reality, T7 RNA polymerase is usually employed for the synthesis of gRNAs (beginning with the 20-nt protospacer sequence) by in vitro transcription. Since the T7 polymerase requires at least the first transcribed nucleotide to be G for efficient transcription, the target site or the protospacer sequence should begin with a G, and the practical formula for a target site should therefore be modified as 5′-G-$(N)_{19}$-NGG-3′, or further simplified as 5′-G-$(N)_{20}$-GG-3′(*see* **Note 1**).
2. To delete a large fragment of chromosomal sequence, one needs to design two gRNAs targeting the 5′- and 3′-ends of the region to be deleted, respectively. Individual Cas9/gRNA target site can be selected manually using the above formula or identified by using web tools, such as CRISPR Design Tool (http://crispr.mit.edu/) [8] or ZiFiT Targeter (http://zifit.partners.org/ZiFiT/CSquare9Nuclease.aspx) [9]. The Cas9/gRNA target sites can be designed to locate in either exonic, intronic, or intergenic regions. Unless the purpose is to delete a defined genomic element (i.e., not much choice for the target sites), the target sites within exons are preferred since they have much less sequence polymorphism.
3. (Optional) For a given sequence, usually there are quite a few choices for the Cas9/gRNA target sites. To simplify the evaluation of targeting efficiency of single gRNA site (and screening for indel mutations in other applications), the target sequence containing a unique restriction enzyme site is usually preferred, which is necessary for restriction enzyme (RE)-resistance assay (*see* **Note 2**).
4. Ensure that each selected target site is unique and highly specific in the zebrafish genome (*see* **Note 3**). Some web tools or software are available to help search for potential off-target sites in the zebrafish genome [8, 10, 11].
5. For each target site, design a pair of primers for the amplification of the genomic region spanning the target site. Choose the primers which are at least 100 bp away from the target site and which give rise to the PCR product of less than 500 bp. In addition, since we have to use a particular pair of primers, one from the upstream target site and the other from the downstream

target site, to detect the large deletions, try to choose the two pairs of primers for the two target sites to have similar annealing temperature.

6. PCR amplify the genomic DNA from parental fish and make sure only one single band is visible after agarose gel electrophoresis. Since mismatches in the target sites will affect the mutagenesis activity, to exclude polymorphisms in the target sites, it is important to confirm their sequences from your in-house fish stock by PCR and direct sequencing, and use the confirmed zebrafish to collect embryos for the gene-targeting experiments.
7. (Optional) Establish RE-resistance assay for the evaluation of targeting efficiency: If one plans to use RE-resistance assay to determine the targeting efficiency, PCR amplify the genomic sequence of the parental fish with the selected primer pairs, perform RE digestion, and evaluate by agarose gel electrophoresis. The RE digestion should be complete and the digestion products should be easily identifiable and distinguishable with the undigested PCR product.

3.2 Preparation of zCas9 mRNA by In Vitro Transcription

1. We recommend to use zebrafish codon-optimized Cas9 (zCas9) rather than other versions of Cas9 (e.g., hCas9, the human codon-optimized Cas9) since zCas9 gives higher targeting efficiency [6]. Our zCas9 plasmid (pGH-T7-NLS-zCas9-NLS) contains a T7 promoter upstream to the *NLS-zCas9-NLS* coding region and can be linearized for making mRNA through in vitro transcription. For linearization, 10 μg zCas9 plasmids are digested by XbaI in a 10 μL system overnight at 37 °C. To monitor the extent of linearization, load 0.5 μL of the reaction mixture to 0.8% agarose gel and examine by electrophoresis (*see* **Note 4**).
2. When the digestion is complete, use a DNA purification kit to purify the linearized plasmids and elute with 20 μL nuclease-free water (*see* **Note 5**). Determine the concentration of the linearized plasmid by a spectrophotometer. (Optional: The linearized plasmid can be stored at −20 °C and used later.)
3. Prepare capped zCas9 mRNA by using the mMessage mMachine T7 Kit (or mMessage mMachine T7 Ultra Kit) (*see* **Notes 6** and **7**). Reaction mixture: 1 μg linearized DNA from the above step, 10 μL 2× NTP/CAP (or NTP/ARCA), 2 μL 10× reaction buffer, and 2 μL T7 enzyme mix, and supplement the volume to 20 μL with nuclease-free water; mix well by pipetting. Incubate the mixture at 37 °C for 2–3 h. To monitor the mRNA synthesis, load 0.5 μL reaction mixture to 1% agarose gel and examine by electrophoresis (*see* **Note 8**).

4. If the transcription is successful, add 1 μL TURBO DNase I supplied by the kit and incubate at 37 °C for 15 min to remove the DNA template.
5. Purify the zCas9 mRNA. Option I: According to the manual of the kit, stop reaction by adding 30 μL LiCl and 30 μL nuclease-free water provided by the kit. Mix well and store at −20 °C for at least 30 min. Then centrifuge at 4 °C for 15 min at top speed. The RNA pellet should be visible at the bottom of the Eppendorf tube. Remove the supernatant and wash with 1 mL cold 70% ethanol. Centrifuge at 4 °C for 10 min at top speed. Remove the 70% ethanol, air-dry the pellet, and add 30–50 μL nuclease-free water to dissolve the pellet. Option II: use RNeasy kit to purify the zCas9 mRNA following the manufacturer's instructions. After purification, determine the concentration of mRNA by a spectrophotometer (typically 500–800 ng/μL; *see* **Note 9**), and aliquot them into small volumes (e.g., 5 μL). Store the aliquots at −80 °C for later use and long-term storage (*see* **Note 10**).

3.3 Preparation of gRNAs by In Vitro Transcription

1. We use purified PCR product as the in vitro transcription template for making gRNAs. Any plasmid containing a full gRNA scaffold sequence can be used as the PCR template (e.g., Addgene plasmid pX459-V2.0 #62988 [7]; *see* **Note 11**). Synthesize (by order from company service) a 57-nt oligo for each gRNA with the sequence 5′-TAATACGACTCACTATA**GXXXXXXXXXXXXXXXXXXX** GTTTTAGAGCTAGAAATAGC-3′, where the nucleotides framed in bold represent the 20-nt protospacer sequence of each particular gRNA, preceded by a 17-nt T7 promoter sequence at the 5′-upstream; and synthesize a universal reverse primer oligo with the sequence 5′-AAAAAAGCACCGACTCGGTGCCAC-3′. PCR reaction mixture: 2 ng plasmid template (*see* **Note 12**), 4 μL 10 μmol/L oligos each, and 20 μL 2× hi-fidelity DNA polymerase master mix, and supplement the volume to 40 μL with nuclease-free water. PCR program: 95 °C 5 min, then 45 cycles of (95 °C 20 s, 55 °C 20 s, 72 °C 30 s), and then 72 °C 6 min. The PCR products can be examined by electrophoresis, the length of which should be 119 bp (17 bp T7 promotor + 102 bp gRNA template).
2. Purify the PCR product by ethanol (EtOH) precipitation (*see* **Note 13**). Adjust the volume by water to 150 μL, add 2.5× volume (375 μL) EtOH and 0.1× volume (15 μL) 3 mol/L NaOAc, and mix well. Store at −20 °C for at least 1 h. Then centrifuge at 4 °C for 10 min at top speed. The pellet of PCR product should be visible at the bottom of the tube. Remove the supernatant and wash with 1 mL cold 70% ethanol. Centrifuge at 4 °C for 10 min at top speed. Remove the 70% ethanol, air-dry the pellet, and add 30 μL nuclease-free

water to dissolve the pellet. Determine the concentration of gRNA DNA template by a spectrophotometer (typical 200–300 ng/μL) (Optional: The PCR product can be stored at −20 °C and used later.).

3. Prepare gRNAs by using the MEGAshortscript T7 Kit or other T7 in vitro transcription system (Fig. 1; *see* **Note 6**). Reaction mixture: 1 μg DNA template from the above step, 2 μL each ATP/CTP/GTP/UTP, 2 μL 10× reaction buffer, and 2 μL T7 enzyme mix, and supplement the volume to 20 μL with nuclease-free water; mix well by pipetting. Incubate the mixture at 37 °C for 2–3 h. Add 1 μL TURBO DNase I supplied by the kit and incubate at 37 °C for 15 min to remove the DNA template.
4. Purify the short gRNAs with mirVana miRNA Isolation Kit, following the manufacturer's instruction. Elute the gRNAs with 50 μL elution buffer. Determine the concentration of gRNA by a spectrophotometer (typically varies from 300 to >1000 ng/μL; *see* **Note 9**).

3.4 Generation of Founder Fish by Microinjection of zCas9 mRNA and gRNA Pairs into Zebrafish Embryos

1. The parental zebrafish intended to give rise to founder embryos for gene targeting should be genotyped to confirm the sequence of the target sites (*see* **Note 14**). One day before the injection day, set up mating tanks and put one pair of zebrafish in each tank and separate by genders with a divider. In the day of injection, remove the divider from one tank each time and collect embryos after spawning.
2. Prepare 3–5 μL injection mixture: ~300 ng/μL zCas9 mRNA and 20–50 ng/μL of each gRNA, add some phenol red to a final concentration of no more than 0.05 %, and supplement the final volume to 20 μL with nuclease-free water. Use disposable capillaries to calibrate the injection volumes by injecting several drops into it. The total volume can be measured by the length of the liquid in the capillary by using a ruler. Then the volume of each drop can be calculated by the inner diameter of the capillary, the length of liquid column made by injection, and the count of drops.
3. Inject 2 nL mixture into the cytoplasm of one-cell stage zebrafish embryos, i.e., ~600 pg zCas9 mRNA and 40–100 pg of each gRNA per embryo (*see* **Notes 15** and **16**). When using different batches of zCas9 mRNA, we recommend reevaluating the optimal injection dosages (*see* **Note 17**).
4. After injection, incubate the embryos in E3 embryo buffer at 28.5 °C. Save some uninjected sibling embryos as a control and process them the same as the injected ones for the determination of the efficiency of individual gRNA. The dead and deformed embryos are counted and removed at 5–6 h post-fertilization (hpf) and 1-day post-fertilization (dpf).

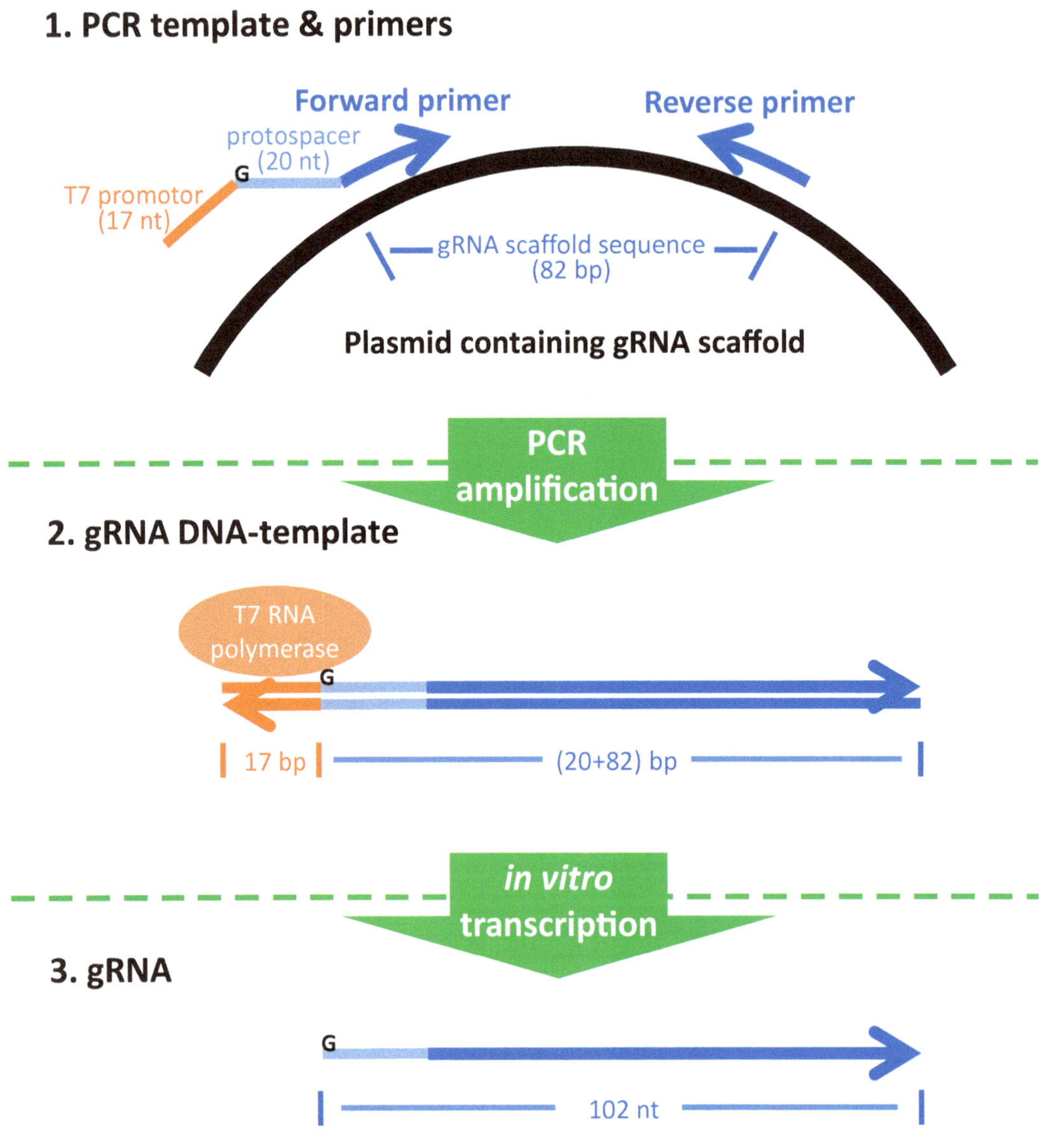

Sequence of the gRNA DNA-template of Step 2:

5’-TAATACGACTCACTATAGXXXXXXXXXXXXXXXXXXXGTTTTAGAGCTAGAAATAGC
AAGTTAAAATAAGGCTAGTCCGTTATCAACTTGAAAAAGTGGCACCGAGTCGGTGCTTTTTTT-3’
(Underlines: Primers used in Step 1)

Fig. 1 The schematic diagram of the steps to prepare gRNAs

3.5 (Optional) Evaluation of Targeting Efficiency of Individual gRNA in Founder Embryos

Evaluation of individual gRNA for its targeting efficiency can help in choosing proper gRNAs for the chromosomal deletion experiments or troubleshooting if the deletion is not successful. We generally use RE-resistance assay to evaluate the efficiency of individual gRNA (Option I) [12]. Other methods can also be used, such as Surveyor (CEL-I) or T7E1 assay (Option II), melting curve assay, and direct sequencing.

1. The genomic DNA of control (uninjected siblings) and injected embryos is extracted using NaOH lysis method [13]. Five to ten normally developing 2–4 dpf embryos are pooled into one PCR tube. Remove extra buffer and add 50 μL 50 mmol/L NaOH. The embryos are lysed by heating for 10–30 min at 95 °C using a PCR machine; then cool down to 4 °C. Vortex the tubes briefly once or twice to break up the embryos. Then add 5 μL Tris–HCl (pH 8.0) to neutralize NaOH and centrifuge for 5 min at 14,000 ×*g*. The supernatant contains crude genomic DNA and is ready to be used as PCR templates.
2. Take 1 μL crude genomic DNA extract as template and assemble a 10 μL PCR reaction (*see* **Note 18**). Recover the PCR products.
3. (Option I) RE-resistance assay: Digest 2 μL PCR products with the proper RE and buffer (*see* **Note 19**). Load reaction mixture to 2–3 % agarose gel and examine the digestion by electrophoresis, using undigested PCR products as a reference. If the DNA from control embryos is digested completely, the efficiency of the gRNA can be estimated with the percentage of the resistant (undigested) band by measuring their intensity. The resistant band can be extracted from the gel and verified by sequencing after cloning into TA-cloning vectors.
4. (Option II) Surveyor (CEL-I) or T7E1 assay: Mix ~250 ng PCR product (from control or Cas9/gRNA injected embryos, respectively) with 1 μL NEB Buffer 2; supplement the volume to 9.5 μL with nuclease-free water. Slowly anneal the DNA in a thermocycler with the following program: 95 °C, 5 min; 95–25 °C at −0.1 °C/s; hold at 4 °C. Then add 0.5 μL Surveyor or T7E1 enzyme on ice. Mix well; incubate in 37 °C for 45 min. Add 1 μL 0.5 mol/L EDTA to stop reaction. Then load reaction mixture to 2–3 % agarose gel and examine the digestion by electrophoresis. If the DNA from control embryos is not digested and the DNA from injected ones is, that means the Cas9/gRNA is functional and small indels have been generated (*see* **Note 20**). The efficiency of mutated allele can be estimated by the corresponding formula [7].

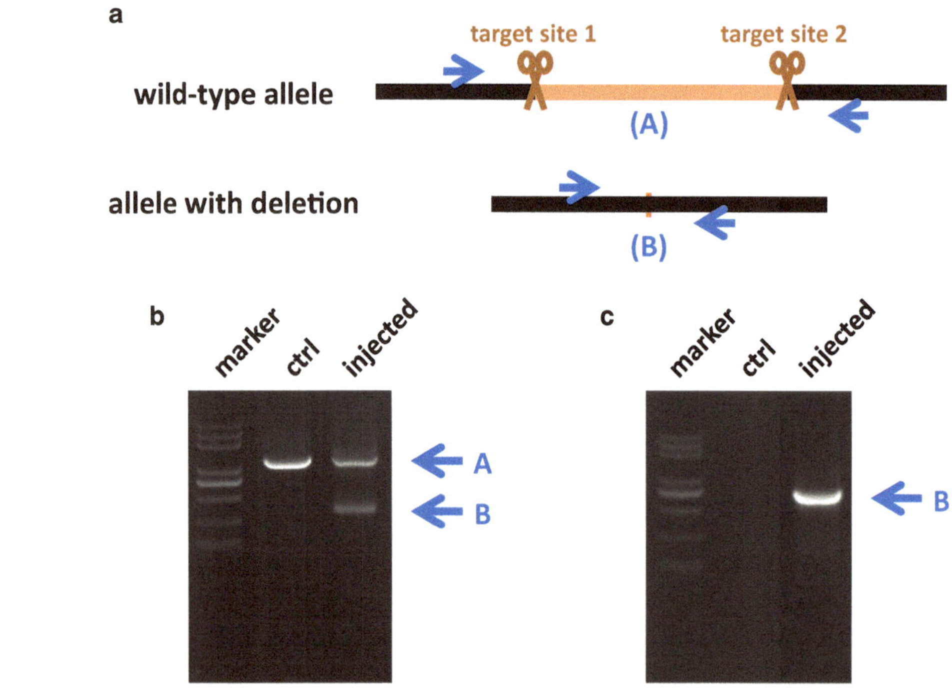

Fig. 2 Detection of chromosomal deletions induced by two Cas9/gRNAs. (**a**) The schematic diagram of the target region and PCR primers used to amplify the genomic fragment. (**b**) In wild-type embryos, if the genomic fragment between the two target sites is short enough to be amplified by PCR, one single PCR product (*band A*) will be obtained from the control (uninjected) embryos. In the injected embryos, an additional and shorter PCR product (*band B*), which corresponds to the deletion allele, should be clearly visible if the two Cas9/gRNAs are functional. (**c**) If the genomic fragment between the two target sites in wild-type embryos is too long to be amplified, no PCR product will be obtained from the control embryos, while only one single band could be detected from the injected embryos if the two Cas9/gRNAs are functional. *Ctrl* control

3.6 Detection and Evaluation of Chromosomal Deletion Efficiency in Founder Embryos

1. To examine the success of the chromosomal deletion, the genomic regions containing the target sites are amplified by PCR with the primers outside the region to be deleted (Fig. 2a). If the distance of the two primers at the chromosome is short enough, the DNA from control embryos can be amplified with a single amplicon, and DNA from injected embryos may have two amplicons: one with the same length as the control and the other should be shorter, which represents the alleles bearing deletions (Fig. 2b). If the distance of the two primers is too long, there may be no amplicon from control DNA, and only a short amplicon from injected embryos can be detected (Fig. 2c). Recover the shorter band and verify the deletions by sequencing after cloning into TA-cloning vectors.
2. The general PCR is not accurate in revealing the exact efficiency of chromosomal deletions [5]. To accurately evaluate the deletion efficiency, design quantitative real-time PCR

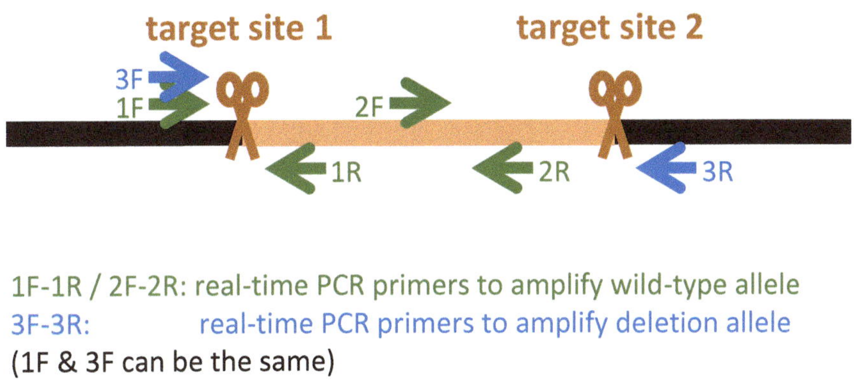

Fig. 3 The schematic diagram of the real-time PCR primers to quantify the ratio of wild-type and deletion alleles

primers and probes for wild-type only alleles (one primer located outside the deletion region and the other located inside) and deletion alleles (primers located at each sides out side the deletion region) (Fig. 3). Use the same amount of DNA template and evaluate the relative ratio of different alleles by ΔCt approach [14] (*see* **Note 21**).

3.7 Evaluation of Germline Transmission Efficiency of Chromosomal Deletions and Screen for Mutants

1. If the chromosomal deletion efficiency is acceptable (e.g., >1 %; *see* **Note 22**), raise enough amount of the same batch of evaluated injected embryos to adulthood as founder (F_0) fish for screening of heritable mutations.
2. Out-cross the mosaic F_0 fish with wild-type zebrafish. After breeding, each F_0 fish is placed and raised separately in a single tank until the F_1 embryos are evaluated. Collect F_1 embryos from each individual F_0, prepare the genomic DNA, and detect the chromosomal deletion with the same strategy as described above. We normally screen 50–100 F_1 embryos (five to ten PCR tubes of ten pooled embryos) for each F_0 fish (*see* **Note 23**).
3. If heritable deletion alleles were identified and confirmed by sequencing, breed the positive F_0 fish to get more F_1 offspring. F_1 zebrafish heterozygous for the chromosomal deletion mutations are identified by genotyping of the genomic DNA from fin clips from each individual, with the same strategy as described above.
4. Homozygous zebrafish mutants can be obtained by in-cross of pairs of heterozygous carriers. They can be verified again with the same process.

4 Notes

1. The promoter sequence for the T7 RNA polymerase is 5′-TAATACGACTCACTATAGGG-3′, where the last triple G (GGG) will also serve as the transcription start site and be incorporated into the RNA transcript; however, one or two G's (i.e., GNN or GGN) is also acceptable for the normal function of T7 RNA polymerase. Alternatively, if the protospacer sequence in the target site does not begin with G, one can simply replace its first nucleotide with G for the synthesis of gRNA with T7 polymerase. The mismatch between the gRNA and the target sequence for the first nucleotide does not seem to significantly compromise the targeting activity of the Cas9/gRNA system (Unpublished observation).
2. The Cas9 protein theoretically cleaves the target sequence between the third and fourth nucleotides upstream of the PAM. The RE recognition and cleavage sites used for the RE-resistance assay should not be too far from this Cas9 cleavage position, since indel mutations may only affect a small region around this position.
3. Since the chromosomal deletion strategy introduces more than one type of gRNA simultaneously in the same embryo, one should be more cautious about the potential off-targeting effects of the Cas9/gRNA system.
4. Make sure all the circular plasmids are digested completely. Use the original (undigested) plasmids as a control in electrophoresis, and there should be no visible bands at the same position as the original plasmid in the lanes loaded with the linearized plasmids. Insufficient linearization will lead to low yield of mRNA from in vitro transcription.
5. Follow the supplier's protocol of the purification kit. Avoid loading too much linearized plasmids into the column, or it may be overloaded and lead to low yield of purified product.
6. All the reagents and consumables used for in vitro transcription should be free of RNase.
7. Both the T7 Kit and T7 Ultra Kit from Ambion are acceptable. The T7 Ultra Kit uses Anti-Reverse Cap Analog (ARCA) instead of the general Cap in the T7 Kit. We found the zCas9 mRNA prepared by using the ARCA Kit showed higher targeting efficiency [6].
8. The size of mRNA might not be accurately revealed by the agarose gel electrophoresis and may show up as multiple bands, possibly due to the formation of secondary structures of the mRNA.

9. If the concentration of zCas9 mRNA or gRNA is not high enough, open the cap of the tubes and carefully put the tubes in a 37 °C incubator (avoid RNase contamination) for a few hours to concentrate the liquid. If the original concentration is too low, check the troubleshooting section in the manual of the kit and make new reactions.
10. We suggest using a new aliquot of mRNA each time for microinjection, since it may degrade if thawing and freezing frequently.
11. Most of the popular Addgene Cas9 plasmids from Feng Zhang Lab (pX330, pX458, pX459-V2.0, etc. [7, 15]) can be used as the PCR template for making gRNA in vitro transcription template, since they contain the gRNA (or called chimeric RNA) scaffold sequence. Avoid using the first-generation Cas9 plasmids (such as pX260), since they do not contain the gRNA scaffold, but use crRNA and tracrRNA instead.
12. Avoid adding too much plasmid template; otherwise it will remain as the main component in the final PCR product.
13. Most PCR purification kits are not designed to recover small fragments. Read the manual carefully and make sure it is suitable for recovering the 119-bp gRNA template if you want to use a kit.
14. Mismatches in the Cas9/gRNA target sites may dramatically decrease targeting efficiency. Due to the strong sequence polymorphisms of the zebrafish genome, we strongly recommend to confirm the actual sequence of each selected target site in each individual fish before using their offspring to perform the mutagenesis experiments.
15. Adjust the parameters of pressure and duration time of the microinjection machine to calibrate the volume of each drop to be 2 nL. 1 nL is also acceptable; however, larger injection volumes make relatively smaller variations between each shot. The injection needle should be recalibrated after any change in the balance pressure, eject pressure, duration time, or re-breaking of the tips of the needles. Avoid plugging of the needle tips, and occasionally try to inject several drops into the medium to see whether the needle still work normally.
16. The dosage of the zCas9 mRNA and gRNAs injected into the embryo varies by different target sites and needs to be optimized. We always choose the dosage which gives no less than 50% embryo survival rate at 2 dpf. We suggest better to inject the RNAs directly into the cytoplasm instead of the yolk.
17. For different batches of zCas9 mRNA, even though the concentrations of mRNA might be the same, the activity may still be different due to variations of the capping efficiency during in vitro transcription.

18. The crude lysate obtained from this fast genomic DNA extraction method is sufficient for the amplification of small fragment (<500 bp).
19. The RE needs to be tested in advance to make sure it can completely digest the PCR product from control embryos (can be done at the step of target sequences confirmation). It is even better if the RE can function with the unpurified PCR system, which can save time by omitting purification steps.
20. The ratio of PCR products and enzymes is important for a successful assay. If the PCR products from control (uninjected) embryos are digested and showed a smear pattern, the concentration of the enzymes should be reduced (particularly for T7E1).
21. Designing good real-time PCR primers and probes for a specific genomic region is an important but elaborate process. Many different types of PCR probes are available; refer to the supplier's instructions or tools for choosing and preparing good probes.
22. The efficiency of large genomic deletions depends on both the activity of individual Cas9/gRNA and the size of deletion and also other undefined factors. We have successfully generated deletions with the length up to 122 kb with paired customized endonucleases and found ~1/10 of the founder fish can transmit the mutations to offspring with an average mosaicism of 14%.
23. Since the purpose of this detection approach is to get a positive PCR signal, more embryos (up to 50 as we have tested) can be pooled in one PCR tube to reduce the amount of samples. One should ensure that all the embryos are broken up completely when preparing the genomic DNA.

Acknowledgment

We thank Zhanxiang Wang, Da Liu, and other members in our lab for their efforts on optimizing the CRISPR/Cas applications in zebrafish. This work was partially supported by the National Natural Science Foundation of China (31110103904, 81371264), the 973 Program of the Ministry of Science and Technology of China (2012CB945101, 2015CB942803), and the Seeding Grant for Medicine and Life Sciences of Peking University (2014-MB-06).

References

1. Huang P, Zhu Z, Lin S et al (2012) Reverse genetic approaches in zebrafish. J Genet Genomics 39(9):421–433. doi:10.1016/j.jgg.2012.07.004
2. Xiao A, Wu Y, Yang Z et al (2013) EENdb: a database and knowledge base of ZFNs and TALENs for endonuclease engineering. Nucleic Acids Res 41(Database issue):D415–D422. doi:10.1093/nar/gks1144
3. Hisano Y, Ota S, Kawahara A (2014) Genome editing using artificial site-specific nucleases in zebrafish. Dev Growth Differ 56(1):26–33. doi:10.1111/dgd.12094
4. Bedell VM, Ekker SC (2015) Using engineered endonucleases to create knockout and knockin zebrafish models. Methods Mol Biol 1239:291–305. doi:10.1007/978-1-4939-1862-1_17
5. Xiao A, Wang Z, Hu Y et al (2013) Chromosomal deletions and inversions mediated by TALENs and CRISPR/Cas in zebrafish. Nucleic Acids Res 41(14):e141. doi:10.1093/nar/gkt464
6. Liu D, Wang Z, Xiao A et al (2014) Efficient gene targeting in zebrafish mediated by a zebrafish-codon-optimized cas9 and evaluation of off-targeting effect. J Genet Genomics 41(1):43–46. doi:10.1016/j.jgg.2013.11.004
7. Ran FA, Hsu PD, Wright J et al (2013) Genome engineering using the CRISPR-Cas9 system. Nat Protoc 8(11):2281–2308. doi:10.1038/nprot.2013.143
8. Hsu PD, Scott DA, Weinstein JA et al (2013) DNA targeting specificity of RNA-guided Cas9 nucleases. Nat Biotechnol 31(9):827–832. doi:10.1038/nbt.2647
9. Sander JD, Maeder ML, Reyon D, Voytas DF, Joung JK, Dobbs D (2010) ZiFiT (Zinc Finger Targeter): an updated zinc finger engineering tool. Nucleic Acids Res 38(Web Server issue):W462–W468. doi:10.1093/nar/gmk319
10. Xiao A, Cheng Z, Kong L et al (2014) CasOT: a genome-wide Cas9/gRNA off-target searching tool. Bioinformatics. doi:10.1093/bioinformatics/btt764
11. O'Brien A, Bailey TL (2014) GT-Scan: identifying unique genomic targets. Bioinformatics 30(18):2673–2675. doi:10.1093/bioinformatics/btu354
12. Huang P, Xiao A, Tong X et al (2014) TALEN construction via "Unit Assembly" method and targeted genome modifications in zebrafish. Methods 69(1):67–75. doi:10.1016/j.ymeth.2014.02.010
13. Meeker ND, Hutchinson SA, Ho L et al (2007) Method for isolation of PCR-ready genomic DNA from zebrafish tissues. Biotechniques 43(5):610, 612, 614
14. Zheng Q, Cai X, Tan MH et al (2014) Precise gene deletion and replacement using the CRISPR/Cas9 system in human cells. Biotechniques 57(3):115–124. doi:10.2144/000114196
15. Cong L, Ran FA, Cox D et al (2013) Multiplex genome engineering using CRISPR/Cas systems. Science 339(6121):819–823. doi:10.1126/science.1231143

Chapter 6

Testing of *Cis*-Regulatory Elements by Targeted Transgene Integration in Zebrafish Using PhiC31 Integrase

Yavor Hadzhiev, Irene Miguel-Escalada, Darius Balciunas, and Ferenc Müller

Abstract

Herein we present several strategies for testing the function of *cis*-regulatory elements using the PhiC31 integrase system. Firstly, we present two different strategies to analyze the activity of candidate enhancer elements. Targeted integration of candidate enhancers into the same genomic location circumvents the variability-associated random integration and position effects. This method is suitable for testing of candidate enhancers identified through computational or other analyses a priori. Secondly, we present methodology for targeted integration of BACs into the same genomic location(s). By using additional reporters integrated into a BAC, this enables experimental testing whether *cis*-regulatory elements are functional in the sequence inserted in the BAC.

Key words Zebrafish, PhiC31 integrase, Transgenesis, Microinjection

1 Introduction

The PhiC31 integrase is a sequence-specific recombinase of the bacteriophage PhiC31. The PhiC31 integrase catalyzes the recombination between two sites attB and attP. In the presence of the PhiC31 enzyme, an attB-containing donor plasmid can be integrated into a target genomic location containing an attP site irreversibly [1, 2]. Utilizing this property of the PhiC31 integrase, we have developed a system based on two components: recipient transgenic lines containing docking attP site in the genome and targeting plasmids containing attB site. To facilitate screening for precise site-specific integrations, we designed a selection method based on fluorescent reporter color change in the lens [3]. The recipient transgenic lines contain a lens-specific crygc/GFP cassette with an attP site placed between the lens-specific *Xenopus laevis* gamma-crystalline promoter and GFP [4]. To monitor site-specific integration, our targeting (donor) vectors contain an attB site followed by either a red fluorescent reporter (mRFP or mCherry) or Cerulean

Koichi Kawakami et al. (eds.), *Zebrafish: Methods and Protocols*, Methods in Molecular Biology, vol. 1451,
DOI 10.1007/978-1-4939-3771-4_6, © Springer Science+Business Media New York 2016

CFP. Site-specific integration into crygc/attP-GFP docking site is expected to produce a crygc/attR-Red/Cyan recombinant site, which can be scored by red or cyan fluorescence in the lens. Random integration of these targeting vectors into the genome is extremely unlikely to result in lens-specific RFP or CFP expression.

Herein we present several strategies for identification and functional validation of regulatory elements using the PhiC31 integrase system. Firstly, we present two different strategies to analyze the activity of candidate enhancer elements. Targeted integration of candidate enhancers into the same genomic location circumvents the variability-associated random integration and position effects. This method is suitable for testing of candidate enhancers identified through computational or other analyses a priori. Secondly, we present methodology for targeted integration of BACs into the same genomic location(s). By using additional reporters integrated into a BAC, this enables experimental testing what (if any) regulatory elements are present within the BAC.

The two different strategies for testing of potential *cis*-regulatory elements are as follows:

1. An "enhancer/promoter/reporter/polyA-signal" cassette is introduced upstream of the attB site of the attB-Red targeting vector (Fig. 1a). In this case after successful integration, a lens color change (green to red) will be observed and a reporter expression driven by the enhancer/promoter cassette. The GFP reporter in the docking line will become inactive.
2. The "enhancer/promoter" cassette (without a reporter and polyA-signal) can be introduced immediately upstream of the attB site of the attB-Red targeting vector (Fig. 1b). Upon successful target integration, a green to red lens color change will be observed; the enhancer/promoter cassette will be placed in front of the GFP of the docking line, thus driving its expression.

Both designs have some advantages and disadvantages. Inclusion of independent reporter in the first design provides more robustness in ensuring reporter expression in comparison to the second, which utilizes the GFP reporter of the docking line, where it will be an additional 84 bp sequence (the attR site) between the GFP reporter and the promoter. This may lead to out-of-frame reporter transcripts with certain promoter designs.

On the other hand, a PhiC31 integrase independent, random integration in the case of the first design, will contribute to the pattern of the targeted integration, which may require additional outcrossing of the founder fish to be able to separate the two patterns. The utilisation of the GFP reporter of the docking line, upon successful targeted integration, facilitates transient transgenic assays in case of the second design.

Targeted integration of BACs relies on a similar lens color switch scheme to select for targeted integration events. Use standard BAC

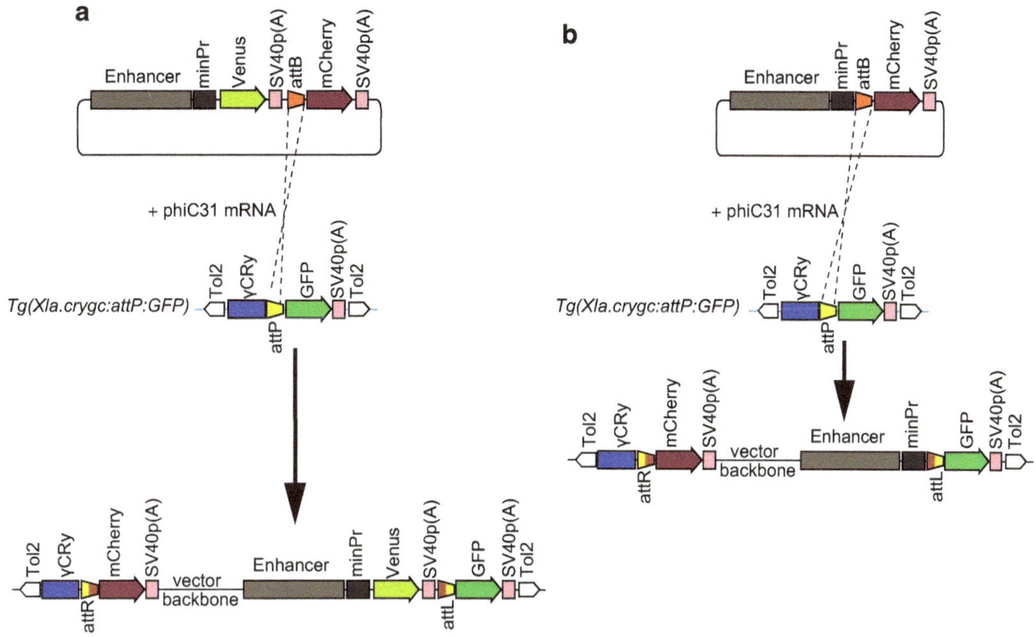

Fig. 1 Schematic representation of two experimental designs for *cis*-regulatory element testing by targeted transgene integration. (**a**) The enhancer/promoter cassette of interest is linked to a reporter Venus (*yellow-green arrow box*) followed by the integration marker mCherry (*dark red arrow box*) after a polyA-signal (*pink box*). Upon successful recombination between the attB and attP sites (*orange* and *yellow trapezium,* respectively), mCherry will be placed in front of the gamma-crystalline promoter (*blue box*) resulting in *green* to *red color* change. The enhancer/promoter cassette will drive the expression of the Venus reporter, and GFP reporter (*green arrow box*) in the docking line will be inactivated. (**b**) The enhancer/promoter cassette of interest relies for expression on the GFP reporter of the docking line. It is placed immediately upstream of attB site and the integration marker mCherry, thus driving its expression prior to targeted integration. After successful integration/recombination, the mCherry will be placed in front of the gamma-crystalline promoter leading to a color change in the lens. The enhancer/promoter cassette will be placed in front of the GFP, thus driving its expression. *γCRY* gamma-crystalline, *p(A)* poly (A) signal, *minPr* minimal promoter

recombineering techniques [5] to insert the attB/cerulean or attB/RFP cassette into the BAC (Grajevskaja et al., under revision). This modified BAC can be directly used for complementation testing. To test for presence of regulatory elements, recombineer a reporter (fluorescent protein or Gal4-VP16) into the gene of interest in the BAC. Since both BACs (as well as any other derivative BACs) will be integrated into the same genomic location, subject to the same (if any) position effects, direct conclusions about requirement for expression in specific subdomains can be drawn from these experiments. Furthermore, candidate *cis*-regulatory elements can be later tested by integration into the same genomic location.

Although the PhiC31 integrase works effectively in zebrafish, high levels of protein are toxic to the early zebrafish embryos. To overcome this problem and make the system more robust to implement, we took advantage of recently published nanos1 3′UTR version of the PhiC31 integrase mRNA [6], which leads to localization of the injected mRNA predominantly in the primordial germ cells (PGCs). This leads to improved germ line transmission and significant increase in embryo survival. The nanos1 3′UTR does not lead to 100 % localization of the PhiC31 integrase mRNA into the PGCs, i.e., there still will be some mRNA in the other cell types which allows the lens color change to be used for prescreening of the injected embryos, even if PhiC31-nanos1-3′UTR is being used. The PhiC31-nanos1-3′UTR does not substantially reduce PGCs survival [3].

2 Materials

2.1 PhiC31 mRNA Synthesis

1. Expression vector containing PhiC31 ORF (with or without *nanos1* 3′UTR) downstream of a phage RNA polymerase promoter (SP6, T7, T3) suitable for in vitro mRNA synthesis, e.g., *pCS2+PhiC31* or *pCS2+PhiC31o-nos1-3′ UTR* [3].
2. In vitro mRNA transcription kit, e.g., mMESSAGE mMACHINE® (Life Technologies).
3. Appropriate restriction enzyme for linearization of the expression vector prior to mRNA synthesis, e.g., NotI or Acc65I/KpnI for *pCS2+PhiC31* and *pCS2+PhiC31o-nos1-3′ UTR* vectors.
4. Phenol/chloroform/isoamyl alcohol 25:24:1 (e.g., Sigma, P3803; for vector template extraction).
5. Acid phenol/chloroform 5:1 (e.g., Sigma P1944, or Life Technologies, AM9720; for mRNA extraction).
6. Chloroform (e.g., Sigma, C2432).
7. 2-Propanolol (isopropanol), molecular biology grade (e.g., Sigma, I9516).
8. 100 and 75 % Ethanol, molecular biology grade.
9. 3 M sodium acetate pH 5.2 (e.g., Sigma, S7899).
10. ***Optional***: PCR (gel) purification kit (e.g., Qiagen, 28104 or equivalent; alternative to phenol/chloroform extraction of the vector template).
11. ***Optional***: MEGAclear™ Transcription Clean-Up Kit (Life Technologies, AM1908) or RNeasy™ RNA Clean-Up Kit (Qiagen 74104) are alternatives to phenol/chloroform extraction of the synthesized mRNA.

2.2 Microinjection and Culturing of the Injected Embryos

1. Docking zebrafish line (stable transgenic zebrafish line containing PhiC31 attP docking site), preferably homozygous.
2. Dissecting stereomicroscope such as Nikon SMZ1500.

3. Pico-injector (e.g., Tritech Research, MINJ-1, or Harvard Apparatus PLI-90/100).
4. Borosilicate capillaries, thin wall with filament, 1.0 mm OD, 0.78 mm ID, 75 mm (e.g., Harvard Apparatus 300040 or World Precision Instruments 1B100F-4).
5. Micropipette Puller (e.g., Sutter Instrument, P-87/97).
6. Micrometer scale slide.
7. Mineral oil.
8. Petri dishes (60 or 90 mm diameter).
9. E3 embryo media, supplemented with 50 μl/ml gentamicin sulfate.
10. 0.5 % phenol red solution in PBS (e.g., Sigma, P0290).
11. Targeting plasmid, containing an attB and the transgene of interest.
12. PhiC31 (or PhiC31-nanos1-3′UTR) integrase mRNA.
13. Small net or tea strainer (for embryo collection).
14. Pasteur pipettes (wide bore and fine tip), forcipes, scalpels.
15. Suitable embryo incubator at 28 °C.
16. Fluorescent stereomicroscope or an upright microscope with a long working distance objective such as Zeiss Fluar 5×/0.25.

RNase-free water for dilution of nucleic acids, e.g., Ambion AM9937. We do not recommend using DEPC-treated water.

Optional: agarose microinjection plates (https://zfin.org/zf_info/zfbook/chapt5/5.1.html)

3 Methods

3.1 PhiC31 mRNA Synthesis

The mRNA in vitro transcription is performed according the instruction manual of the mMESSAGE mMACHINE kit. The main steps are surmised below.

3.1.1 Vector Template Preparation

1. Linearize the plasmid containing the PhiC31 ORF (*see* **Note 1**). About 10 μg of circular plasmid should be digested in the appropriate restriction buffer for ~2 h at the recommended for the restriction enzyme incubation temperature.
2. Purify the linearized plasmid template by using a spin column purification kit (*see* **Note 2**).

3.1.2 In Vitro mRNA Transcription Reaction

1. Assemble the in vitro transcription reaction as described in the mMESSAGE mMACHINE kit manual by suing ~1 μg linearized and purified plasmid template. Optional: add 0.5 μl of RiboLock RNase inhibitor (Thermo Scientific EO0381) to transcription reaction.

2. Incubate at 37 °C for 2 h.
3. Remove the plasmid DNA template by Turbo DNase treatment.
4. Purify the transcribed mRNA by phenol/chloroform extraction or using the MEGAclear™ Transcription Clean-Up Kit or RNeasy RNA clean-up kit.
5. Quantify the mRNA on a NanoDrop (or other type of spectrophotometer) and assay the quality by agarose gel electrophoresis.

3.1.3 Preparation of Donor Plasmid DNA

1. Prepare the donor plasmid DNA with anion-exchange chromatography kits like QIAGEN tip 20, 100 or 500 kits or equivalent (*see* **Note 3**).

3.1.4 Preparation of Donor BAC DNA

1. Prepare BAC DNA using Qiagen Plasmid Maxi Kit (Qiagen 12164).

3.2 Microinjection

3.2.1 Preparation for Microinjection

1. Prepare the microinjection needles from borosilicate microcapillaries using a needle puller instrument (*see* **Note 4**).
2. Prepare the injection solution containing the target plasmid DNA, PhiC31 mRNA, and injection dye marker, e.g., phenol red, allowing to distinguish the injected from non-injected embryos. Injection volumes between 1.5 and 3.0 nl at plasmid DNA concentration between 8 and 25 ng/μl will work well (*see* **Note 5**).

   Example recipe for injection solution (10 μl total volume):

Target plasmid DNA:	*X* μl (5–25 ng/μl final concentration)
PhiC31 mRNA:	*Y* μl (15–30 ng/μl final concentration)
Phenol red (0.5 %):	2 μl (0.1 % final concentration)
Nuclease-free H_2O:	up to 10 μl

3. Alternatively, prepare aliquots of PhiC31-nanos1-3′UTR mRNA in advance. Dilute the PhiC31-nanos1-3′UTR RNA to 42 ng/μl (5× of final concentration), prepare 2 μl aliquots, and store them in a −80 °C freezer. Before injection, dilute donor plasmid DNA to 10.4 ng/μl and add 8 μl of diluted plasmid to the 2 μl aliquot of integrase mRNA. This will result in 25 pg of both DNA and RNA being injected in 3 nL volume. For BAC injections, dilute BAC DNA to 18.3 ng/μl. Add 8 μl of diluted BAC to the 2 μl aliquot of integrase RNA.
4. Set crosses of the docking zebrafish line in the evening before the injection day, according to the procedures in your zebrafish facility (*see* **Note 6**).

3.2.2 Microinjections Procedure

Each zebrafish laboratory has established its own microinjection procedures and equipment. Achieving a good microinjection technique and speed requires initial training by an experienced person and subsequent practicing. Here we summarize the procedures used in our laboratories and some general points which should be considered.

1. Measure and adjust the injection volume by using a micrometer scale or a microcapillary.
2. The micrometer method:

   Put a drop of mineral oil exactly on top of the micrometer scale's division marks, and place it under the injection stereomicroscope.

   Inject solution into the oil drop. The aqueous solution will form a perfect sphere in the hydrophobic oil drop.

   Measure the diameter of the sphere using the micrometer scale.

   The sphere diameter corresponding to the desired injection volume can be calculated using the following formula:

   $$d = 100\sqrt[3]{\frac{6V}{\pi}}$$

   where *d* is the sphere diameter in *micrometers* (μm), *V* is the desired injection volume in *nanoliters* (nl), and π is the pi constant (~3.14).

   For injection volumes between 1.5 and 2.0 nl, the sphere diameter should be approximately 150 μm.

   Adjust the injection pressure and/or time until you get a sphere with the desired diameter.

   The adjustment of the volume should be done every time after a needle change.
3. The microcapillary method:

   After breaking the tip of the needle, perform ten injections into Drummond microcapillary (Cat. No. 1-000-0010) and measure the size of the water column. One millimeter of capillary corresponds to 30 nl volume. Adjust injection time so that ten injections would produce 1 mm water column. For example, if the water column is 2 mm long injection time should be reduced by half.
4. Collect the embryos by pouring the water, from the vessel in which the eggs are being laid, through a small net or tea strainer. Transfer them into a Petri dish with fish water (water from your fish system).
5. Take an empty Petri dish and transfer no more than 200 embryos into it using a wide-bore Pasteur pipette leaving as less as possible water. Aspirate the leftover liquid completely by using a fine-tip Pasteur pipette. Use the same pipette to gently arrange the embryos in a monolayer (not to have embryos on top of each other) if necessary. The complete removal of resid-

ual liquid is very important in order to prevent the embryos from moving/sliding during injection.

6. Alternatively, embryos can be loaded onto an agarose plate with grooves (https://zfin.org/zf_info/zfbook/chapt5/5.1.html) (see also Balciuniene and Balciunas JoVE) (http://www.jove.com/video/50113/gene-trapping-using-gal4-in-zebrafish). Remove excess water to allow surface tension to stabilize embryos in the grove during injection, no cover is necessary.
7. Preform microinjection manually by hand or using a micromanipulator under a stereomicroscope (*see* Fig. 11.4 in [7]) (http://link.springer.com/protocol/10.1007/978-3-642-20792-1_11) (*see* **Note 7**).
8. After injecting all the embryos, add appropriate amount of E3 media (with 50 μl/ml gentamicin sulfate) into the Petri dish and place in incubator adjusted to 28 °C. Alternatively, embryos can be incubated in a simple 1 g/L Instant Ocean salt solution.

3.3 Embryo Culturing, Prescreening, and Screening

1. Grow the injected embryos until 72–96 hpf before lens color change can be observed. The culturing of the embryos is performed at 28 °C in E3 media supplemented with 50 μg/ml gentamicin sulfate or instant salt solution described above. The embryo media needs to be cleaned (by removing any debris and dead/abnormal embryos) and exchanged with fresh media daily. This will prevent the development of fungal/bacterial infections.
2. Prescreen the embryos at 72–96 hpf for lens color change (green to red or cyan), using a fluorescent microscope (stereo or upright) with appropriate filter set (filter cubes optimized for detection of GFP and mCherry are recommended). The gamma-crystalline promoter is highly active starting after 30 hpf, but it takes time for the florescent reporter protein to accumulate in amounts sufficient for detection. This usually occurs by 72–96 hpf (*see* **Note 8**).
3. Grow all injected and normally developing embryos (including embryos negative for mCherry) to adulthood and screen for germ line transmission. A successful transgene integration could have happened in the germ cells but not in the lens, especially if the nos1-3′UTR version of the integrase is used.
4. For BAC transgenesis, it is recommended to prescreen injected embryos for lens color conversion (GFP to CFP) as well as any reporters integrated into the BAC and only select embryos positive for one or both fluorescent markers.
5. After reaching sexual maturity, screen the founder fish for individuals transmitting the target-integrated transgene to their offspring (*see* **Note 9**).

6. For BAC transgenesis, screening by incrossing is recommended because transgenesis efficiency is lower (around 5 %, Grajevskaja et al., under revision).

4 Notes

1. Using circular template will result in long heterogeneous transcripts. Ideally uniquely present restriction enzyme, cutting close after the end of the poly A signal should be used. We recommend using either NotI or Acc65I to linearize pCS2-based PhiC31 and PhiC31-nanos1-3′UTR vectors.
2. After digestion, the linearized plasmid template needs to be purified. If the starting plasmid DNA is of high purity and quality, e.g., prepared with anion-exchange chromatography column (most commercial maxi-, midi-, and some miniprep kits), the purification of the restriction digest reaction can be done using a standard spin column PCR/gel purification kit. The resulting plasmid DNA may have a residual RNase contamination if silica-membrane spin column miniprep kit has been used. In this case a phenol/chloroform extraction is highly recommended to ensure complete removal of RNases.
3. Plasmid DNA extracted with most if not all commercially available plasmid minipreparation kits, based on silica-membrane spin column, will often be contaminated with low amounts of RNase, due to the high amounts of RNase A in the resuspension solution. This can cause degradation of PhiC31 RNA during microinjection, resulting in failure of the experiment. For this reason we recommend using plasmid DNA prepared with anion-exchange chromatography kits like QIAGEN tip 20 (cat#12123), 100 (cat#12143), or 500 (cat#12163) or equivalent. If spin column miniprep kit is used, additional phenol/chloroform extraction of the plasmid DNA is recommended to ensure RNase removal.
4. The type (e.g., diameter, wall thickness, etc.) of the microcapillaries used depends on the type of the injection instrument you are using, in particular the needle holder. Check the specification of your micro-injector before purchasing microcapillaries to ensure that they will fit into the needle holder. It is recommended to use microcapillaries with an inside filament. They are easier to fill with injection solution and less likely to retain air bubbles, which can result in clogging of the injection needle. The optimal pulling parameters on the puller instrument need to be determined empirically. The resulting tip of the needle shouldn't be too long, which will cause them to bend during injection and making the penetration of the embryo's chorion difficult. Too short and/or thick needle tip

on the other side will result in a wide opening and difficulties to adjust/control the injection volume and could damage the embryos during injections. Read the instruction manual of your needle puller instrument to know how changing different parameters/settings affects the length and the thickness of the needle's tip.

5. The optimal concentration of the PhiC31 mRNA requires being determined empirically, due to the toxicity of the intra-state protein. In our experiments we consider a given concentration to be working when minimum 25–30 % of the injected embryos develop normally and around 5–10 % of them show a lens color change (a marker for successful integration event in our experimental design [3]).
6. In brief, a one or more pairs of zebrafish should be put in a small cage containing insert with grid/mesh bottom, filled with system fish water. It is recommended to separate the female from the male fish using a divider. This will prevent them from mating. On the injection day the dividers can be removed when desired, allowing time control of the egg production.
7. The optimal time for injection is within the first 10–15 min. after laying, before the formation of the animal pole; thus the embryo can be injected anywhere (not necessary to target the cytoplasm), and the DNA/RNA will be brought into the cytoplasm by the cytoplasmic streams forming the animal pole. Aim for yolk/cytoplasm interface in late one-cell embryos, and avoid deep injection into the yolk at late one-cell stage to prevent trapping of the injected volume in the yolk cell after cytoplasmic streaming.
8. Due to the mosaic nature of zebrafish transgenesis, a complete color conversion should not be expected, i.e., injected embryos, which are positive for mCherry in the lens, will also express GFP. If the injections are successful, at least 5 % of the injected and normally developing embryos should be positive for mCherry/CFP in the lens. The lens color change should be observed with both versions the wild type and the nos1-3′UTR containing PhiC31 integrase. This is due to the fact that the nos1-3′UTR version is not exclusively localized in the PGCs, and there is sufficient amount of mRNA left in the other cell types for targeted integration to occur.
9. For analysis of *cis*-regulatory elements, it is recommended the screening crosses to be an outcross to wild type fish. Upon successful target integration, the founder embryos should show complete conversion of the lens color from GFP to mCherry, correlating 100 % with the presence of reporter expression from the introduced transgene. As mentioned above the opti-

mal time to screen for lens color conversions is at 72–96 hpf. The stage for detecting reporter activity from the introduced transgene will depend on the properties of the regulatory elements of that transgene. The identified positive embryos can be grown to adulthood in order to establish stable transgenic lines. The expected percentage of founder fish, with germ line transmission is around 10% [3].

Acknowledgments

We thank Shannon Fisher for the pCS2+PhiC31-nos1-3 UTR plasmid. Work in the Mueller laboratory was supported by *ZF-Health* Integrating Project of the Framework7 program of the European Commission.

References

1. Kuhstoss S, Rao RN (1991) Analysis of the integration function of the streptomycete bacteriophage phi C31. J Mol Biol 222(4):897–908
2. Smith MC, Till R, Brady K, Soultanas P, Thorpe H, Smith MC (2004) Synapsis and DNA cleavage in phiC31 integrase-mediated site-specific recombination. Nucleic Acids Res 32(8):2607–2617. doi:10.1093/nar/gkh538
3. Roberts JA, Miguel-Escalada I, Slovik KJ, Walsh KT, Hadzhiev Y, Sanges R, Stupka E, Marsh EK, Balciuniene J, Balciunas D, Muller F (2014) Targeted transgene integration overcomes variability of position effects in zebrafish. Development 141(3):715–724. doi:10.1242/dev.100347
4. Davidson AE, Balciunas D, Mohn D, Shaffer J, Hermanson S, Sivasubbu S, Cliff MP, Hackett PB, Ekker SC (2003) Efficient gene delivery and gene expression in zebrafish using the Sleeping Beauty transposon. Dev Biol 263(2):191–202
5. Warming S, Costantino N, Court DL, Jenkins NA, Copeland NG (2005) Simple and highly efficient BAC recombineering using galK selection. Nucleic Acids Res 33(4):e36. doi:10.1093/nar/gni035
6. Hu G, Goll MG, Fisher S (2011) PhiC31 integrase mediates efficient cassette exchange in the zebrafish germline. Dev Dyn 240(9):2101–2107. doi:10.1002/dvdy.22699
7. Geurts A, Balciunas D, Mates L (2011) Vertebrate transgenesis by transposition. In: Pease S, Saunders TL (eds) Advanced protocols for animal transgenesis, Springer protocols handbooks. Springer, Berlin, pp 213–236. doi:10.1007/978-3-642-20792-1_11

Chapter 7

Fluorescence-Activated Cell Sorting and Gene Expression Profiling of GFP-Positive Cells from Transgenic Zebrafish Lines

Hideyuki Tanabe, Masahide Seki, Mari Itoh, Ailani Deepak, Pradeep Lal, Terumi Horiuchi, Yutaka Suzuki, and Koichi Kawakami

Abstract

Gene expression profiling is a useful approach for deeper understanding of the specificity of cells, tissues, and organs in the transcriptional level. Recent development of high-throughput next-generation sequence (NGS) allows the RNA-seq method for this profiling. This method provides precise information of transcripts about the quantitation and the structure such as the splicing variants. In this chapter, we describe a method for gene expression profiling of GFP-positive cells from transgenic zebrafish by RNA-seq. We labeled specific cells in the brain with GFP by crossing a Gal4 driver line with the UAS:GFP line, isolated those cells by fluorescence-activated cell sorting (FACS), and analyzed by RNA-seq.

Key words Tol2 transgenesis, Gene trap, Enhancer trap, Gal4-UAS system, GFP, FACS, RNA extract, RNA-seq, Next-generation sequence, Gene expression analysis

1 Introduction

We developed Tol2-mediated transgenesis, enhancer trap, gene trap, and Gal4-UAS methods [1–4], and, to date, we generated more than 1000 transgenic lines that expressed Gal4FF in temporally and spatially restricted patterns. These transgenic fish are valuable resources for the study of developmental biology, organogenesis, and neuroscience and can be viewed on the zTrap database (http://kawakami.lab.nig.ac.jp/ztrap/) [5] (Fig. 1). By using these transgenic fish, the specific cell types are labeled by GFP. Also, because of the ease of transgenesis, numerous transgenic zebrafish have been created that express GFP in specific cells, tissues, and organs.

Recent development of massive parallel sequencing, known as next-generation sequence (NGS), allows a new method for the gene expression profiling termed RNA-seq [6, 7]. In this strategy, RNA is converted into a library of cDNA fragments with adaptor

Koichi Kawakami et al. (eds.), *Zebrafish: Methods and Protocols*, Methods in Molecular Biology, vol. 1451,
DOI 10.1007/978-1-4939-3771-4_7, © Springer Science+Business Media New York 2016

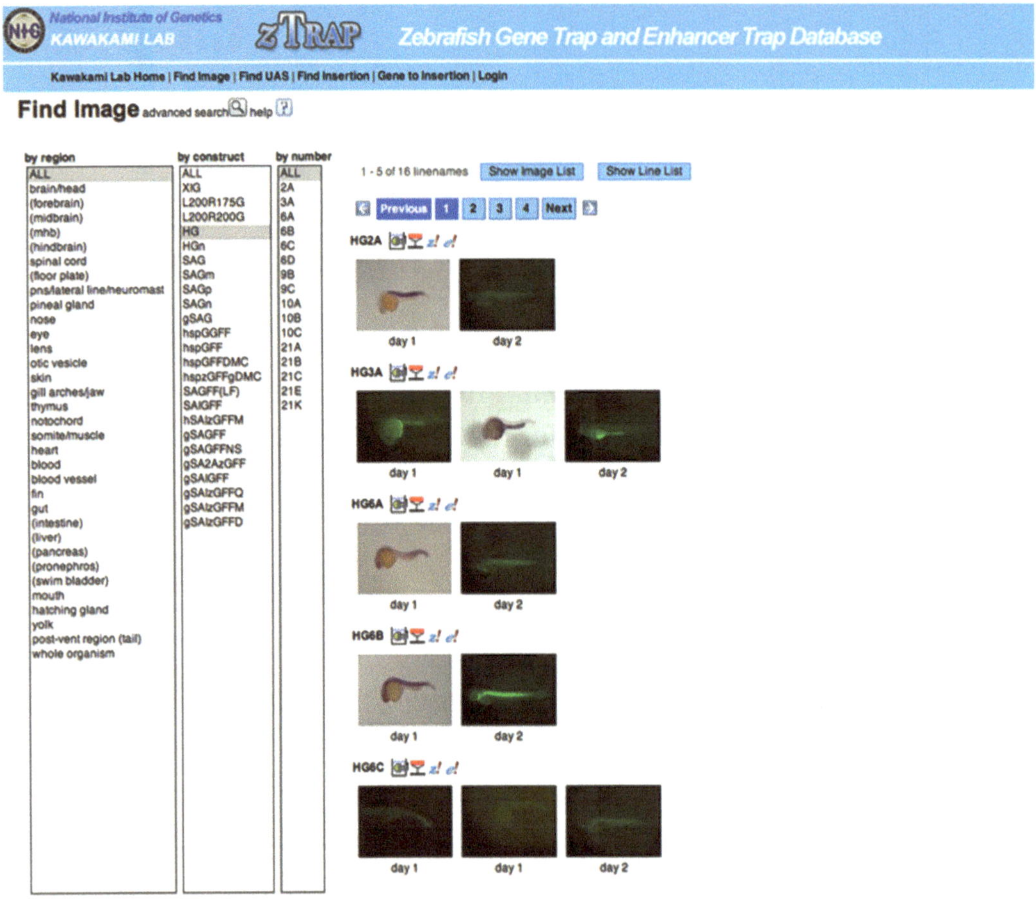

Fig. 1 zTrap database. Top page of the zebrafish gene trap and enhancer trap database (zTrap) that contains the information about expression patterns and genomic loci where transposon insertions are landed. URL: http://kawakami.lab.nig.ac.jp/ztrap/

molecules, and these fragments are sequenced in a high-throughput manner. For the data analysis, these sequence reads are either aligned to the reference genome or de novo alignment of these reads. In this method, the gene expression levels are estimated by counting the number of the sequence reads [6, 7].

This RNA-seq method shows several advantages over the existing hybridization-based approach, DNA microarray [8, 9]. For example, in DNA microarray, the detectable genes and splicing variants are limited in which probes are designed previously. Furthermore, the dynamic range of the expression level is also limited owing to the high background and the saturation of signals. On the other hand, RNA-seq doesn't have the limitation in detection of novel genes and splicing variants, and the dynamic range is broader than the microarray. This quantitative information of RNA-seq has been shown to be highly accurate and reproducible [10–12]. In addition, it is also an advantage of the RNA-seq that less RNA sample is enough for the analysis.

In this chapter, we introduce a protocol to combine the GFP-expressing transgenic fish resource with gene expression profiling by RNA-seq. We collected a brain from transgenic zebrafish, which expressed GFP in specific cells in the brain, and isolated GFP-positive cells by using fluorescence-activated cell sorting (FACS). Then we extracted a small amount (~10 ng) of total RNA from the isolated cells and analyzed gene expression by using RNA-seq.

2 Materials

2.1 Fluorescence-Activated Cell Sorting of GFP-Positive Cells from Brain

1. Transgenic zebrafish with GFP expression in the brain (telencephalon) and wild-type zebrafish.
2. PBS (137 mM NaCl, 2.7 mM KCl, 10 mM Na_2HPO_4, 2 mM KH_2PO_4).
3. Glucose solution (4 mg/mL glucose in PBS).
4. 0.025 % Tricaine in system water.
5. Forceps (DUMOSTAR 55) × 2 (Dumont, Switzerland).
6. 3.5 cm Petri dish coated by silicon (silicon plate, *see* **Note 1** and Fig. 2a).
7. Solution A (1 mg/mL BSA, 1.8 mg/mL glucose in PBS).
8. Papain solution (0.5 mg/mL papain in solution A).
9. PBS with 1 % FBS.
10. CellTrics filter (50 μm) (Sysmex, Japan).
11. Polystyrene round-bottom tube (Falcon).
12. JSAN desktop cell sorter (Bay bioscience Corp. Japan).
13. Sonicator.
14. Milli-Q water.
15. FACS Rinse Solution (BD Bioscience, USA).
16. FACS Sheath Solution (BD Bioscience, USA).

2.2 RNA Extraction and cDNA Synthesis

1. TRIzol Reagent (Thermo Fisher Scientific, USA).
2. Chloroform/isoamyl alcohol 24:1 (Sigma, USA).
3. 2 mg/mL Glycogen solution.
4. Isopropanol.
5. 75 % Ethanol (v/v).
6. Agilent RNA 6000 Pico kit (Agilent, USA).
7. 2100 Bioanalyzer (Agilent, USA).
8. SMARTer Ultra Low RNA kit for Illumina Sequencing (Clontech, USA).
9. Advantage 2 PCR Kit (Clontech, USA).

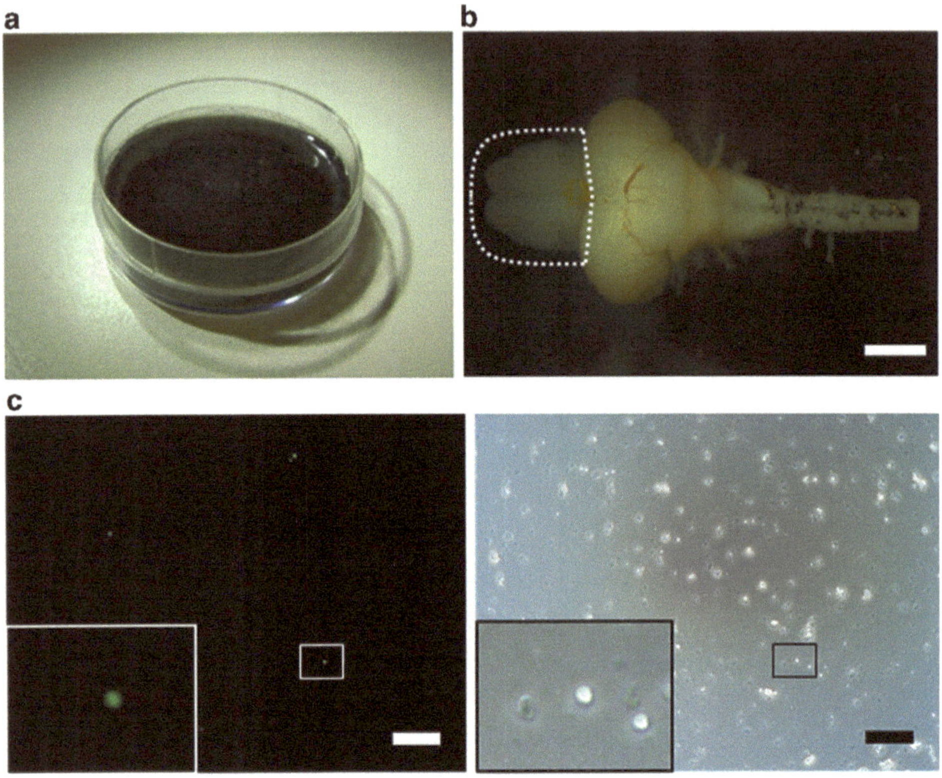

Fig. 2 Dissection of the brain. (**a**) Silicon plate. (**b**) The top view of the adult brain. The *dots* indicate the telencephalon. Scale bar: 1 mm. (**c**) Microscopic images of dissociated cells from the telencephalon. (*Left*) GFP fluorescence. (*Right*) Bright field. *Insets* show magnified views. Scale bar: 100 μm

10. IsoFreeze rack (precooled at −20 °C).
11. AMPure XP Beads (Beckman Coulter).
12. DynaMag-2 (Thermo Fisher Scientific, USA).
13. 80 % ethanol (fresh).
14. Ultrapure water (Milli-Q).
15. Agilent High Sensitivity DNA kit (Agilent, USA).

2.3 Library Preparation and Sequencing

1. Covaris S2 (Covaris, USA).
2. TruSeq DNA Sample Preparation Kits (Illumina, USA).
3. Thermal cycler.
4. Agilent DNA 7500 kit (Agilent, USA).
5. 0.1 M NaOH.
6. 10 mM Tris-Cl (pH 8.5) with 0.1 % Tween 20.
7. TruSeq PE Cluster Kit v3-cBot-HS (Illumina, USA).
8. cBot (Illumina, USA).
9. TruSeq SBS Kit v3-HS (Illumina, USA).
10. HiSeq 2500 (Illumina, USA).

3 Methods

3.1 Fluorescence-Activated Cell Sorting of GFP-Positive Cells from the Brain

1. Anesthetize the adult fish with 0.025 % Tricaine.
2. Cut the head and transfer into cooled PBS in a silicon plate (Fig. 2a; *see* **Note 1**).
3. By using forceps, remove the soft tissue of the ventral side of head and eyes by cutting optic nerve.
4. Remove the skull (*see* **Note 2**) and put the brain sample (telencephalon) (Fig. 2b) into cooled glucose solution. We used 40 transgenic fish. Five wild-type fish were used to set up FACS system.
5. Transfer five brain samples into one 1.5 mL tube, and gently remove glucose solution.
6. Add 800 μL of papain solution to the tube. Dissociate the brain by pipetting ten times (cut the end of a tip). Incubate at 28 °C for 40 min. Debris precipitates at the bottom.
7. Gently remove excess solution and add 800 μL of Solution A. Dissociate the brain by pipetting 10–20 times (cut the end of a tip). Incubate at 28 °C for 10 min.
8. Collect cell suspension into a new 15 mL tube, and keep it on ice.
9. Repeat **step 8**, until no debris remains at the bottom (about five times).
10. Centrifuge the 15 mL tube at 800 × *g* for 10 min at 4 °C, and remove supernatant.
11. Add 3 mL of cooled PBS with 1 % FBS. Centrifuge at 800 × *g* for 10 min at 4 °C. Remove the supernatant. Repeat this step once.
12. Add 500 μL of cooled PBS with 1 % FBS to the tube. Filtrate the cell suspension with CellTrics filter (50 μm), and transfer it to a new polystyrene round-bottom tube (FACS tube). Keep it on ice, until you perform the FACS sorting (Fig. 2c).
13. Start up the main unit of JSAN.
14. Create an experiment file, an FSC-SSC plot, and the histogram plot of GFP expression (*X*-axis: Fl1 log) (Fig. 3a, b) (*see* **Note 3**).
15. Set wild-type sample on the sample station, and run. Adjust flow rate within the range of 500–1000 events/s. Adjust the FSC and SSC voltages to place your population of interest on scale within *X*-axis or *Y*-axis. Adjust histogram plot to place cells within the range of 10^0 and 10^1 in *X*-axis.
16. Set GFP-positive samples on the sample station, and run. Create gate on the GFP-positive region in the histogram and populated area of GFP cells in the FSC-SSC plot. If you can't

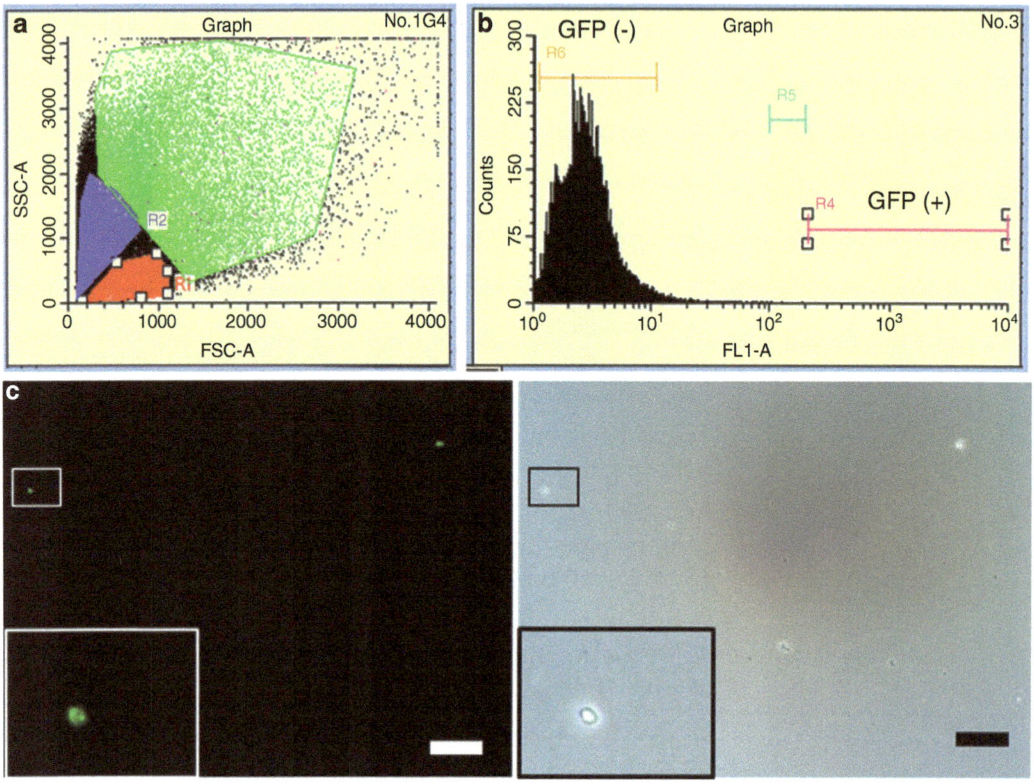

Fig. 3 FACS of the GFP-positive cells. (**a**) Characterization of cells dissociated from the brain (FSC-SSC plot). We tentatively divided cells into three groups (*blue*, *red*, and *green*). These three groups of cells contain GFP-positive cells similarly. (**b**) Histogram plot of all of the cells dissociated from the brain. *X*-axis and *Y*-axis show the intensity of GFP fluorescence and the number of cells, respectively. We set the gate at 2×10^2 for GFP-positives and at 10^0–10^1 for GFP-negatives. (**c**) Microscopic images of cells from GFP-positive fraction. (*Left*) GFP fluorescence. (*Right*) Bright field. *Insets* show magnified views. Scale bar: 100 μm

find the populated area in FSC-SSC plot, skip the FSC-SSC process. We created a gate on the histogram. The range of GFP-positive is higher than 2×10^2 in *X*-axis, and the range of the GFP-negative is between 10^0 and 10^1 (Fig. 3a, b).

17. Set up cell sorter (*see* **Note 4**), and set tubes on the tube stand. Put 500 μL of PBS with 1 % FBS into the tubes. Cool the tubes on ice.
18. Collect sorted cells both from GFP-positive and the GFP-negative fractions (Fig. 3c). Transfer tubes with sorted cells immediately on ice. Supply the cooled tubes as new tubes.

3.2 RNA Extraction and cDNA Synthesis

3.2.1 RNA Extraction

1. Transfer GFP-positive cells (in our experiments, 6.0×10^3 cells and 2.0×10^6 cells were obtained from GFP-positive and GFP-negative fractions, respectively) into new 1.5 mL tubes. Centrifuge at 1200 × *g* for 15 min at 4 °C.

2. Remove the supernatant. Add 800 μL of TRIzol Reagent and homogenize on ice.
3. Add 160 μL of chloroform/isoamyl alcohol. Mix by shaking, and incubate for 3 min at room temperature. Centrifuge at 18,000 × *g* for 15 min at 4 °C.
4. Transfer upper layer into a new tube. Add 5 μL of 2 mg/mL glycogen and 400 μL of isopropanol, mix by tapping the tube, and incubate for 10 min at room temperature.
5. Centrifuge at 18,000 × *g* for 15 min at 4 °C. Remove the supernatant.
6. Wash precipitates with 75 % ethanol (200 μL). Centrifuge at 18,000 × *g* for 5 min at 4 °C, and remove the supernatant. Dry for 5 min. Do not dry too much.
7. Add 10 μL of RNase free water. The sample can be stored at −80 °C.
8. Evaluate RNA by using the Agilent RNA 6000 Pico kit and the 2100 Bioanalyzer (Fig. 4a).

3.2.2 cDNA Synthesis

1. Perform cDNA synthesis using SMARTer Ultra Low RNA kit for Illumina sequence
2. Prepare reaction buffer by mixing 1 μL of RNase inhibitor with 19 μL of dilution buffer.

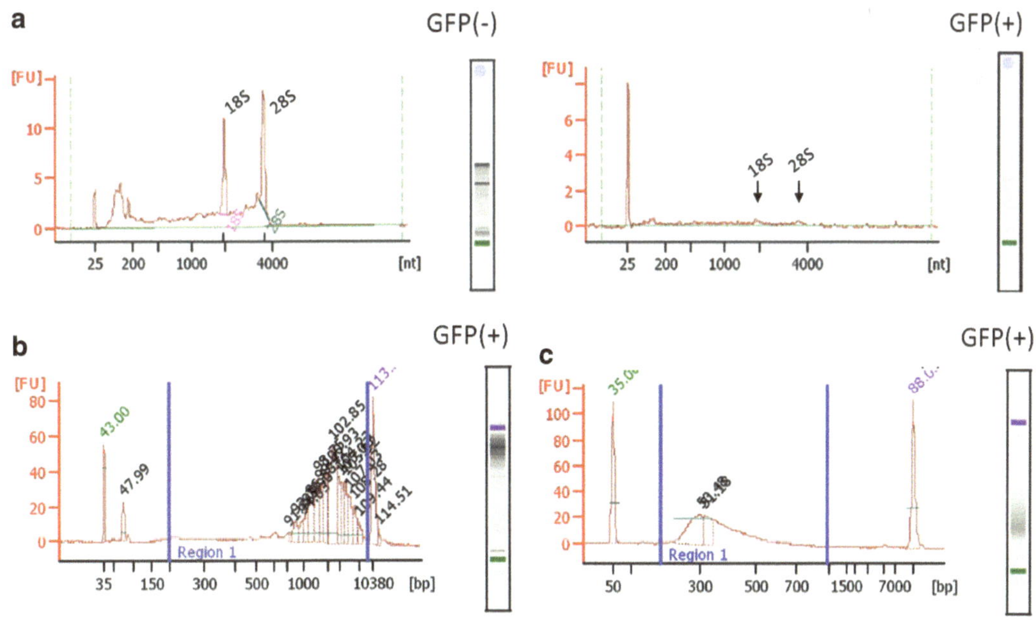

Fig. 4 Analysis of RNA and cDNA samples by Agilent 2100 Bioanalyzer. (**a**) Analysis of the RNA quality. (*Left*) Electrogram of RNA from GFP-negative cells. (*Right*) Electrogram of RNA from GFP-positive cells. Small peaks of 18s and 28s rRNA can be seen in the *right* and this RNA sample was used for cDNA synthesis. (**b**) Analysis of cDNA. The average size is about 2350 bp. (**c**) Analysis of fragmented library cDNA. The average size is about 400 bp

3. Gently mix 1 μL of RNA samples with 2.5 μL of reaction buffer. Place on IsoFreeze rack. Add 1 μL of 3′ SMART CDS Primer IIA (12 μM).
4. Incubate samples at 72 °C for 3 min in thermal cycler. Then transfer to the IsoFreeze rack immediately.
5. Prepare Master Mix: 2 μL of 5× First Strand Buffer, 0.25 μL of DTT (100 mM), 1 μL of dNTP Mix (10 mM), 1 μL of SMARTer IIA Oligonucleotide (12 μM), 0.25 μL of RNase inhibitor (40 U/μL), 1 μL of SMART Scribe Reverse Transcriptase (100 U/μL).
6. Mix the sample with Master Mix gently. Incubate at 42 °C for 90 min.
7. Terminate reaction by heating at 70 °C for 10 min.
8. Vortex AMPure XP Beads, and add 25 μL of the beads to the cDNA sample (*see* **Note 5**). Mix by pipetting ten times. Incubate at room temperature for 15 min.
9. Place samples on the magnetic stand (DynaMag-2) at room temperature, until the liquid becomes completely clear (about 5 min).
10. Remove the supernatant by pipette while sample is on the magnetic stand.

3.2.3 Amplification of cDNA by Long-Distance PCR

1. Prepare PCR Master Mix (Advantage 2 PCR Kit): 5 μL of 10× Advantage 2 PCR Buffer, 2 μL of dNTP Mix (10 mM), 2 μL of IS PCR Primer (12 μM), 2 μL of 50× Polymerase Mix, 39 μL of nuclease-free water.
2. Add 50 μL of Master Mix to each sample. Run PCR: 95 °C, 1 min; 15 cycles of 95 °C 15 s, 65 °C 30 s, 68 °C 6 min; 72 °C 10 min followed by 4 °C.
3. Vortex AMPure XP Beads, and add 90 μL of the beads to the PCR sample, and mix by pipetting ten times. Incubate at room temperature for 15 min.
4. Place on the magnetic stand for 5 min. Remove the supernatant.
5. Add 200 μL of 80% ethanol, while samples are on the magnetic stand. Incubate for 30 s and carefully remove the supernatant. Repeat this step once. Remove ethanol, and dry the samples at room temperature for 15 min. Do not dry the beads too much.
6. Add 50 μL of Milli-Q water. Remove the samples from the magnetic stand, and incubate at room temperature for 2 min. Mix well and back on the magnetic stand for 1 min.
7. Transfer the clear supernatant to a new tube. The sample can be stored at −20 °C.
8. Evaluate the cDNA by using Agilent High Sensitivity DNA kit and 2100 Bioanalyzer (Fig. 4b).

3.3 Library Preparation and Sequence

Library preparation for Illumina sequencing is done by the TruSeq DNA sample preparation kit and the low sample protocol.

1. In our experiment, we prepared 35.9 ng cDNA for the GFP-positive sample and 130.5 ng for the GFP-negative sample.
2. *cDNA fragmentation.* Turn on the Covaris S2 instrument at least 30 min before use. Start the fragmentation procedure at 6 °C.
3. Make the double-strand cDNA sample to a final volume of 55 μL. Transfer 52.5 μL of the sample to Covaris tubes. Fragmentation conditions: duty cycle 10%; intensity 5.0; bursts per second 200; duration 120 s.
4. *End repair.* Mix the 50 μL of fragmented cDNA with 40 μL of End Repair Mix and 10 μL of ultrapure water. Incubate at 30 °C for 30 min in the thermal cycler.
5. Add 160 μL of AMPure XP Beads and gently mix.
6. Incubate at room temperature for 15 min. Place samples on the magnetic stand for 5 min. And remove the supernatant. Wash beads with 200 μL of 80% ethanol. Incubate for 30 s on the magnetic stand, and remove the supernatant. Repeat this step once. Dry beads on the magnetic stand for 15 min at room temperature for 15 min. Remove the samples from the magnetic stand.
7. Resuspend the dried pellet with 17.5 μL of ultrapure water, and gently mix. Incubate the sample at room temperature for 2 min.
8. Place the sample on the magnetic stand for 5 min. Transfer 15 μL of the clear supernatant to a new tube. The sample can be stored at −20 °C.
9. *Adenylation of 3′ ends.* Mix the sample with 2.5 μL of ultrapure water and 12.5 μL of A-Tailing Mix.
10. Incubate the sample in the thermal cycler at following conditions: 37 °C for 30 min, 70 °C for 5 min, and cool to 4 °C. Proceed immediately to adapter ligation.
11. *Adapter ligation.* Mix 30 μL of samples with 2.5 μL of Ligation Mix, 2.5 μL of DNA adaptor, and 2.5 μL of ultrapure water. Incubate at 30 °C for 10 min.
12. Add 5 μL of Stop Ligation Buffer to the sample.
13. Add 42.5 μL of AMPure XP Beads and gently mix.
14. Repeat **step 6**.
15. Resuspend the dried pellet with 52.5 μL of ultrapure water and gently mix. Incubate the sample for 2 min at room temperature. Place the sample on the magnetic stand for 5 min, and transfer 50 μL of clear supernatant to a new tube. Add 50 μL of AMPure XP Beads, and gently mix.

16. Repeat **step 6**.
17. Resuspend the dried pellet with 22.5 μL of ultrapure water. Incubate for 2 min at room temperature. Place the sample on the magnetic stand for 5 min, and transfer 20 μL of clear supernatant to 0.2 mL PCR tubes.
18. *Enrichment of DNA fragments*. Mix sample with 5 μL of PCR Primer Cocktail and 25 μL of PCR Master Mix.
19. Place samples in a thermal cycler and run PCR: 98 °C 30 s; ten cycles of 98 °C 10 s, 60 °C 30 s, 72 °C 30 s; 72 °C 5 min followed by 10 °C.
20. Add 50 μL of AMPure XP Beads.
21. Repeat **step 6**.
22. Resuspend the dried pellet with 32.5 μL of ultrapure water. Incubate the sample for 2 min at room temperature.
23. Place the sample on the magnetic stand for 5 min, and transfer 30 μL of clear supernatant to a new tube.
24. Validate cDNA library with Agilent DNA 7500 kit and 2100 Bioanalyzer (Fig. 4c).
25. *Cluster generation*. We employed cBot and TruSeq PE Cluster Kit v3-cBot-HS for the cluster generation. Prepare 0.1 M NaOH and 10 mM Tris-Cl (pH 8.5) with 0.1% Tween 20.
26. *Sequencing*. We employed HiSeq 2500 and TruSeq SBS Kit v3-HS (*see* **Note 6**).

3.4 Data Analysis

For gene expression profiling, you align your sequence reads against reference genome (alignment), and count the aligned reads on the gene region of the reference to estimate the gene expression levels (read count). Finally, you perform statistical analysis for differential expression. For this profiling, various types of software are available (*see* **Note 7**). Here, we describe the Eland-based methodology for the quantification. This algorithm is provided by CASAVA, Illumina software for alignment, running on Linux server. The version we used for the analysis is CASAVA-1.8.2:

1. You get sequence information in Bcl format. Convert Bcl files into FASTQ files by using the CASAVA.
2. Prepare the genome reference information, the exon-junction information, and ribosomal RNA information (optional, *see* **Note 8**) of zebrafish. Save the genome information and the ribosomal information in FASTA format and the annotation file in the original format.

   Genome information: danRer7.fa (UCSC) http://hgdownload.cse.ucsc.edu/goldenPath/danRer7/bigZips/

Exon-junction information: refFLAT.txt (UCSC) http://hgdownload.soe.ucsc.edu/goldenPath/danRer7/database/refFlat.txt.gz

Ribosomal RNA information

5S: Ensemble/BioMart/Data set: Ensemble Gene 81, Danio rerio genes/Filters/Gene/Gene type: rRNA; and push Results

18S: http://www.ncbi.nlm.nih.gov/nuccore/FJ915075.1 (NCBI)

28S: http://www.ncbi.nlm.nih.gov/nuccore/AF398343.1 (NCBI)

3. Make configuration text file. You describe here the design of the alignment. Example is shown below.

   ELAND_GENOME/path/to/directory (folder) of danRer7.fa

   ELAND_RNA_GENOME_ANNOTATION/path/to/directory of refFLAT.txt.gz

   ELAND_RNA_GENOME_CONTAM/path/to/directory of ribosomal RNA.fa

   USE_BASES Y100,n100 (Align the first 100 bases of read1)

   ANALYSIS eland_rna (*see* **Note 9**)

   ELAND_FASTQ_FILES_PER_PROCESS 3 (Align FASTQ file for the 3 references in one process)

4. *Alignment.* Make folders to store the alignment files and the unalignment files. Perform the following command by using Linux for alignment: Run CASAVA (configureAlignment.pl), reference the configuration file, and store the output alignment files in the folder.
5. *Read count.* To estimate the gene expression level, you choose the aligned reads suitable for your calculation. Select the aligned reads satisfying the following conditions by using Perl Script: Allow less than two mismatches per 100 bases; form a pair with read1 and read2.
6. Count the read number on the gene region of the reference: between UTR Start (uppermost region of the gene) and UTR End (lowermost region of the gene), including the splice junction (by using Perl Script).
7. Calculate ppm (read numbers × million/proper mapped reads) and RPKM (ppm/mRNA length).
8. *Differential expression analysis.* Perform DE-seq analysis [13]. You can download the application from the Bioconductor web site (http://www.bioconductor.org/packages/release/bioc/html/DESeq.html).

4 Notes

1. Preparation of Silicon plate

   Materials: Silicon (KE-106, Shin-Etsu silicones Co., Japan), Curing catalyst (CAT-RG, Shin-Etsu silicones Co., Japan), Indian ink (Boku eki, Daiso, Japan).

   (a) Add 10 g CAT-RG into KE106 (10% w/w), and mix.

   (b) Add Indian ink and mix until the color changes to gray.

   (c) Centrifuge the mixture to remove the bubble.

   (d) Spread the mixture into 3.5 cm dishes. Incubate overnight at 60 °C.

2. (1) Cut the tip of olfactory bulb and both sides of the telencephalon. (2) Cut the bone covering the telencephalon to remove the skull on it. (3) Cut both sides of the tectum and cerebellum, and gently remove the remaining skull over the brain.

3. If you sort cells in multi-color FACS, you should use dot-plot of FL1–FL2. In this case, you have to calculate the color compensation to prevent the leakage of the fluorescent.

4. Make sure that the drop formation is stable and the break off point is near the center. You can choose the purification level in the sort mode. In our experiment, we used one droplet.

5. AMPure XP Beads are viscous, then suck the entire volume up, and push it out slowly. Before each use, keep beads at room temperature for at least 30 min, and mix well by Vortex to disperse.

6. In RNA-seq, you need to set up read length and format (single-end read or paired-end read). In the Illumina sequence support page (http://support.illumina.com/sequencing/faqs.html), it is recommended as follows:

   For gene profiling (gene-level counts): single-end read for 50 bp.

   For complete transcriptome annotation: paired-end read for 75–100 bp.

7. Alignment algorithms: TopHat [14, 15], MapSplice [16], Eland (Illumina). Annotation: Cufflinks [17], Scripture [18]. Differential expression analysis: edgeR [19], DE-seq [13, 20].

8. To check the validity of polyA selection, we mapped reads on the ribosomal RNA sequence as negative control.

9. eland_rna is one mode of the alignment of Eland v2. You can align your reads against reference genome and also splice junction.

Acknowledgments

We thank Naoyuki Inagaki for the helpful advice about cell dissociation methods. This work was partly supported by the National BioResource Project (to KK), Grant-in-Aids for Scientific Research on Innovative Areas "Genome Science" (221S0002 to YS), and Grant-in-Aids for Scientific Research (A) (23241063 and 15H02370 to KK) from the Ministry of Education, Culture, Sports, Science and Technology of Japan.

References

1. Kawakami K, Takeda H, Kawakami N, Kobayashi M, Matsuda N, Mishina M (2004) A transposon-mediated gene trap approach identifies developmentally regulated genes in zebrafish. Dev Cell 7:133–144
2. Nagayoshi S, Hayashi E, Abe G, Osato N, Asakawa K, Urasaki A, Horikawa K, Ikeo K, Takeda H, Kawakami K (2008) Insertional mutagenesis by the Tol2 transposon-mediated enhancer trap approach generated mutations in two developmental genes: tcf7 and synembryn-like. Development 135:159–169
3. Scott EK, Mason L, Arrenberg AB, Ziv L, Gosse NJ, Xiao T, Chi NC, Asakawa K, Kawakami K, Baier H (2007) Targeting neural circuitry in zebrafish using GAL4 enhancer trapping. Nat Methods 4:323–326
4. Asakawa K, Suster ML, Mizusawa K, Nagayoshi S, Kotani T, Urasaki A, Kishimoto Y, Hibi M, Kawakami K (2008) Genetic dissection of neural circuits by Tol2 transposon-mediated Gal4 gene and enhancer trapping in zebrafish. Proc Natl Acad Sci U S A 105:1255–1260
5. Kawakami K, Abe G, Asada T, Asakawa K, Fukuda R, Ito A, Lal P, Mouri N, Muto A, Suster ML et al (2010) zTrap: zebrafish gene trap and enhancer trap database. BMC Dev Biol 10:105
6. Wang Z, Gerstein M, Snyder M (2009) RNA-Seq: a revolutionary tool for transcriptomics. Nat Rev Genet 10:57–63
7. Ozsolak F, Milos PM (2011) NIH public access. Nat Rev Genet 12:87–98
8. Sultan M, Schulz MH, Richard H, Magen A, Klingenhoff A, Scherf M, Seifert M, Borodina T, Soldatov A, Parkhomchuk D et al (2008) A global view of gene activity and alternative splicing by deep sequencing of the Human Transcriptome. Science 321:956–960
9. Zhao S, Fung-Leung WP, Bittner A, Ngo K, Liu X (2014) Comparison of RNA-Seq and microarray in transcriptome profiling of activated T cells. PLoS One 9:e78644
10. Carninci P, Kasukawa T, Katayama S, Gough J, Frith MC, Maeda N, Oyama R, Ravasi T, Lenhard B, Wells C et al (2005) The transcriptional landscape of the mammalian genome. Science 309:1559–1563
11. Cloonan N, Forrest ARR, Kolle G, Gardiner BBA, Faulkner GJ, Brown MK, Taylor DF, Steptoe AL, Wani S, Bethel G et al (2008) Stem cell transcriptome profiling via massive-scale mRNA sequencing. Nat Methods 5:613–619
12. Mortazavi A, Williams BA, McCue K, Schaeffer L, Wold B (2008) Mapping and quantifying mammalian transcriptomes by RNA-Seq. Nat Methods 5:621–628
13. Anders S, Huber W (2010) Differential expression analysis for sequence count data. Genome Biol 11:R106
14. Trapnell C, Pachter L, Salzberg SL (2009) TopHat: discovering splice junctions with RNA-Seq. Bioinformatics 25:1105–1111
15. Kim D, Pertea G, Trapnell C, Pimentel H, Kelley R, Salzberg SL (2013) TopHat2: accurate alignment of transcriptomes in the presence of insertions, deletions and gene fusions. Genome Biol 14:R36
16. Wang K, Singh D, Zeng Z, Coleman SJ, Huang Y, Savich GL, He X, Mieczkowski P, Grimm SA, Perou CM et al (2010) MapSplice: accurate mapping of RNA-seq reads for splice junction discovery. Nucleic Acids Res 8:1–14
17. Trapnell C, Williams BA, Pertea G, Mortazavi A, Kwan G, van Baren MJ, Salzberg SL, Wold BJ, Pachter L (2010) Transcript assembly and quantification by RNA-Seq reveals unannotated transcripts and isoform switching during cell differentiation. Nat Biotechnol 28:511–515
18. Guttman M, Garber M, Levin JZ, Donaghey J, Robinson J, Adiconis X, Fan L, Koziol MJ, Gnirke

A, Nusbaum C et al (2010) Ab initio reconstruction of cell type-specific transcriptomes in mouse reveals the conserved multi-exonic structure of lincRNAs. Nat Biotechnol 28:503–510

19. Robinson MD, McCarthy DJ, Smyth GK (2009) edgeR: a bioconductor package for differential expression analysis of digital gene expression data. Bioinformatics 26:139–140

20. Love MI, Huber W, Anders S (2014) Moderated estimation of fold change and dispersion for RNA-seq data with DESeq2. Genome Biol 15:550

Chapter 8

Construction of the Inbred Strain

Minori Shinya

Abstract

Genetically homogeneous populations such as inbred strains are valuable experimental tools in various fields of biomedical analyses. In many animals, inbred strains are established by consecutive sib-pair mating for a minimum of 20 generations. As the generation proceeds, fitness of the population reduces usually. Therefore, in order to establish inbred strains, the important point is the selection of pairs in good condition at each generation. Here, I describe the procedure and tips for generating inbred strains in zebrafish.

Key words Zebrafish, Inbred strain, Genetic homogeneity, Polymorphisms, Sib-pair mating, Inbreeding depression

1 Introduction

A species usually retains genetic variations in a population, and these genetic variations are partly, at least, responsible for the phenotypic diversity of the species. Recent remarkable progress of the sequencing technology enables us to determine the genomic sequence of individuals, providing enormous information about genetic variations, or polymorphisms in a whole genome. Taking this accumulation of knowledge, researches have become active in the fields connecting "diversity" of various phenotypes with genotypes, for instance, which genotypes relate to the differences in the effectiveness and the side effects of a medicine and how genetic polymorphisms affect to the differences in our thought or personality.

An inbred strain is a population of particular species in which individuals are nearly identical to each other in genotype. Therefore, from the point of genetic diversity, inbred strains are quite peculiar and artificial populations in contrast to the populations in nature. This genetic uniformity, however, is a great advantage to analyze various kinds of biological and medical subjects, such as immunology, cancer, and genetics of complex traits. Because inbred strains provide highly reproducible results and also make possible to dissect genetic and environmental effects, they are essential tools to investigate the genetic background of phenotypic diversity [1, 2].

Koichi Kawakami et al. (eds.), *Zebrafish: Methods and Protocols*, Methods in Molecular Biology, vol. 1451, DOI 10.1007/978-1-4939-3771-4_8, © Springer Science+Business Media New York 2016

Thus, inbred strains have been established in many of the so-called model organisms. In both mice and medaka, over ten inbred strains have been generated by continuous sib-pair mating for more than 20 generations [3, 4]. The repetition of sib-pair mating decreases the genetic diversity of the family, and continuous full sib-pair mating for 20 generations will result in, on average, each individual homozygous in 98.6% of the genome [5].

In zebrafish, there are few inbred strains which are comparable to those in mice and medaka. One of the reasons for that may be inbreeding depression, the reduced survival and fertility of offspring of related individuals. As the genetic variations decrease in a population, the biological fitness also reduces. In fact, one of the two zebrafish strains which had started to be inbred through sib-pair mating, failed to thrive and was lost after 13 generations [6]. Even though the inbreeding depression in zebrafish seems to be stronger than that in medaka, it is possible to establish an inbred strain in zebrafish. The other strain inbred, IM strain, has been passed to generation 16 [6] and is successfully maintained for more than 20 generations by continuous sib-pair mating (unpublished data). Furthermore, another strain is at generation 17 as of May, 2016 and looks quite healthy (unpublished data). The procedure to generate an inbred strain itself is quite simple, just mating and breeding fish repeatedly. What you have to care is the inbreeding depression. A key to overcome the inbreeding depression and to obtain an inbred is simply how much good pair is selected in each generation.

2 Materials

2.1 Fish Mating, Collecting Eggs, and Maintenance of Embryos and Fish

1. Fish tanks (*see* **Note 1**).
2. Cages for pair-mating as many as the numbers of pair.
3. 10-cm plastic petri dish.
4. 15-cm plastic petri dish.
5. Water for egg: Autoclave tap water (*see* **Note 2**). Incubate the water at 28.5 °C before use.
6. Embryo transfer pipette: disposable plastic pipette (3 mL) or Pasteur pipette with a wide opening (cut and flame polished).
7. Brine shrimp.
8. Dry flake food, e.g., TetraMin (Tetra, Melle, Germany).

2.2 *Tetrahymena thermophila* Culture

1. Culture bottles: narrow necked 30–50 mL-bottles with an air permeable stopper on the top, as such shown in Fig. 1.
2. Culture medium: 1% Yeast Extract (w/v, BD Biosciences, San Jose, CA, USA), 2% Proteose Peptone No. 3 (w/v, BD

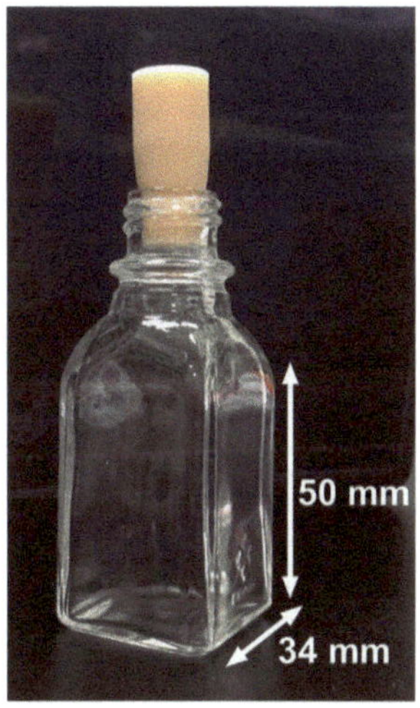

Fig. 1 A bottle example for the culture of *Tetrahymena thermophila*. This is a 30 mL bottle with 34 mm × 34 mm bottom

Biosciences, San Jose, CA, USA), and 0.6% Glucose (w/v). Dispense 10 mL each of the medium in the culture bottle. Autoclave and store at 4 °C.

3. 1.5 mL tubes.
4. 15 mL tubes.

3 Methods

3.1 Inbreeding a Strain

Fish are maintained in a usual condition for zebrafish at 27–28 °C with a light cycle consisting of 14 h of light and 10 h of dark. You should take maximum care not to contaminate fish or embryos during inbreeding (*see* **Note 3**). Still, however much careful you may be, contamination can occur accidently. Thus, it is better to check genetic homogeneity of the inbreeding strain at some or every generation by scale transplantation or genotyping polymorphic markers [6] (*see* **Note 4**). The genetic monitoring also provides the information about how homozygosity proceeds in the inbreeding population (Fig. 2). In the case of IM strain, 42 out of 50 markers were polymorphic at generation 0 [6], and then, the number of polymorphic markers reduced to only one at generation 17 (unpublished data).

1. Fix a rule for identifiers (IDs) of pairs (pair IDs) and individual fish of the pairs (individual IDs) at each generation. It is helpful for handling fish that the IDs contain the information of generation and/or sex.
2. Place a pair of fish (one female and one male) from the population which you want to inbreed, in a mating cage to obtain eggs.

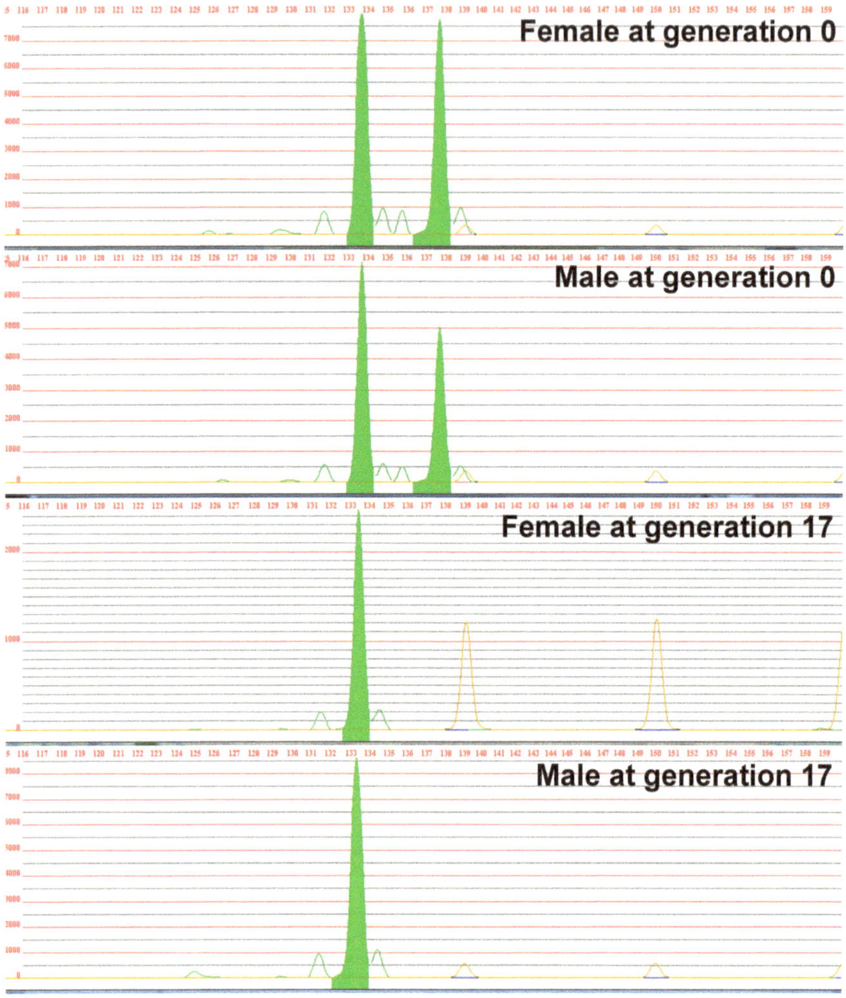

Fig. 2 Examples of genotyping a genetic marker (Z21055) in the IM strain. *Greened peaks* represent alleles. Two alleles were observed in both female and male of the best pair at generation 0, that is, they were heterozygotes. Only the *left peak* was inherited to the best pair at generation 17, indicating loss of polymorphisms at this locus

At the very beginning, the population above is a heterogeneous original strain, and in the progress of generating inbred strain, it is a sibling fish from one pair of parental fish. Set two or more pairwise mating (*see* **Note 5**).

3. Several hours after the lights come on the next day, transfer the pair in a tank temporally, and collect all eggs from the pair in a plastic petri dish. Carry out this work for all the pairs with eggs. Be careful not to lose the connection of which dish containing eggs derives from which pair.
4. Count unfertilized and fertilized eggs of each pairs, respectively. Select embryos from one or more pairs to be raised to the adults

(*see* **Note 6**), judging from the overall condition of the eggs from each pair: the better are more fertilized eggs, higher ratio of the fertilized eggs and less abnormal-looking embryos.

5. Assign pair IDs to the parental pairs of which embryos are selected to be raised, and also, assign individual IDs to each fish of the pairs. Label their embryos with the parental pair IDs.
6. Record the numbers of unfertilized and fertilized eggs of the selected pairs, according to their IDs.
7. Discard the unfertilized eggs. Clean the embryos right away by rinsing them several times with water for egg. Remove feces and other residue with an embryo transfer pipette. Incubate them at 28.5 °C.
8. Put the pairs in fish tanks separately (one fish per a fish tank), and maintain in a usual way for adult fish. Do not keep a pair together in the same tank, as one fish chases the other and weaker fish may die (*see* **Note 7**).
9. Count and remove the dead embryos of each dish once per day for 3 days after fertilization (dpf). Put the data on the record. Check also their development carefully, and if there are abnormal developments found, make a comment on the record.
10. Breed the embryos to the adults as described in Subheading 3.2. Usually, 40–80 embryos per pair are to be bred (*see* **Note 8**). If the embryos are too many, select proper numbers of healthier embryos at 3 dpf. Write down on the record anything noticed.
11. Cross the maintained pairs once in 2–4 weeks, and collect the eggs, as described above (**steps 2–4**, and 7). Record the numbers of unfertilized and fertilized eggs and of dead embryos per day until 3 dpf for each pairs (**steps 6** and **9**). Discard all of them after 3 dpf, if the embryos are not necessary to be raised to the adult. These data are collected in order to assess the fertility of each pairs.
12. When the embryos described above (**step 10**) are matured, determine which parental pair is the best pair basically from the point of view of fertility. This pair is the best pair at generation 0. To elucidate each pair, use the following criteria. The children from the best pair will be used for the next generation pairing.
    (a) Average number of eggs laid.
    (b) Fertilization efficiency.
    (c) Average number of embryos surviving until 3 dpf.
    (d) Survival rate of embryos at 3 dpf.
    (e) Normal development in the most embryos.
    (f) Normal growth into healthy adult fish in the most larvae.
    (g) Sex ratio.

13. Repeat **steps 2–12** for the children from the best pair 19 times more (*see* **Note 9**), so that the best pair at the 19th repeat is the best pair at generation 19. The children from the best pair at generation 19 now belong to an inbred strain.
14. To maintain an established inbred strain, repeat also **steps 2–12**, basically (*see* **Note 10**).

3.2 Breeding the Embryos to the Adults

Here, I describe a standard protocol which we use for breeding the IM strain. You may feel this protocol time-consuming. Therefore, you may first try the protocol which you use normally, and in the case that the protocol doesn't work well for your inbreeding strain, you can change the breeding protocol to ours.

1. Obtain embryos and take care of them as described in **steps 2–7** of Subheading 3.1.
2. Keep clean and change the water as much as possible with fresh water for egg once a day until 3 dpf (*see* **Note 11**). Incubate at 28.5 °C.
3. At 3 dpf, replace 40 larvae per 10-cm plastic petri dish.
4. For 4 and 5 dpf larvae, remove the dead larvae and change 10–25% of the water in a dish with the system water (the water maintaining adult fish). Feed 0.75 mL of *Tetrahymena thermophila* per 10-cm plastic petri dish once a day (*see* **Note 12**). *See* Subheading 3.3 for the preparation of *Tetrahymena thermophila*.
5. At 6 dpf, remove the dead larvae and change the 10-cm plastic petri dish to a 15-cm plastic petri dish. Add the system water enough to cover the bottom of the dish. Feed 1.5 mL of *Tetrahymena thermophila* and some brine shrimps (*see* **Note 13**).
6. After 7 dpf, care larvae as follows daily.
   (a) Remove the dead larvae and the leftovers.
   (b) Change about 20–30% of the water with the fresh system water (*see* **Note 14**). Increase the water depth day by day, e.g., from around 5 mm depth at 7 dpf to 13 mm at 10 dpf. If the bottom of the dish becomes quite dirty, change the dish also.
   (c) Feed 1.5 mL of *Tetrahymena thermophila* and some brine shrimps. Increase the amount of brine shrimps as the larvae grow up and eat more (*see* **Note 15**).
7. When most larvae begin to catch and eat brine shrimps (*see* **Note 16**), transfer them with the water to a 1.5 L tank (or in one or more tanks at the density of about 25 larvae/L). Put it in the fish housing system, and start water dropping (*see* **Note 17**). Hereafter, feed brine shrimps twice a day.
8. At 4 weeks after fertilization, replace fry at the density of 8 fry/L. Continue dropping water supply.

9. At 7 weeks after fertilization, replace fish again at the density of four fish/L. Change the water supply from dropping to a flow, a usual way for the adult fish. Hereafter, feed brine shrimps and some dry flake food twice a day (*see* **Note 18**).
10. At 13 weeks after fertilization, replace fish at the density of three fish/L. Feed brine shrimps and dry flake food to fry twice a day. Fish are usually matured at this time.

3.3 Preparation of *Tetrahymena thermophila*

To avoid contamination, it is better to make passages and to harvest in a sterile tissue culture hood. However, it is possible to perform all the steps of this section in a usual laboratory condition.

1. Add 30 μL of *Tetrahymena thermophila* stock (a procedure to make stock is described in **step 5** in a culture bottle containing the culture medium.
2. Lay the culture bottle on its side in order to keep a good water surface to air ratio, and incubate at 23–25 °C. *Tetrahymena thermophila* will ready to use in 4–7 days. Firstly, the culture is clear and then gets cloudy as days go by (*see* **Note 19**). You can see *Tetrahymena thermophila* as small specks.
3. For the next passage, add 10–30 μL of the above culture in a new medium, and repeat **step 2** (*see* **Note 20**).
4. To harvest *Tetrahymena thermophila*, take 1.5 mL of the cultured medium into a 1.5 mL tube, and spin down for several seconds at about 2000 × *g*. Discard the supernatant immediately (*see* **Note 21**). Rinse *Tetrahymena thermophila* with the water for egg. Spin down again and discard the supernatant. Add water for egg up to 1.5 mL, and now it is ready for feeding.
5. To generate a stock, prepare the culture of *Tetrahymena thermophila* as described in **steps 1** and **2**. Transfer the cloudy cultured medium to 15 mL tube. Seal the tube cap, and put it in a cool dark place. In this way, *Tetrahymena thermophila* can survive a half of year or more (sometimes a year), unless it is contaminated (*see* **Note 22**).

4 Notes

1. To keep only one fish in a tank, a small tank (e.g., 1-L tank) is better. Flat and not too big tanks are better for baby fish such as 1.5–3-L tanks with 5–10 cm deep in order to keep a good water surface to air ratio. Both tanks shown in Fig. 3 are 3-L tanks, for example, but the right tank is better to breed baby fish. Deep-water tanks can save the space, but they are sometimes not suitable for baby fish.
2. Whatever water usually used for eggs in your laboratory is no problem. Autoclaved tap water can be used for embryos when

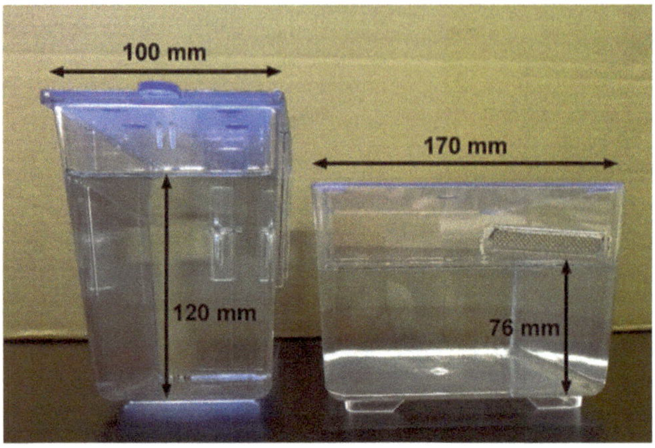

Fig. 3 3 L tanks. The *left* has higher water depth than the *right*. To save the space, the *left* is better, but fry kept in the *right tank* grow better

the quality of the tap water is good enough to keep embryos healthy. If the quality is not suited for embryos, use egg water [7], 1/3 Ringer's solution [8], or 0.3× Danieau's solution (1× Danieau's solution: 58 mM NaCl, 0.7 mM KCl, 0.4 mM $MgSO_4 \cdot 7H_2O$, 0.6 mM $Ca(NO_3)_2 \cdot 4H_2O$, 5 mM HEPES, pH 7.6).

3. Contaminations can often occur more easily than you assume. Don't put tanks of fish from different parents with the opened lid next to each other. Adult fish jumps out through a small crevice of lid, even the crevice width is less than 1 cm. Eggs or embryos may be mixed more often without awareness through fish nets, embryo transfer pipettes, plastic petri dish, or even your hands. Therefore, make sure no live eggs or embryos are left in fish nets, tanks, cases, and all other items used for mating and breeding fish before use. I describe some efficient ways to avoid contaminations of eggs or embryos as follows:
   (a) Use new items. This is a firm and easy but expensive way. It is applicable for transfer pipettes and plastic petri dish.
   (b) Autoclave items. This is also reliable for heat-resistant items, such as tanks, mating cases, and fish nets.
   (c) Immerse items in hot water (70 °C or more), EtOH, or bleach water at least for a few minutes. This is an inexpensive way and most items can be applied. However, it takes more time and work to collect hot water or to wash out EtOH or bleach. Besides, you have to pay attention to the temperature of the hot water, the concentration of EtOH, or the titer of the bleach water.
4. A procedure for scale transplantation is described elsewhere [6]. Scale transplantation costs less, but the accuracy is rather low. In medaka, the transplantation was successful between

individuals from sib-pair mating for only eight generations [9]. Considering this result in medaka, contamination of the different zebrafish strain can be detected by the scale transplantation technique. However, it may be hard to assess contamination among the families which belong to the same inbreeding strain. For this case, more confidence can be obtained by genotyping the best pair with genome-wide polymorphic markers.

5. The more pairs you test, the better pairs you find, but the more work you need to do. Set the appropriate number of pairs for your condition.
6. Embryos from at least two pairs are recommended to be brought up to the adult. Raising embryos from more pairs provides you more choice for the next generation, resulting in the increased guarantee to success generating and maintaining an inbred line. However, it takes more space and work instead. Therefore, you should decide the proper number of pairs on which embryos are raised, considering your space, time, and the condition of the inbreeding line. If the line is getting weak, it is better to keep embryos from more pairs.
7. If the space for fish breeding is quite limited, it may be possible to put back the pairs to the original population soon after collecting the eggs. In this case, omit **step 11** in Subheading 3.1. However, this reduces the sureness of the data for **steps 12a–d** of Subheading 3.1, as you can't assess the variation among batches.
8. As the sex is not strictly determined by the Mendelian genetic factor in zebrafish [10], sex ratio is often biased toward one sex during inbreeding. Then, it can easily happen in a small population that all embryos grow up as male or female. Therefore, a certain population size is desirable. In our experience, it is 40–80, but it depends on the condition of your fish facility.
9. During inbreeding, you may experience various troubles [6]. Basically, fish become weaker and weaker as inbreeding generation proceeds. However, it is not always getting worse. The worse condition in some time, the better another time. Thus, a key to success generating an inbred strain is how you manage the hard time. Fish around generation 2 or 3 tend to suffer the first difficulty. The total number of eggs often decreases, and besides, fertilization efficiency is lowered, so that the population size becomes smaller, e. g., less than 40 normal embryos at 3 dpf. In such a case, test additional pairs and try to find pairs with better fertility. This selective force on several generations may recover the fertility. Even after generation 10 or more (possibly after passing through generation 20, that is a generation to define an inbred strain), there can be hard times to breed. The difficulties observed in the IM strain (an inbred

strain generated by method described here) were biased sex ratio, slow growth, and bad mating success, for example [6]. Furthermore, both embryos and adults in the IM strain appear weak in low dissolved oxygen (unpublished observations). The inbred strain is also sensitive to the environmental changes, so that sudden changes in water quality including temperature and pH sometimes lead the fish to death [6]. On the other hand, those fish may endure gradual changes. In fact, the IM fish could survive at 22 °C when the temperature gradually lowered and kept good health after the temperature was returned to 27 °C. As there are not so many trials to establish an inbred strain in zebrafish, you may observe a wide variety of traits which is described nowhere. Whenever a hard time comes, select the best pair giving weight to the problem at that time.

10. If the inbred strain is in a bad condition and it is hard to obtain embryos by pair-mating, you may breed embryos from mass-mating in one or two generations. However, maintenance by continuous mass-mating will increase the polymorphisms in the strain. If too many polymorphisms are accumulated, the strain is no more an inbred strain. Therefore, avoid maintaining the strain by mass-mating as much as possible, and when you unwillingly use mass-mating for the next generation, record the fact in order to notify people who use the strain for their experiments.
11. As genetically homogeneous embryos are often sensitive to low dissolved oxygen, keep less than 100 embryos in a 10-cm plastic petri dish. Otherwise, their growth may be retarded severely and many embryos may die.
12. Instead of *Tetrahymena thermophila*, you can use paramecia as food for zebrafish larvae. Method for raising paramecia is described elsewhere [7]. *Tetrahymena thermophila* is smaller and moves more slowly than paramecia, so that it seems better for the larvae with a small mouth, such as the IM larvae.
13. Don't feed too many brine shrimps. Ten shrimps or so are enough in this step, as most larvae at 6 dpf can't eat brine shrimps yet. Too many dead brine shrimps cause fungus and bacteria to grow in the dish and the water condition becomes bad for larvae.
14. Be careful not to replace too much water. As inbreeding proceeds, fish are getting sensitive to sudden changes of water quality. For an extreme example, when most water (probably 80–90%) had been changed at once, almost all larvae have been found dead in the next day.
15. Too many brine shrimps are not good for larvae as described in **Note 13**, but larvae grow better when they are fed so many

that some (not too many) leftovers are found in the next day, especially in the early days such as 7–10 dpf.

16. In the case of the IM strain, it is about 14 dpf. You can easily check whether larvae eat brine shrimps or not, by looking them from the side of the dish. When a larva eats brine shrimps, a part of its body, where the stomach is, becomes orange (a color of brine shrimps).
17. Water is better to be supplied by dropping. This makes the surface of water wave constantly, so that a film doesn't form on the surface. Larvae seem to grow better without the film on the water surface.
18. Young fish has a smaller mouth than the adult. So, make the dry flake food smaller and feed that as much as they can finish eating within a few minutes. As the young fish don't eat so much dry flake, be careful too much leftovers which make the water condition bad.
19. In the case that the cultured medium becomes cloudy in short time, like the next day of the passage, it is often contaminated. Contaminated medium smells sour, and specks in the medium are smaller, so that brief spin down yields less or no pellet. Discard the contaminated medium and start a passage from a clean culture or stock.
20. If you do not use a sterile tissue culture hood for passages and/or harvests, use a clean cloudy culture as a seed culture, which has never been opened since its culture started. Otherwise, it is quite easily contaminated.
21. As *Tetrahymena thermophila* move around in the medium, its pellet starts to scatter, and a boundary between pellet and supernatant becomes unclear soon after the spin down. Therefore, it is not necessary to discard all the supernatant strictly. From the 1.5 mL of 4–7 days cultured medium, about 50 μL or more pellet can be obtained usually. If the volume is less, add one more 1.5 mL of cultured medium and spin down.
22. Dead *Tetrahymena thermophila* goes down and forms debris at the bottom of the tube. The living can be observed as small dots in the clear medium near the surface. As long as the living are there, you can use it as a stock.

Acknowledgment

I thank Dr. Noriyoshi Sakai and his lab members in the National Institute of Genetics for their valuable assistance in breeding zebrafish. I am also grateful to Ms. Hideko Utsumi, Dr. Taijiro Yabe, and Prof. Shinji Takada for their kind support in the maintenance of the IM strain and Drs. Atsuko Shimada, Kiyoshi Naruse, Yasuhiro Kamei, Tomonori Deguchi, and Koichi Kawakami for their helpful support and discussions. The work was supported partly by Grant-in-Aid for Young Scientists (B) from Japan Society for the Promotion of Science, by the Center for the Promotion of Integrated Sciences (CPIS) of Sokendai, and by NIBB Collaborative Research Program (c-12-351, b-13-202).

References

1. Kimura T, Shimada A, Sakai N et al (2007) Genetic analysis of craniofacial traits in the medaka. Genetics 177:2379–2388
2. Frankel WN (1995) Taking stock of complex trait genetics in mice. Trends Genet 11:471–477
3. Naruse K, Hori H, Shimizu N et al (2004) Medaka genomics: a bridge between mutant phenotype and gene function. Mech Dev 121:619–628
4. Peters LL, Robledo RF, Bult CJ et al (2007) The mouse as a model for human biology: a resource guide for complex trait analysis. Nat Rev Genet 8:58–69
5. Falconer DS (1989) Introduction to quantitative genetics. Longman, Essex, UK
6. Shinya M, Sakai N (2011) Generation of highly homogeneous strains of zebrafish through full sib-pair mating. G3 (Bethesda) 1:377–386
7. Westerfield M (1993) The zebrafish book: a guide for the laboratory use of zebrafish (*Danio rerio*). University of Oregon Press, Eugene, OR
8. Koshida S, Shinya M, Mizuno T et al (1998) Initial anteroposterior pattern of the zebrafish central nervous system is determined by differential competence of the epiblast. Development 125:1957–1966
9. Hyodo-Taguchi Y, Egami N (1985) Establishment of inbred strains of the medaka *Oryzias latipes* and the usefulness of the strains for biomedical research. Zoolog Sci 2: 305–316
10. Wilson CA, High SK, McCluskey BM et al (2014) Wild sex in zebrafish: loss of the natural sex determinant in domesticated strains. Genetics 198:1291–1308

Part II

Disease Models and Mechanism

Chapter 9

Zebrafish as a Model for the Study of Solid Malignancies

Genevieve C. Kendall and James F. Amatruda

Abstract

Zebrafish cancer models have provided critical insight into understanding the link between aberrant developmental pathways and tumorigenesis. The unique strengths of zebrafish as compared to other vertebrate model systems include the combination of fecundity, readily available and efficient transgenesis techniques, transparency that facilitates in vivo cell lineage tracing, and amenability for high-throughput applications. In addition to early embryo readouts, zebrafish can develop tumors at ages ranging from 2 weeks old to adulthood. Tumorigenesis is driven by genetically introducing oncogenes using selected promoter/tissue-specific expression, with either mosaic expression or with the generation of a stable transgenic line. Here, we detail a research pipeline to facilitate the study of human oncogenes in zebrafish systems. The goals of this approach are to identify conserved developmental pathways that may be critical for tumor development and to create platforms for testing novel therapies.

Key words Transgenesis, Functional genomics, Solid malignancies, Histopathology, Gateway cloning, Tol2, Zebrafish tumor model

1 Introduction

Zebrafish are a powerful model system to study cancer given their size, fecundity, and that they are an easily genetically manipulated vertebrate system. Normally, zebrafish rarely develop tumors, but can present with a range of spontaneous tumors with various incidences covering the majority of human neoplasias, including the most frequent, seminomas, to other varieties such as sarcomas, epithelial carcinomas, and hemangiomas [1, 2]. These solid tumors usually occur after 1–2 years of age without genetic modification or chemical exposure. The first example of zebrafish being chemically manipulated for cancer research was in the 1960s, when Dr. Mearl Stanton at the National Cancer Institute exposed wild-type strains to diethylnitrosamine and observed liver neoplasias [3]. Since then, the implementation of a variety of chemical or genetic methods has facilitated the use of zebrafish to study cancer, by interrogating known genes and pathways, introducing human genes that drive oncogenesis, or performing forward genetics to identify novel genetic modifiers [4–19].

Koichi Kawakami et al. (eds.), *Zebrafish: Methods and Protocols*, Methods in Molecular Biology, vol. 1451,
DOI 10.1007/978-1-4939-3771-4_9, © Springer Science+Business Media New York 2016

With the rise of high-throughput sequencing and human genetic analyses, zebrafish uniquely meet the need to quickly determine the functional effects of putative oncogenes or gene mutations in a vertebrate system. Zebrafish are easily genetically manipulated, and approximately 82 % of known human disease genes have a zebrafish counterpart, emphasizing the applicability of the system [20–24]. By implementing a variety of genetic strategies, such as forward genetics, reverse genetics, or facile transgenesis techniques, many zebrafish cancer models have been successfully developed. For example, the TILLING technique (ENU mutagenesis followed by re-sequencing of candidate genes) was used to recover the zebrafish *tp53* M214K missense mutation. These *tp53* mutants exhibit defective apoptosis and also develop malignant peripheral nerve sheath tumors (MPNSTs) at high penetrance [6]. Gain-of-function models introduce known mammalian oncogenes under the control of a variety of promoters to interrogate their functions and model specific cancers. Using these methods, investigators have created zebrafish models of leukemia [7–10], pancreatic neuroendocrine carcinoma [11], melanoma [12–14], embryonal rhabdomyosarcoma [15], Ewing sarcoma [16], liposarcoma [17], and hepatocellular carcinoma [18]. Forward genetic screens have identified novel genes involved in tumor development, with an example from our own group being that nonsense mutations in the bone morphogenetic protein receptor *bmpr1bb* result in 100 % penetrance of testicular germ cell tumors [19]. Taken together, these data highlight the utility of zebrafish for understanding the integral ties between developmental signaling pathways and how and when deregulated they can result in cancer predispositions.

Compared to the most common leukemias, for which survival rates have steadily increased in the past two decades, the outcomes for solid tumors in advanced stages have remained very poor [25]. The generation of a variety of animal models, each with complementary strengths and weaknesses, will facilitate understanding the basic biology and the discovery of novel therapies for these tumors. Zebrafish models can be applied to both of these contexts. For example, a common theme for many solid tumors that can be addressed with zebrafish is identifying the tumor cell of origin. In this strategy, oncogenes can be directly labeled and tracked to determine susceptible cell types, or even the migration or incorporation of these cells into solid tumors. Zebrafish also represent an important vertebrate translational screening tool. Historically, zebrafish have been used in high-throughput screening applications to identify genetic mutants. However, this has evolved into applying these genetic mutants or transgenic reporter lines as the basis for screens to identify chemical inhibitors of cancer processes or even metastasis. In this context, the advantage of zebrafish is that their external embryonic phenotypes can translate into

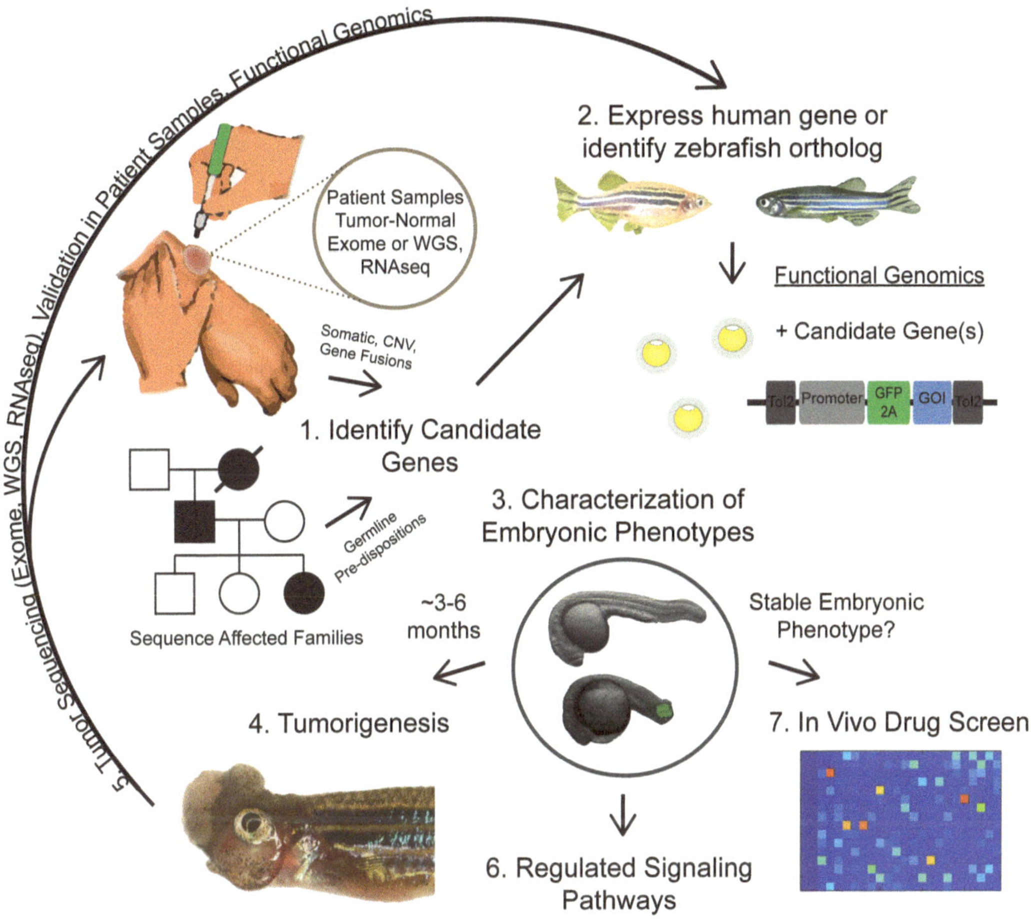

Fig. 1 Zebrafish as a functional genomics and translational medicine model for the study of solid tumors. Zebrafish models contribute to the study of solid tumors by two primary means: understanding the underlying biology of how solid tumors develop and using these models for translational and clinical applications. In this process, candidate genes are first identified by sequencing affected family pedigrees for identification of germline variants, or tumor-normal whole-exome or whole-genome sequencing (WGS), to identify likely somatic drivers. Candidate genes can be prioritized for modeling based on the identification of a zebrafish ortholog and/or the overall conservation of the relevant signaling pathways. Transgenes and mutations can be easily incorporated into the zebrafish system using a variety of techniques, and the expression of these transgenes tracked and monitored by the addition of a GFP Viral2A sequence, or another fluorescent marker. Zebrafish expressing a potent oncogene could exhibit tumors at ~3–6 months old, if not earlier. These tumors can be resected and analyzed by exome/WGS or RNAseq, in addition to histology to validate the tumor type. Such analyses will identify additional cooperating mutations that can be further incorporated into a functional genomics model, and then tumor latency evaluated for earlier onset. During this 3–6 month interim, the embryonic phenotypes can be quantified and are especially valuable if they are specific and reliable. The study of these early embryos could reveal developmental signaling pathways relevant to the pathogenesis of the oncogene. If these phenotypes are consistent and penetrant, they can be utilized to perform a high-throughput drug screen to determine in vivo chemical or genetic inhibitors of the oncogene's activity. WGS whole-genome sequencing

predictors of tumorigenesis. Previous success with embryonic high-throughput chemical screening in zebrafish models is seen in applications ranging from the study of muscle-specific diseases to cancer, namely, the identification of phosphodiesterase inhibitors as a potential therapy for Becker and Duchenne muscular dystrophy, and the identification of prostaglandin E2 as an enhancer of hematopoietic stem cell numbers, which has completed Phase I safety trials in adults and is now being expanded to pediatric patients [26–28]. Given that human oncogenes have similar functions in zebrafish, and their expression initiates and drives tumor formation that recapitulates their human counterparts, the rationale is sound to proceed with zebrafish as a high-throughput screening tool for solid tumor therapies [29]. Finally, the economy with which large numbers of tumor-bearing zebrafish can be produced makes possible the tumor-normal numbers required to observe somatic mutations driving rare solid tumor disease subsets, which sequencing cohorts to date may not have the statistical power to detect [30]. These advantages could realize the intersection of human genetics and high-throughput functional genomics in a vertebrate model.

Several points are important to develop a cancer model. These include the oncogene or mutation being assessed, strategies for expression or induction in a zebrafish system, and readouts that will be considered informative. Here, we outline the schematic for initiating the study of solid cancers, including the development of transgenic models with embryonic phenotypes that form solid tumors, and detail the process for analyzing resulting tumors (Fig. 1). Although the initial strategy includes a generalized and ubiquitous expression of the human oncogene, the approach can easily be adapted to probe the effects of an oncogene on a particular cell or tissue of interest.

2 Materials

2.1 Gateway Plasmid Constructs and Cloning Reagents [31]

1. *See* Table 1 and the Tol2Kit website for further information: http://neuroweb.neuro.utah.edu/tol2kitwiki/index.php/Main_Page
2. High-fidelity Taq polymerase (e.g., Platinum *Pfx* DNA polymerase, Thermo Fisher).
3. Gateway donor vectors (select donor vectors with chloramphenicol/kanamycin), and Gateway destination vectors (select destination vectors with chloramphenicol/ampicillin).
4. Gateway BP Clonase II enzyme mix (Thermo Fisher).
5. Gateway LR Clonase II Plus enzyme mix (Thermo Fisher).
6. Gel purification kit (e.g., QIAGEN or Thermo Fisher).

Table 1
Available Tol2 Kit constructs (reproduced from the Tol2Kit website) [31]

Plasmid ID #	Name	Insert	Size (bp)	Made by	Ver	MTA?
5′ entry clones, attL4-R1 (kan resistant)						
299	p5E-bactin2	5.3 kb beta-actin promoter (ubiquitous)	7950	KMK	1	
380	p5E-h2afx	1 kb histone2A-X promoter (quasi-ubiquitous)	3604	KMK	1	
382	p5E-CMV/SP6	1 kb CMV/SP6 cassette from pCS2+	3704	KMK	1	
222	p5E-hsp70l	1.5 kb hsp70l promoter for heat-shock induction	4163	BM	1	
327	p5E-UAS	10× UAS element and basal promoter for Gal4 response	3127	DSC	1	
228	p5E-MCS	Multiple-cloning site from pBluescript	2810	BM	1	
381	p5E-Fse-Asc	Restriction sites for 8-cutters FseI and AscI	2663	EF	1	
Middle entry clones, attL1-L2 (kan resistant)						
383	pME-EGFP	EGFP	3327	KMK	1	
384	pME-EGFPCAAX	Membrane-localized (prenylated) EGFP; fused to 21 aa of H-ras	3345	KMK	1	
385	pME-nlsEGFP	Nuclear-localized EGFP	3342	KMK	1	
386	pME-mCherry	Monomeric red fluorophore mCherry	3261	KMK	1	
232	pME-mCherryCAAX H80D	Prenylated mCherry (deleterious H80D mutation; superseded by 450)	3321	KMK	1	
233	pME-nlsmCherry	Nuclear-localized mCherry	3288	KMK	1	
234	pME-H2AmCherry	mCherry fused to zebrafish histone H2A.F/Z	3651	KMK	1	

(continued)

Table 1
(continued)

Plasmid ID #	Name	Insert	Size (bp)	Made by	Ver	MTA?
387	pME-Gal4VP16	Gal4 DNA binding domain fused to VP16 transactivation domain	3204	EF	1	
237	pME-MCS	Multiple-cloning site from pBluescript	2765	BM	1.1	
450	pME-mCherry CAAX	Prenylated mCherry (some preps contam w/ H80D, superseded by #550)	3321	KMK/ SYC	1.2	
455	pME-EGFP no stop	EGFP, no stop (to make N-terminal fusions)	3324	KMK	1.2	
456	pME-mCherry no stop	mCherry, no stop (to make N-terminal fusions)	3258	KMK	1.2	
550	pME-mCherryCAAX	Prenylated mCherry	3321	KMK/ SYC	1.2	
3′ entry clones, attR2-L3 (kan resistant)						
302	p3E-polyA	SV40 late polyA signal	2838	KMK	1	
229	p3E-MTpA	6× myc tag for C-terminal fusions, plus SV40 late polyA	3151	BM	1	
366	p3E-EGFPpA	EGFP for C-terminal fusions, plus SV40 late polyA	3634	MEH	1	
388	p3E-mCherrypA	mCherry for C-terminal fusions, plus SV40 late polyA	3586	MEH	1	
389	p3E-IRES-EGFPpA	IRES driving EGFP plus SV40 late polyA	4219	KMK	1	
390	p3E-IRES-EGFPCAAXpA	IRES driving prenylated EGFP plus SV40 late polyA	4250	KMK	1	
391	p3E-IRES-nlsEGFPpA	IRES driving nuclear EGFP plus SV40 late polyA	4248	KMK	1	
Destination vectors, attR4-R3 (amp/chlor resistant; grow in ccdB-tolerant cells)						
394	pDestTol2pA2	pDestTol2pA with ~2 kb extraneous sequence removed	5883	EF	1	KK

(continued)

Table 1 (continued)

Plasmid ID #	Name	Insert	Size (bp)	Made by	Ver	MTA?
395	pDestTol2CG2	pDestTol2CG2 with ~2 kb extraneous sequence removed	7796	EF	1	KK
Donor vectors (kan/chlor resistant; grow in ccdB-tolerant cells)						
218	pDONR221	Middle donor vector; attP1-P2 flanking chlor/ccdB cassette		Invitrogen	1.1	
219	pDONRP4-P1R	5′ donor vector; attP4-P1R flanking chlor/ccdB cassette		Invitrogen	1.1	
220	pDONRP2R-P3	3′ donor vector; attP2R-P3 flanking chlor/ccdB cassette		Invitrogen	1.1	
Transposase clones						
396	pCS2FA-transposase	For in vitro transcription of capped Tol2 transposase mRNA	6034	KMK	1	KK

Notes: # is the Chien lab stock number, made by shows who made the construct; *CG* Clemens Grabher, Look lab; *BM* Ben Mangum, Chien lab; *DSC* Doug Campbell, Chien lab; *EF* Esther Fujimoto, Chien lab; *KMK* Kristen Kwan, Chien lab; *MEH* Melissa Hardy, Chien lab; *SYC* Seok-yong Choi, Chitnis lab. ver shows the version of the Tol2kit in which this construct first appeared. MTA? indicates whether an MTA is associated with each clone. *KK* Koichi Kawakami

7. TE Buffer pH 8.
8. Competent cells (e.g., Top10 chemically competent cells, Thermo Fisher).
9. Kanamycin (for entry clones).
10. Ampicillin (for expression clones).
11. LB broth and LB agar.
12. QIAprep Spin Miniprep Kit (QIAGEN).

2.2 Synthesis of Transposase mRNA (Refer to [23] for Details)

1. *NotI* restriction enzyme.
2. mMessage mMachine SP6 Transcription Kit (Thermo Fisher).
3. Spin columns for purification (e.g., QIAGEN or Roche).

2.3 Mosaic Transgenesis of Oncogene

1. Wild-type strain of zebrafish (AB, TL, WIK, etc.).
2. *tp53*$^{zdf1/zdf1}$ strain containing a *tp53*^{M214K} missense mutation (ZIRC ID, ZL1057).
3. Fish tanks and system for maintenance of fish.

4. Fish food.
5. Mating chambers.
6. Mating chamber dividers.
7. Petri dishes (100 mm × 15 mm deep).
8. 6-well zebrafish injection molds (Adaptive Sciences).
9. Agarose (dissolve 1.5 % in E3 for injection molds).
10. Standard glass capillaries, 4 in., 1/0.58 OD/ID, filament/fire polished (World Precision Instruments).
11. Micropipette needle puller (Sutter Instrument Company).
12. Micropipette holder.
13. Microinjector.
14. Dissecting light microscope.
15. E3 embryo buffer.
16. 1 % phenol red.
17. Ultrapure RNase-/DNase-free distilled water.
18. 30× Danieau's buffer.
19. ddH_2O.

2.4 Screening Materials

1. Fluorescent stereoscope.
2. 50 mg/mL pronase for dechorionation (Roche).
3. 0.02 % tricaine (for screening).

2.5 Characterizing Embryonic Phenotypes

2.5.1 Phenotype Analysis

1. Dissecting light microscope.
2. 50 mg/mL pronase for dechorionation (Roche).

2.5.2 FACS Analysis (See Ref. 32 for Details)

1. Deyolking buffer.
2. 2 %FBS in Danieau's.
3. FACSMax Cell Dissociation Solution (Genlantis).
4. Cell strainers, 40 μm (Fisher).
5. 1/2 mL BD Lo-Dose™ U-100 insulin syringe (VWR).
6. Petri dish, 60 mm in diameter.
7. 1× PBS.
8. 5 mL Falcon round-bottom FACS tubes (Fisher).
9. MoFloXDP Cell Sorter (Beckman-Coulter).

2.5.3 Total RNA Isolation

1. Liquid nitrogen.
2. 1.5 mL Eppendorf tubes, RNAse/DNase-free.
3. RNeasy Microkit (QIAGEN).

2.5.4 Microscopy and Live Cell Imaging

1. Fluorescence microscope with time lapse capabilities.
2. Glass slides.
3. 2 % methylcellulose.
4. Glass pipettes and bulb.
5. Glass cover slips.
6. Low-melt agarose.
7. Tricaine in E3.
8. Vacuum grease (e.g., Dow Corning).

2.6 Tumor Analysis in Adults

2.6.1 Pathology of Embryos or Resected Tumors

1. 0.01 % tricaine diluted in fish water.
2. Fine forceps.
3. Razor blades.
4. Thermo Scientific Shandon Biopsy Processing/Embedding Cassettes II, small pore (Fisher).
5. Paraformaldehyde, granular (resuspended in 1× PBS to 4 %).
6. PBS.
7. 0.5 M EDTA for decalcifying [33].
8. TissuePrep-2 Embedding Media (Fisher Scientific).
9. H&E staining [2].

2.6.2 DNA Isolation (of Embryos, Tumors, and Tail Clips)

1. DNA isolation kit for tissues or blood (QIAGEN).
2. 1.5 mL RNAse/DNAse-free microcentrifuge tubes.

2.6.3 Total RNA Isolation from Tumor Samples

1. Liquid nitrogen to snap freeze.
2. Homogenizer.
3. RNeasy Microkit (QIAGEN).

3 Methods

3.1 Identifying a Gene of Interest (GOI) to Study its Tumorigenic Capacity in a Zebrafish System

There are numerous gene mutations, amplifications, deletions, or fusion oncogenes implicated in the initiation or development of solid tumors. Choosing which genes to begin with is subject to the goal of the particular study. Here we have outlined our strategy for the introduction, or expression, of human fusion oncogenes using the beta-actin or heat-shock promoter and Tol2 transposon system [16, 23, 24, 31] (Fig. 2, *see* **Note 1**). However, it should be noted that this can be applied in a variety of scenarios, including gene overexpression, or for the introduction of dominant negative mutations. First, it is important to identify that the gene and its signaling or developmental pathways are conserved in the zebrafish system. In addition, the relevant target tissue should be present,

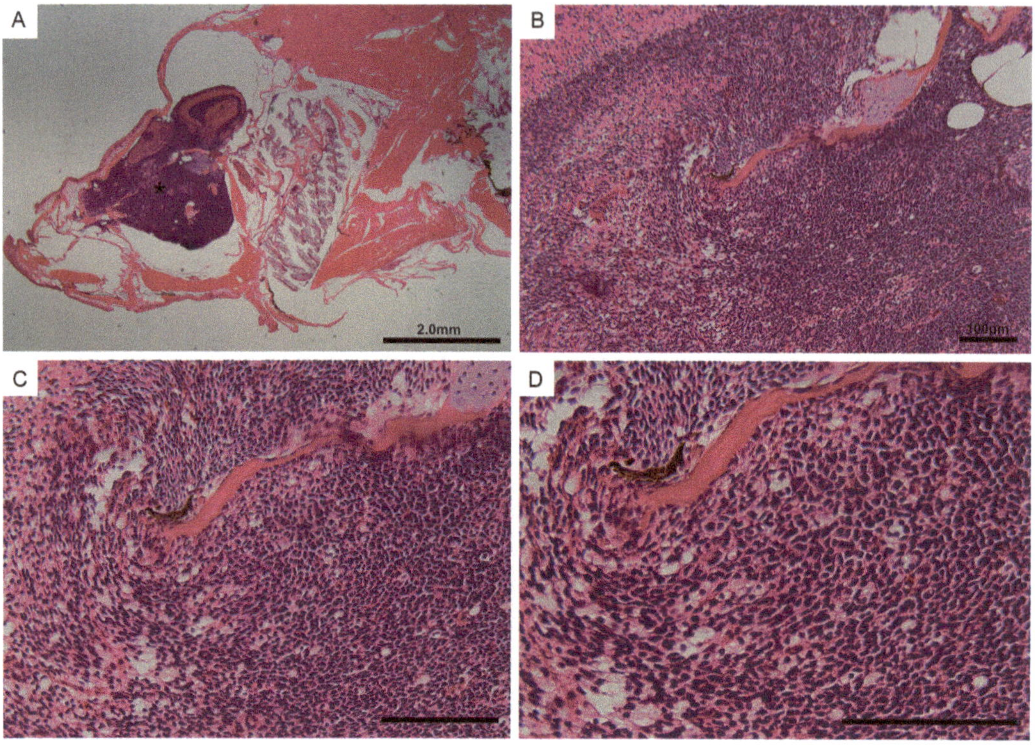

Fig. 2 Tol2 transgenesis of human EWS-FLI1 in zebrafish initiates tumor formation that recapitulates the human disease. (**a**) Sagittal section of an adult zebrafish injected with EWS-FLI1 under the beta-actin or heat-shock hsp70l promoter using the Tol2 system initiates the formation of a solid tumor indicated by the asterisk. Scale bar, 2 mm. (**b–d**) The histology of the EWS-FLI1 induced zebrafish tumor is consistent with a solid, small, round, blue cell tumor and what would be expected for Ewing Sarcoma. 20×, 40×, and 60×, respectively. Scale bar, 100 μm

i.e., skeletal muscle or the liver, or be represented by a homologous structure. Further insight could be gained for the study of a fusion oncogene, or dominant negative forms, by overexpressing the non-oncogenic form of the gene(s) and contrasting these differences. Such a strategy may highlight phenotypes specific to the oncogene, but these results should be interpreted with care. A simple control is to express the chosen fluorophore used to tag the oncogene under the same promoter.

The choice of promoter (tissue specific vs. ubiquitous or inducible) may have important effects on developmental phenotypes and on tumor latency, penetrance, and tissue distribution. While there are exceptions, in general the expression of an oncogene from a tissue-specific promoter will result in fewer developmental abnormalities and will cause tumors to arise in the target tissue, with two examples being the *rag2:mMyc* and *mitfa:BRAFV599E* transgenics, which develop T-cell leukemia and melanoma, respectively [7, 13]. An alternative approach is to use the UAS/GAL4 binary expression system to restrict expression of the oncogene. UAS/GAL4 transgenic repositories, such as zTRAP from Koichi Kawakami's lab or the Zebrafish Enhancer Trap Database from Harold Burgess' lab, contain GAL4 lines that have been developed, characterized, and made available to research groups [34–36]. In some cases, investigators may wish to produce high-level mosaic expression in embryos to reveal developmental pathways engaged by specific oncogenes. In other cases, the relevant promoter may not be available, or the cell of origin of the human cancer may not be known. In these situations, expression from a ubiquitous promoter such as ubiquitin or beta-actin can be helpful [31, 37]. Low tumor incidence can be overcome by injecting large numbers of fish, which is reasonable given that housing zebrafish is a low cost as compared to other vertebrate systems. Finally, a variety of secondary sensitizing mutations can be anticipated, and incorporated into the model (*see* **Note 2**), such as the knockout of tumor suppressors using CRISPR/Cas9, or the introduction of missense/activating mutations in known genes with a relevant human context [21, 22].

3.2 Cloning Your GOI into a Gateway Expression System for Tol2 Transgenesis

3.2.1 Creating 3′ Entry GOI and Control Plasmid Constructs

1. Design primers that amplify the coding sequence of your gene of interest for cloning into pDONRP2R-3, the 3' donor vector. Add an attb2r sequence to your forward primer and an attb3 sequence to your reverse primer (see below). Include a stop codon in the coding sequence of your reverse primer.

   FWD: 5′ GGGGACAGCTTTCTTGTACAAAGTGGCC... your GOI sequence 3′

   REV: 5′ GGGGACAACTTTGTATAATAAAGTTGC... your GOI sequence 3′
2. Amplify using a high-fidelity Taq polymerase.
3. Gel purify PCR product.
4. Set up a BP reaction with 50 fmol of pDONR and 50 fmol of gel purified PCR product, add TE buffer, pH 8 to bring the volume up to 8 μL. Add 2 μL of BP Clonase II enzyme mix.
5. Incubate overnight at 25 °C.
6. Add 1 μL proteinase K × 10 min at 37 °C.
7. Transform into chemically competent cells (e.g., Top10).
8. Plate on kanamycin (50 μg/mL) selective LB agar plates, and incubate overnight at 37 °C.
9. Screen clones by colony PCR or restriction digest.
10. Miniprep clones that are positive by colony PCR or restriction digest, and confirm the correct clone by Sanger sequencing.

3.2.2 Creating an Expression System with Your GOI and a Relevant Control

Set up two reactions, one with your GOI in the 3′ entry and one reaction with the polyA 3′entry as a control.

1. Set up LR reactions with 20 fmol of DEST vector and 10 fMol of 5′entry, middle entry (GFPViral2A or mCherryViral2A),

and 3′entry. Add TE buffer, pH 8 to bring the volume up to 8 μL.

2. Add 2 μL of LR Clonase and incubate at 25 °C overnight.
3. Add 1 μL proteinase K × 10 min at 37 °C.
4. Transform into chemically competent cells (e.g., Top10). Plate on AMP selective (100 μg/mL) LB agar plates, and incubate overnight at 37 °C.
5. Screen clones by colony PCR or restriction digest.
6. Miniprep clones that are positive by colony PCR or restriction digest, and confirm the correct clone by Sanger sequencing.

3.3 Injection and Mosaic Expression of Your GOI in Zebrafish Embryos

1. The day prior to injections, set up two males and two females per mating chamber, separated by dividers. There will need to be enough fertilized embryos for three injection groups: uninjected controls, GFP or mCherry fluorescent injected controls, and the oncogene-injected groups.
2. Injection mixes can be made the morning of injections, or the night prior and frozen at −80 °C. Make injection mixes with titrations of plasmid DNA containing your GOI. Suggested beginning concentrations are 25, 50, and 100 ng/μL plasmid DNA, with 50 ng/μL of Tol2 transposase mRNA, 1 μL of 3× Danieau's buffer, and 1 μL of 1% phenol red in a 10 μL total volume. Add nuclease-free water to bring the volume up to 10 μL.
3. Inject 300 or more 1–2 cell stage wild type, or *tp53*$^{zdf1/zdf1}$ embryos, into the cell body for mosaic expression of the transgene.
4. Raise embryos at 28.5 °C. Remove dead embryos and change E3 daily following injections.
5. Repeat this process on multiple days of injections (3–5 days total).

3.4 Screening Injected Embryos from 24hpf to 5dpf

1. At 24hpf, note the number of dead embryos in all groups and record that number.
2. Dechorionate embryos for easier observation by adding eight drops of pronase to 20 mL of E3 per petri dish.
3. Incubate at 25 °C for 10 min without agitation.
4. Rinse 3 times with E3. Remove embryos to a new petri dish. This should remove the majority of the chorions, if a few remain dechorionate those by hand with forceps.
5. Screen embryos for GFP fluorescence under the dissecting microscope.
6. Separate the GFP-positive embryos from the negative embryos. Record the number of GFP positive and negative embryos.
7. Discriminate phenotypes. Record these. Quantify across groups and days to determine consistency.

8. Daily, document mortality and the percentage of GFP-positive embryos in each group.
9. At 5 days, divide into GFP-positive and GFP-negative groups, before placing them on the nursery and raising them according to each system's specific guidelines. There will likely be high mortality in oncogene-injected embryos as compared to controls; if this is a significant issue, *see* **Note 3** and **Note 4** for alternative approaches and resources.

3.5 Continue Monitoring Larvae and Adults for Cell Masses or Tumors

1. Screen larvae for GFP-positive masses of cells, or tumors, under the dissecting fluorescent microscope at 2 weeks and 30 days.
2. If nothing is seen at 30 days, continue monitoring the status of these fish by checking gross morphology weekly, and screening for GFP fluorescence at 1–2 month intervals. The tumor onset will depend on the oncogene and genetic background, and reports have ranged from as early as 7 days post fertilization to a tumor latency of 2 years [38, 39]. For example, in a *tp53*^{M214K} mutant sensitized background, spontaneous tumors such as malignant peripheral nerve sheath tumors will begin forming around 1 year of age, making it potentially difficult to sort out what is due to the oncogene or the genetic background [6]. Hence, it is useful to fluorescently tag the oncogene to directly track its expression and incorporation into the tumor mass. Typically, tumors will form by 3–6 months of age if the transgene is successfully integrated, expressed, and oncogenic in a zebrafish system. Effort can be devoted during this period to characterizing the embryonic morphologic effects of a given oncogene. Such insight may pinpoint the developmental pathways that are modulated by the transgene and are potentially required for tumor development.

3.6 Analysis of Embryonic Phenotypes: Overall Survival, Phenotypes, GFP Tissue-Specific Tolerance and Temporal Expression Kinetics, RNA Expression Analysis, and Live Cell Lineage Tracing

3.6.1 Understanding How Oncogenes Affect Early Zebrafish Development

Oncogenes associated with human cancers typically exhibit strong effects on signaling pathways and gene expression programs associated with growth, survival, angiogenesis, and other hallmarks of cancer. Not surprisingly, widespread mis-expression of these same genes in zebrafish embryos also frequently affect early developmental pathways resulting in observable phenotypes. These early phenotypes are one of the strengths of the zebrafish system, especially for drug development or high-throughput genetic modifier studies, and are of value to study and characterize in parallel to developing a tumor model. Understanding the early developmental pathways that are targeted and modulated by oncogenes in vivo likely translates to their importance in driving tumor development. Below we have detailed a strategy for characterizing early phenotypes due to mosaic expression of transgenes. For maximum efficiency and specificity, as is required for high-throughput drug screening, most investigators would want to establish stable lines with 100% penetrance of the developmental phenotype. These early studies will guide the development of such a stable line.

3.6.2 Embryonic Analysis: Overall Survival, Phenotypes, and GFP-GOI Expression Patterns

These phenotypes may be unique to the specific oncogene that is expressed and can be quantified in order to provide insight into targeted cell type(s) and likely regulated developmental pathways.

1. Compare the overall survival of early embryos across groups such as uninjected, control groups injected with fluorophores alone, if appropriate the normal form of the gene, and then the mutation or fusion that makes it oncogenic.
2. Monitor how the GFP expression patterns change over time across these groups, both in terms of their restriction to specific tissues, or if expression is lost during the course of development. These details may be important to define tolerant cell types and guide the design of a refined experiment with more specific transgene expression. Although these are rough metrics, they are simple to quantify and valuable to distinguish relevant expression patterns.

3.6.3 Embryonic Analysis: Isolation of GFP-GOI-Positive Cell Populations for Expression Analysis

Understanding the signaling pathways that are modulated in cells expressing the oncogene is greatly enhanced when this cell population can be experimentally isolated. FACS sorting and isolation of GFP-positive cells from stable transgenic lines has been well described. However, we have also shown that it is useful in the study of embryos with mosaic expression of transgenes that co-express fluorophores such as GFP or mCherry. For example, we have isolated GFP-positive cell populations from 24hpf embryos injected in the cell body with BetaActin-GFP-2A-pA using the Tol2 system. After disaggregation into single cell suspension, using the Manoli and Driever [32], protocol with modifications, we found that on average, 25–35% of the total cells were GFP positive, suggesting this method is useful for the study of individual cells or cell populations. Injecting large numbers of embryos facilitates purification of enough GFP-positive cells to allow isolation of high-quality total RNA, for use in subsequent downstream applications, including gene expression studies with microarray or RNAseq.

3.6.4 Embryonic Analysis: Microscopy and Live Cell Imaging

Zebrafish embryos are transparent and can be manipulated to eliminate early development of pigment, facilitating their use for the tracking of cell lineages. Live cell imaging of oncogenes that are tagged with GFP or mCherry viral 2A constructs allows for studies such as determining tolerated cell types and the movement of cell populations in a dynamic environment. This strategy could even provide early insight into metastasis profiles and how various gene additions, or deletions, could modify metastasis. In order to visualize early developing embryos, resuspend zebrafish in low-melt agarose and place on a cover slip. Let the agarose solidify, and then surround the embedded embryo with vacuum grease. Add a small amount of E3 embryo water with 0.02% tricaine to minimize movement. Secure the embryos in place by putting another cover slip on top of the setup. Place the embryo on the microscope stage, and capture fluorescent and bright-field stacks as is required to determine the migration of oncogenic cells, and where these cells are localizing in a dynamic system in real time.

3.7 Analysis of Adult Zebrafish Tumors

Once a tumor has been identified, a few factors need to be taken into account prior to processing. First, note the time of tumor onset, and evaluate whether it is life-threatening or causing distress to the fish or alternatively can be allowed to continue to grow to collect more tissue material for histology, RNA, and DNA isolation. Secondly, the location and orientation of the tumor dictates how the tumor is collected and poised for histology (Figs. 3 and 4). After these decisions are made, image the fish/tumor under a dissecting microscope to record its location and fluorescence patterns. Before sacrificing the zebrafish, have everything in place to harvest for histology, DNA, and RNA so that you can work quickly. Proceed according to the tumor location and number. Refer to Subheadings 3.8–3.10, and Figs. 3 and 4 as a guide for each approach.

3.8 Approach for Analysis of Adult Zebrafish Tumors: Suspected Tumor

If there is a suspected tumor, but its location is not clear, it may be useful to sacrifice and observe if there are any structural abnormalities, or tumor masses, that are not externally visible (*see* Fig. 3a).

1. Sacrifice adult fish in an overdose of tricaine solution (2 g/L stock; dilute to 0.1% in fish water for euthanasia) for up to 5 min.
2. Image gross morphology and fluorescence patterns under a dissecting fluorescence microscope.
3. Remove the tail fin and freeze in liquid nitrogen for analysis of germline DNA.
4. Prepare the fish for histology. It is important to open the ventral surface of the fish with a pair of dissecting scissors, to ensure that fixative penetrates to the internal organs. Place in histology cassette, and submerge in 4% paraformaldehyde in 1× PBS in an Erlenmeyer flask. Fix in 4% PFA for 48 h at 4 °C on a gentle rocker.
5. Place the histology cassette in 0.5 M EDTA at 25 °C and decalcify for 3–5 days [33].
6. Process and embed in paraffin. Prepare sagittal sections as follows [2]. Refer to Fig. 4a, b as a guide for sagittal sections.
   - (a) Level 1—waste paraffin until reaching the middle of the eye. Prepare 1 H&E and 1–2 unstained sections.
   - (b) Level 2—waste paraffin until reaching the point just medial to the eye. Prepare 1 H&E and 1–2 unstained sections.

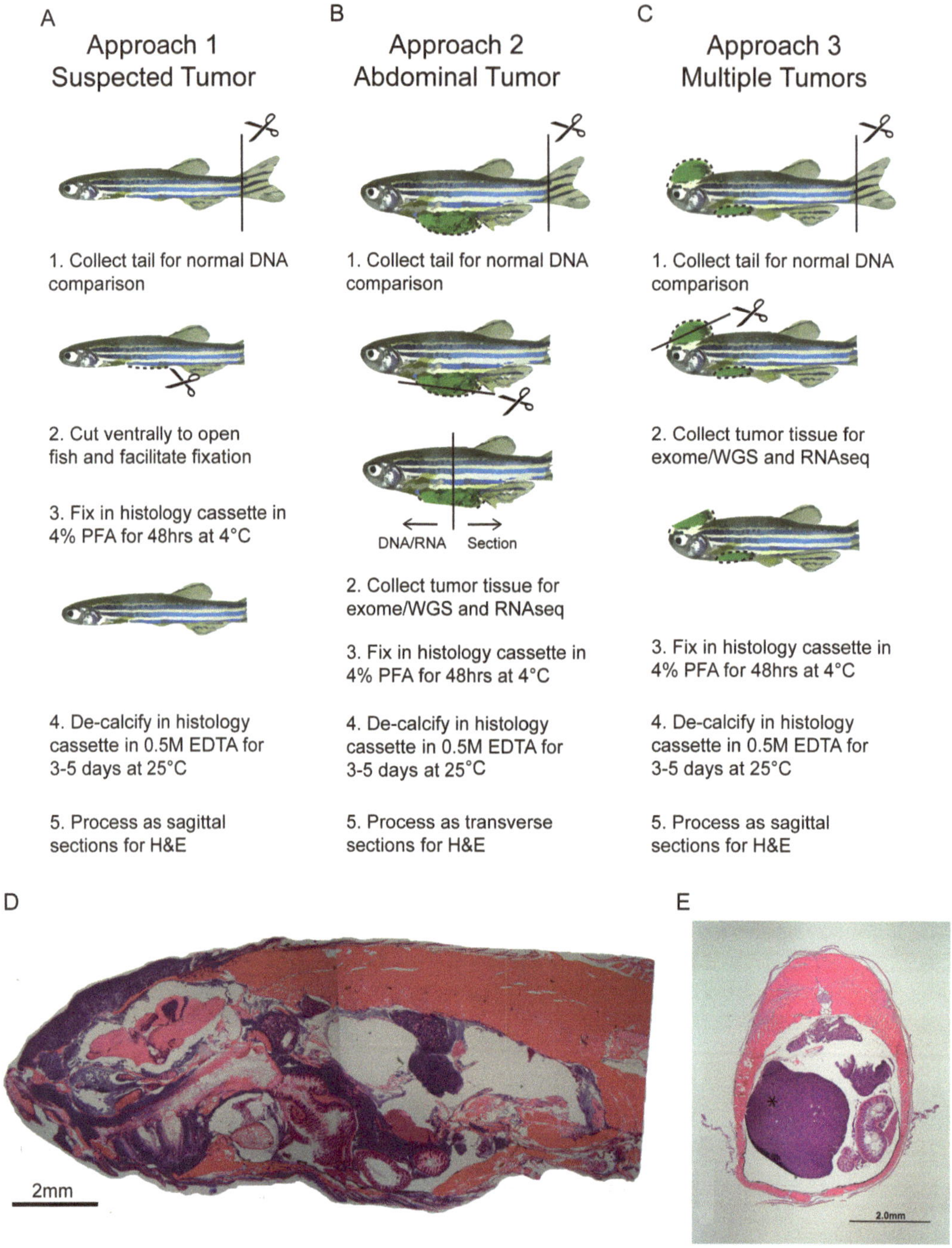

Fig. 3 Strategy for harvesting zebrafish tumors for analysis of DNA, RNA, and histology. The overall approach for processing tumors depends on the presentation, position, and size. (**a**) If a zebrafish has been injected with an oncogene and has a suspected tumor, but its location is not clear, it may be useful to sacrifice and observe if there are any structural abnormalities, or tumor masses, that are not externally visible. In this case, the best approach is to perform sagittal sections. This will increase the chances of observing tumor masses throughout the body, if they are present. Additional details are available in the text. (**b**) Many zebrafish tumors present in the abdomen. In these cases, it is best to cut the zebrafish in half at the tumor, assuming that the oncogene is labeled with a GFP viral 2A sequence, which acts as a guide when harvesting. This allows for half of the tumor to be utilized for DNA/RNA and the other half for histology. (**c**) In some cases the tumor will not be best represented by transverse sections, an example being if there are multiple masses that are visible by GFP fluorescence. If this is true, harvest any external tumors for analysis of RNA/DNA, and perform sagittal sections to

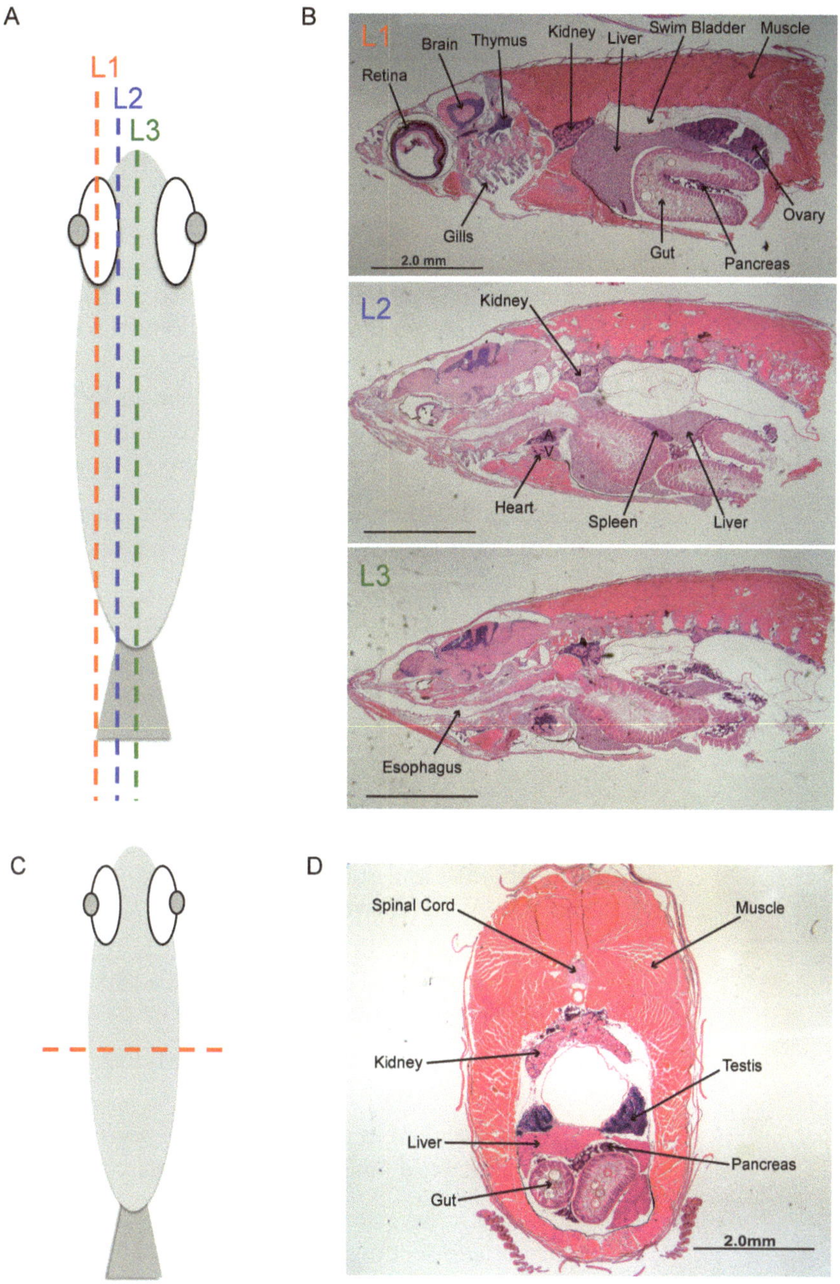

Fig. 4 Sagittal and transverse sectioning strategy demonstrated in normal zebrafish. (**a**) Dorsal view of a schematic of an adult zebrafish prepared for sagittal step sectioning. Level 1 (L1) is at the middle of the eye. Level 2 (L2) is just medial to the eye. Level 3 (L3) is at the midline of the fish. (**b**) Representative H&E staining from each Level 1–3 from a normal adult zebrafish. (**c**) Dorsal view of a schematic of an adult zebrafish prepared for transverse sectioning. (**d**) Representative H&E staining from a transverse section from a normal adult zebrafish

Fig. 3 (continued) capture the histology of remaining malignancies. (**d**) Sagittal H&E stain of an example of a leukemia that was identified in a zebrafish with a suspected tumor. Scale bar, 2 mm. (**e**) A H&E stain of a transverse section reveals a large germ cell tumor, denoted by the asterisk, in a bmpr1bb homozygous mutant. Scale bar, 2 mm

(c) Level 3—waste paraffin until reaching the midline of the fish. The remaining block can be saved, or recut to generate additional unstained sections.

7. Examine H&E stained sections from L1, L2, and L3 to assess gross anatomy and health [2]. Note unstained sections are critical to collect proactively in case immunohistochemistry or immunofluorescence analysis is required to evaluate the expression of the transgene, or further pinpoint the tumor type.

3.9 Approach for Analysis of Adult Zebrafish Tumors: Abdominal Tumor

Many zebrafish solid tumors present in the abdomen. In the case specified here, the oncogene is labeled with a GFP viral 2A sequence, which acts as a guide when harvesting (*see* Fig. 3b).

1. Sacrifice adult fish in an overdose of tricaine solution (2 g/L stock; dilute to 0.1 % in fish water for euthanasia) for up to 5 min.
2. Image gross morphology and fluorescence patterns under a dissecting fluorescence microscope.
3. Remove the tail fin and freeze in liquid nitrogen for analysis of germline DNA.
4. Cut in the middle of the fluorescent mass under the dissecting microscope and using the fluorescence as a guide. Resect external pieces of the tumor. Remove additional tumor from one half of the fish and snap freeze in liquid nitrogen to save for tumor DNA/RNA analysis.
5. Prepare the fish for histology. Place in histology cassette, and submerge in 4 % paraformaldehyde in 1× PBS in an Erlenmeyer flask. Fix in 4 % PFA for 48 h at 4 °C on a gentle rocker.
6. Place the histology cassette in 0.5 M EDTA at 25 °C and decalcify for 3–5 days [33].
7. Process and embed in paraffin. Prepare transverse sections [2]. Refer to Fig. 4c, d as a guide for transverse sections.

   (a) Collect transverse sections starting at the center of the tumor and sectioning away. Prepare 1 H&E and 1–2 adjacent serial unstained sections [2].

8. Examine H&E stained to assess gross anatomy and health. Note unstained sections are critical to collect proactively in case immunohistochemistry or immunofluorescence analysis is required to evaluate the expression of the transgene, or further pinpoint the tumor type.

3.10 Approach for Analysis of Adult Zebrafish Tumors: Multiple Tumors

In some cases the tumor will not be best represented by transverse sections, an example being if there are multiple masses that are visible by GFP fluorescence (*see* Fig. 3c).

1. Sacrifice adult fish in an overdose of tricaine solution (2 g/L stock; dilute to 0.1 % in fish water for euthanasia) for up to 5 min.

2. Image gross morphology and fluorescence patterns under a dissecting fluorescence microscope.
3. Remove the tail fin and freeze in liquid nitrogen for analysis of germline DNA.
4. Resect pieces of the tumor that are externally visible, then use the fluorescent dissecting microscope as a guide to obtain any remaining tumor sample. Snap freeze in liquid nitrogen for analysis of tumor DNA and RNA.
5. Prepare the fish for histology. Place in histology cassette, and submerge in 4 % paraformaldehyde in 1× PBS in an Erlenmeyer flask. Fix in 4 % PFA for 48 h at 4 °C on a rocker.
6. Place the histology cassette in 0.5 M EDTA at 25 °C and decalcify for 3–5 days [33].
7. Process and embed in paraffin. Prepare sagittal sections as follows [2]. Refer to Fig. 4a, b as a guide for sagittal sections.
   (a) Level 1—waste paraffin until reaching the middle of the eye. Prepare 1 H&E and 1–2 unstained sections.
   (b) Level 2—waste paraffin until reaching the point just medial to the eye. Prepare 1 H&E and 1–2 unstained sections.
   (c) Level 3—waste paraffin until reaching the midline of the fish. The remaining block can be saved, or recut to generate additional unstained sections.
8. Examine H&E stained sections from L1, L2, and L3 to assess gross anatomy and health [2].

Note unstained sections are critical to collect proactively in case immunohistochemistry or immunofluorescence analysis is required to evaluate the expression of the transgene, or further pinpoint the tumor type.

3.11 Histology, Immunohistochemistry, and Immuno-fluorescence

Fix zebrafish tumors for sectioning by incubating in 4 % PFA for 48 h at 4 °C. Embed in paraffin and take serial sections. Perform a H&E stain on one section to assess the pathology of the tumor [2]. At this point it is critical to collaborate with a pathologist or veterinary pathologist in order to determine the correct tumor sub-type (Fig. 2). Depending on the tumor you are modeling, use unstained sections for immunohistochemistry or immunofluorescence to assess expression of markers that are utilized for diagnosis, or are associated with the human disease.

3.12 DNA Isolation of Both Tumor and Germline Tissue Samples

Snap freeze the tail fin and tumor sample in liquid nitrogen for isolation of normal and tumor DNA. Isolate DNA using the QIAGEN DNeasy Blood and Tissue kit. Perform exome or whole-genome sequencing on tumor-normal matched samples to determine somatic mutations and copy number variation.

3.13 RNA Isolation and Expression Analysis

Snap freeze the tumor sample in liquid nitrogen for RNA isolation. Isolate total RNA using the QIAGEN RNeasy Microkit. Validate the quality of the total RNA by running a sample on a bioanalyzer to verify the integrity of the 28s/18s ribosomal subunits. Perform a gene expression analysis using a microarray platform, such as Zebrafish 1.1 ST Array from Affymetrix, or by performing RNAseq.

4 Notes

1. *Cloning.*

   In many cases, the cell type or promoter of interest is not defined. Therefore, we begin analysis of oncogenes by expression ubiquitously with Tol2 kit components, including the beta-actin promoter or the hps70l heat-shock promoter for more controlled expression. In addition, including a fluorescent tag is critical to determine the tissue-specific preference of the oncogene or its tolerance in various cell types. The most efficient system includes cloning either the zebrafish or human gene of interest in the 3′ entry, making this system readily applicable to try in various combinations of promoter constructs and fluorophores.

2. *Implementing sensitizing secondary mutations to promote tumorigenesis.*

   Latency of oncogenes can vary, and the early implementation of secondary mutations could improve the overall penetrance of the tumor phenotype. One strategy is to perform tumor modeling in the $tp53^{zdf1/zdf1}$ strain containing a $tp53^{M214K}$ sensitizing mutation (ZIRC ID, ZL1057; [6]), which has a missense mutation in the DNA binding domain of *tp53*. These zebrafish are readily available through ZIRC and have been described to increase tumor incidence in other zebrafish tumor systems [6, 16]. If secondary cooperating mutations are already known in the human cancer, yet the mechanisms for how they cooperate is not understood, these can be implemented early in the development of a tumor model to understand the relevant underlying biology and potentially increase tumor penetrance and decrease latency.

3. *Alternatives to overcome significant death in early embryos due to oncogene overexpression.*

   There will likely be high mortality in oncogene-injected embryos as compared to controls, and the number of GFP-positive embryos will decrease over time, highlighting the need for a high number of embryos to be injected initially. If mortality is a significant issue, there are a variety of approaches to circumvent this. One possibility is to co-inject with a GFP morpholino targeting the translational start site (Gene Tools, GFP morpholino) to inhibit early expression of the oncogene and facilitate long-term survival. This can be applied in constructs that contain the GFP-2A middle entry clone or in which the oncogene is fused to the GFP coding sequence. Alternatives include using oncogenes fused to the mutated estrogen receptor, which are inducible upon tamoxifen exposure, using a GAL4/UAS system to restrict expression, or a tissue-specific promoter driving the oncogene [34–36, 40].

4. *Transgenic zebrafish resources.*

   ZIRC: https://zebrafish.org/home/guide.php

   zTRAP, Koichi Kawakami Lab: http://kawakami.lab.nig.ac.jp/ztrap/

   Zebrafish Enhancer Trap Database, Harold Burgess Lab: http://burgesslab.nichd.nih.gov/#tableWell

   CreZoo, Michael Brand Lab: http://crezoo.crt-dresden.de/crezoo/

Acknowledgments

GCK is supported by a Cancer Prevention and Research Institute of Texas postdoctoral fellowship through the UTSW Cancer Intervention and Prevention Discoveries training program, a QuadW-American Association for Cancer Research Fellowship for Clinical/Translational Sarcoma Research, and a Young Investigator Grant from Alex's Lemonade Stand Foundation. Supported by grants R01CA135731 from the NIH and RP120685 from the Cancer Prevention and Research Institute of Texas (to JFA). JFA is supported by the Nearburg Family Professorship in Pediatric Oncology Research.

References

1. Smolowitz R, Hanley J, Richmond H (2002) A three-year retrospective study of abdominal tumors in zebrafish maintained in an aquatic laboratory animal facility. Biol Bull 203(2):265–266
2. Kent ML, Spitsbergen JM, Matthews JM, Fournie JW, Murray KN, Westerfield M Diseases of zebrafish in research facilities. In ZIRC Health Services zebrafish disease manual 2012. http://zebrafish.org/health/diseaseManual.php
3. Stanton MF (1965) Diethylnitrosamine-induced hepatic degeneration and neoplasia in the aquarium fish Brachydanio Rerio. J Natl Cancer Inst 34:117–130
4. Pliss GB, Khudoley VV (1975) Tumor induction by carcinogenic agents in aquarium fish. J Natl Cancer Inst 55(1):129–136
5. Feitsma H, Cuppen E (2008) Zebrafish as a cancer model. Mol Cancer Res 6(5):685–694
6. Berghmans S et al (2005) tp53 mutant zebrafish develop malignant peripheral nerve sheath tumors. Proc Natl Acad Sci U S A 102(2):407–412
7. Langenau DM et al (2003) Myc-induced T cell leukemia in transgenic zebrafish. Science 299(5608):887–890
8. Langenau DM et al (2005) Cre/lox-regulated transgenic zebrafish model with conditional myc-induced T cell acute lymphoblastic leukemia. Proc Natl Acad Sci U S A 102(17):6068–6073
9. Sabaawy HE et al (2006) TEL-AML1 transgenic zebrafish model of precursor B cell acute lymphoblastic leukemia. Proc Natl Acad Sci U S A 103(41):15166–15171

10. Zhuravleva J et al (2008) MOZ/TIF2-induced acute myeloid leukaemia in transgenic fish. Br J Haematol 143(3):378–382
11. Yang HW et al (2004) Targeted expression of human MYCN selectively causes pancreatic neuroendocrine tumors in transgenic zebrafish. Cancer Res 64(20):7256–7262
12. Santoriello C et al (2010) Kita driven expression of oncogenic HRAS leads to early onset and highly penetrant melanoma in zebrafish. PLoS One 5(12):e15170
13. Patton EE et al (2005) BRAF mutations are sufficient to promote nevi formation and cooperate with p53 in the genesis of melanoma. Curr Biol 15(3):249–254
14. Dovey M, White RM, Zon LI (2009) Oncogenic NRAS cooperates with p53 loss to generate melanoma in zebrafish. Zebrafish 6(4):397–404
15. Langenau DM et al (2007) Effects of RAS on the genesis of embryonal rhabdomyosarcoma. Genes Dev 21(11):1382–1395
16. Leacock SW et al (2012) A zebrafish transgenic model of Ewing's sarcoma reveals conserved mediators of EWS-FLI1 tumorigenesis. Dis Model Mech 5(1):95–106
17. Gutierrez A et al (2011) Aberrant AKT activation drives well-differentiated liposarcoma. Proc Natl Acad Sci U S A 108(39):16386–16391
18. Nguyen AT et al (2012) An inducible kras(V12) transgenic zebrafish model for liver tumorigenesis and chemical drug screening. Dis Model Mech 5(1):63–72
19. Neumann JC et al (2011) Mutation in the type IB bone morphogenetic protein receptor Alk6b impairs germ-cell differentiation and causes germ-cell tumors in zebrafish. Proc Natl Acad Sci U S A 108(32):13153–13158
20. Howe K et al (2013) The zebrafish reference genome sequence and its relationship to the human genome. Nature 496(7446):498–503
21. Hwang WY et al (2013) Efficient genome editing in zebrafish using a CRISPR-Cas system. Nat Biotechnol 31(3):227–229
22. Hruscha A et al (2013) Efficient CRISPR/Cas9 genome editing with low off-target effects in zebrafish. Development 140(24):4982–4987
23. Suster ML et al (2009) Transgenesis in zebrafish with the tol2 transposon system. Methods Mol Biol 561:41–63
24. Kawakami K, Shima A, Kawakami N (2000) Identification of a functional transposase of the Tol2 element, an Ac-like element from the Japanese medaka fish, and its transposition in the zebrafish germ lineage. Proc Natl Acad Sci U S A 97(21):11403–11408
25. Novakovic B (1994) U.S. childhood cancer survival, 1973–1987. Med Pediatr Oncol 23(6):480–486
26. Martin EA et al (2012) Tadalafil alleviates muscle ischemia in patients with Becker muscular dystrophy. Sci Transl Med 4(162):162
27. Kawahara G et al (2011) Drug screening in a zebrafish model of Duchenne muscular dystrophy. Proc Natl Acad Sci U S A 108(13):5331–5336
28. North TE et al (2007) Prostaglandin E2 regulates vertebrate haematopoietic stem cell homeostasis. Nature 447(7147):1007–1011
29. Amatruda JF et al (2002) Zebrafish as a cancer model system. Cancer Cell 1(3):229–231
30. Lawrence MS et al (2014) Discovery and saturation analysis of cancer genes across 21 tumour types. Nature 505(7484):495–501
31. Kwan KM et al (2007) The Tol2kit: a multisite gateway-based construction kit for Tol2 transposon transgenesis constructs. Dev Dyn 236(11):3088–3099
32. Manoli M, Driever W (2012) Fluorescence-activated cell sorting (FACS) of fluorescently tagged cells from zebrafish larvae for RNA isolation. Cold Spring Harb Protoc 2012(8)
33. Moore JL et al (2002) Fixation and decalcification of adult zebrafish for histological, immunocytochemical, and genotypic analysis. Biotechniques 32(2):296–298
34. Scheer N, Campos-Ortega JA (1999) Use of the Gal4-UAS technique for targeted gene expression in the zebrafish. Mech Dev 80(2):153–158
35. Asakawa K, Kawakami K (2008) Targeted gene expression by the Gal4-UAS system in zebrafish. Dev Growth Differ 50(6):391–399
36. Kawakami K et al (2010) zTrap: zebrafish gene trap and enhancer trap database. BMC Dev Biol 10:105
37. Mosimann C et al (2011) Ubiquitous transgene expression and Cre-based recombination driven by the ubiquitin promoter in zebrafish. Development 138(1):169–177
38. Burger A et al (2014) A zebrafish model of chordoma initiated by notochord-driven expression of HRASV12. Dis Model Mech 7(7):907–913
39. Amsterdam A et al (2004) Many ribosomal protein genes are cancer genes in zebrafish. PLoS Biol 2(5):E139
40. Jungke P, Hans S, Brand M (2013) The zebrafish CreZoo: an easy-to-handle database for novel CreER(T2)-driver lines. Zebrafish 10(3):259–263

Chapter 10

Melanoma Regression and Recurrence in Zebrafish

Sonia Wojciechowska, Zhiqiang Zeng, James A. Lister, Craig J. Ceol, and E. Elizabeth Patton

Abstract

Melanoma is the most lethal form of skin cancer with high mortality rates. Most melanoma cases have activating mutations in BRAF (V600E) and the selective inhibitors of $BRAF^{V600E}$ have been successfully used in patients. However, after initial tumor regression, the majority of patients develop drug resistance resulting in tumor regrowth. It is therefore important to understand the mechanisms underlying these processes. We have recently described the role of the master melanocyte transcription factor MITF in tumor growth, regression, and recurrence. Here, we describe protocols to study regression and recurrence in vivo, as well as for histology and immunohistochemistry, using a temperature-sensitive zebrafish model of human melanoma.

Key words Melanoma, Mitf, Regression, Recurrence, Zebrafish cancer models

1 Introduction

Melanoma is an aggressive and deadly form of skin cancer with rapidly rising mortality rates. The V600E mutation in the BRAF gene has been established as one of the common genetic mutations in melanoma [1], and the BRAF inhibitors are now used as a part of therapy resulting in tumor regression [2]. However, most of the patients develop drug resistance and succumb to the disease after few months of treatment [3]. Therefore, the development of new therapies is an important task. One of the useful and powerful tools to study melanoma regression and recurrence mechanisms is zebrafish, which has been used as a cancer model for over 10 years [4]. Zebrafish tissues and organs share cellular organization with human counterparts, and their genomes are highly conserved with human (zebrafish have orthologs for over 70% of human genes). Importantly, zebrafish develop cancers that share molecular, genetic and histopathological features with human cancers [4–7]. Here we describe a method to investigate melanoma regression and recurrence in a temperature-sensitive zebrafish model of human melanoma *Tg(mitfa:BRAF*^{V600E}*); mitfa*^{vc7} [8].

Koichi Kawakami et al. (eds.), *Zebrafish: Methods and Protocols*, Methods in Molecular Biology, vol. 1451,
DOI 10.1007/978-1-4939-3771-4_10, © Springer Science+Business Media New York 2016

In this model, transgenic fish carrying a human mutated *BRAF*^{V600E} gene expressed under the zebrafish *mitfa* promotor (to restrict expression to melanocytes) [6] are crossed to *mitfa*^{vc7} fish carrying a splice-site mutation in the *mitfa* gene that enables the conditional control of its endogenous activity [9, 10]. MITF (microphthalmia-associated transcription factor) is a highly conserved "master melanocyte regulator" that is known to control the expression of genes involved in melanocyte specification, differentiation, and function, as well as genes involved in cell survival and division [11]. Amplifications as well as gain- and loss-of-function mutations in *MITF* have been identified in melanoma [12–14]. The *mitfa*^{vc7} allele causes a reduction in melanocytes when zebrafish are reared at <26 °C and an almost complete loss of melanocytes at >28 °C. Grown in a permissive temperature, *Tg(mitfa:BRAF*^{V600E}*)*; *mitfa*^{vc7} fish develop melanomas that often display superficial spreading growth pattern with invasion into the muscle and the presence of large, heavily pigmented macromelanophages throughout the tumor. After upshifting into 32 °C to block the *mitf* activity, the melanomas dramatically regress, but once the *mitf* activity is restored, they recur near or in the original tumor site [8].

Here we describe the methods for inducing melanoma development, regression, and recurrence using the temperature-sensitive zebrafish model, as well as protocols for pathology, histology, and immunohistochemistry that can also be applied to other cancer models in zebrafish.

2 Materials

2.1 Zebrafish Strains and Primers

1. *mitfa*^{vc7} genetic line (*see* **Note 1**).
2. *Tg(mitfa:BRAF*^{V600E}*)*; *mitfa*^{vc7} transgenic line or the *mitfa:BRAF*^{V600E} DNA construct (*see* **Note 2**).
3. Primer sequences (all 5′ to 3′) for *BRAF*^{V600E} genotyping: wt, Fwd-TGCTCTTGACCTCAGACTGG and Rev—CCTCAATAAACACCCTACGG, and *BRAF*^{V600E}, Fwd-GAGGCTTTTGTCGAATCGGACCGGTG and Rev-TTGAACAGAGCCTGGCCCGGCT. For *mitfa*^{vc7} genotyping: Fwd - CAAAAGAAGGACAACCACAACCTCtG and Rev GAATTAAAGTCCCCAGCTCTTTAA.

2.2 Zebrafish Maintaining Solutions

1. E-3 media, 60× stock solution: 34.8 g NaCl, 1.6 g KCl, 5.8 g $CaCl_2 \cdot 2H_2O$, and 9.78 g $MgCl_2 \cdot 6H_2O$; dissolve all in H_2O to final volume of 2 l. Mix and adjust pH 7.2 using NaOH. To prepare 1 l of 1× E-3 media, use 16.5 ml of 60× stock, make up to 1 l with H_2O, and add 100 μl methylene blue.
2. Tricaine (3-amino benzoic acid ethyl ester): 400 mg tricaine, 2.1 ml Tris (pH 9), and 97.9 ml distilled H_2O. Dissolve using hot plate and stirrer, and adjust pH to 7.

3. Bleach: 180 µl of bleaching solution (10–13% sodium hypochlorite) with 500 ml system water.

2.3 Genotyping Reagents

1. Tissue lysis buffer (e.g., DirectPCR Ear, Viagen).
2. Proteinase K (10 mg/ml).
3. 10× PCR reaction buffer (e.g., Invitrogen).
4. 50 mM $MgCl_2$.
5. 5 mM dNTPs.
6. 10 µM primer (F+R) mix.
7. Taq DNA polymerase (e.g., Invitrogen).
8. Distilled H_2O.
9. UltraPure Agarose.
10. TBE buffer: prepare 5× stock solution by dissolving 54 g of Tris base, 27.5 g of boric acid, and 20 ml of 0.5 M EDTA (pH 8) in 1 l of distilled H_2O. Dilute with distilled H_2O to get a 1× working solution.
11. DNA gel stain (e.g., SYBR safe or ethidium bromide).
12. Gel loading dye.
13. 1 kb Quick-Load DNA ladder.

2.4 Fixation and Histology Solutions

1. PBS (phosphate-buffered saline): prepare a 10× stock by dissolving 80 g NaCl, 2 g KCl, 14.4 g Na_2HPO_4, and 2.4 g KH_2PO_4 in 800 ml of distilled H_2O. Adjust to pH 7.4 using HCl and make up to final 1 l volume using distilled H_2O. Dilute 1:10 with H_2O to make a 1× working solution.
2. Paraformaldehyde: dilute commercial 16% paraformaldehyde to a 4% working concentration with PBS.
3. EDTA: 0.5 M ethylenediaminetetraacetic acid (pH 7.8).
4. TBS (Tris-buffered saline): prepare a 10× stock by dissolving 61 g Trizma base and 90 g NaCl in 800 ml of distilled H_2O. Adjust to pH 8.4 using HCl and make up to final 1 l volume with distilled H_2O. Dilute 1:10 with H_2O to make a 1× working solution.
5. Citrate buffer: prepare 1 l of 0.01 M solution (pH 6) by dissolving 18 ml 0.1 M citric acid and 82 ml sodium citrate in 900 ml distilled H_2O on a hot plate.
6. Ethanol: a range of percentages (30, 50, 70, 90, 99%).
7. DPX mounting media for histology.
8. Xylene.
9. Hematoxylin solution (basic dye for staining).
10. Eosin solution (acidic dye for staining).
11. Lithium carbonate: 1% solution to blue up fixed sections.

12. Acid-alcohol solution: 1 % HCl in 70 % alcohol.
13. Bleach for IHC: 1 % KOH with 3 % H_2O_2.
14. Protein blocking solution (e.g., DAKO).
15. Antibody diluent (e.g., DAKO).
16. DAB chromogen and substrate (e.g., DAKO).
17. Antibodies.

3 Methods

3.1 Generating the Tg(mitfa:BRAF^{V600E}); mitfa^{vc7} line

1. Cross the two lines into each other (or, alternatively, incross the *mitfa*^{vc7} genetic mutant fish and co-inject the embryos at 1-cell stage with mitfa:BRAF^{V600E} construct with transposase RNA) [15] (*see* **Note 3**), collect the embryos into E-3 media, grow up until 5 dpf in 28.5 °C incubator, and then move into the system (*see* **Note 4**).
2. Once the fish are adults (after 3 months), incross the first generation, select the white embryos (indicating the *mitfa*^{vc7} homozygotes), and grow them up in the system.
3. In order to tail-clip the fish, anesthetize them first (*see* **Note 5**) by placing in a small tank with 4.5 ml tricaine mixed with 100 ml water. Once under anesthesia, place the fish on a Petri dish and cut the small piece of tail fin with a scalpel. Put the fish back to system water to recover and keep isolated until the genotyping results are confirmed (*see* **Note 6**).
4. To extract the DNA, place the tail tissue into a tube, add 25 μl of direct PCR lysis buffer and Proteinase K (10 μl of 10 mg/ml Proteinase K per 1 ml of lysis buffer mix), then incubate on thermocycler in 56 °C for 2 h, followed by 15–20 min in 84 °C, and hold in 10 °C.
5. Dilute the extracted DNA in 25 μl of water and use 1–2 μl of this to set up the PCR reaction.
6. Add 1.25 μl of 10× PCR reaction buffer, 0.38 μl of 50 mM $MgCl_2$, 0.5 μl of 5 mM dNTPs, 0.5 μl of 10 μM primer mix (forward+reverse, separate PCR reaction for wild-type and mutant BRAF primers), and 0.1 μl of Taq polymerase and top up with distilled water to a total volume of 12.5 μl per tube (*see* **Note 7**).
7. Run the PCR for wild-type and mutant *BRAF*. For pigmented fish, you can also run a PCR for *mitfa*^{vc7} to identify the heterozygotes, which can be used for further crosses (*see conditions in* Table 1).
8. Check the product by running 5 μl with 1 μl loading dye on 1 % agarose gel electrophoresis with DNA gel stain (100 V for 15–20 min) and visualize the bands under UV light (*see* **Note 8**) (*see a typical result* Fig. 1).

Table 1
***Tg(mitfa:BRAF^{V600E})* genotyping – PCR conditions**

	Wild-type	*mitfa:BRAF^{V600E}*	*mitfa^{vc7}*
Initial denaturation	95 °C for 2 min		
Number of cycles:	35x	35x	30x
Denaturation	95 °C for 10 s	95 °C for 10 s	95 °C for 10 s
Annealing	56 °C for 20 s	58 °C for 20 s	60 °C for 30 s
Elongation	72 °C for 50 s	72 °C for 30 s	72 °C for 30 s
Final elongation	72 °C for 5 min		
Hold	4 °C		

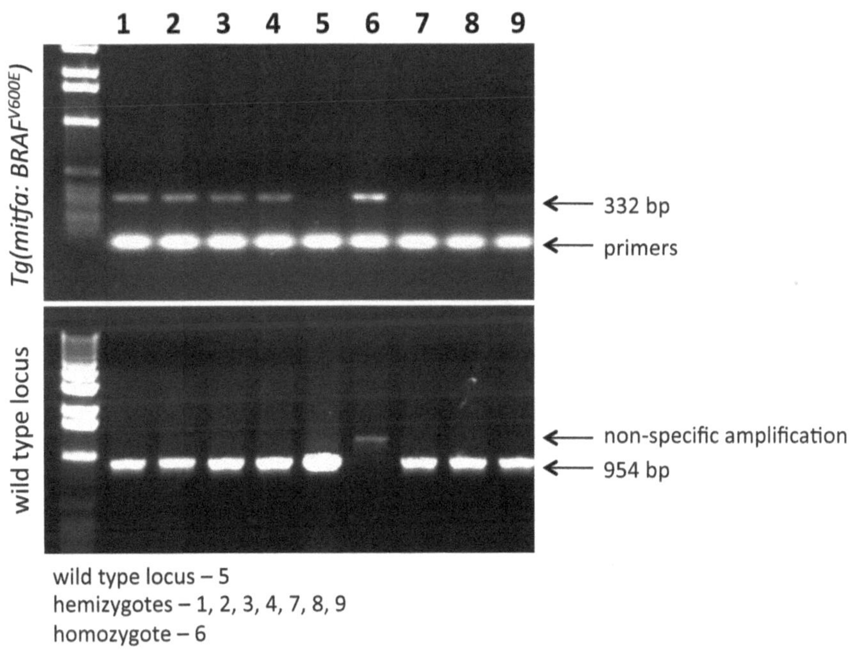

Fig. 1 A typical result of *Tg(mitfa:BRAF^{V600E})* genotyping. An image of 1 % agarose gels showing the results of the two PCR reactions required to genotype the *Tg(mitfa:BRAF^{V600E})* line. Fish with numbers 1, 2, 3, 4, 7, 8, and 9 are hemizygotes, as determined by a PCR product at 954 bp representing the wild-type locus and 332 bp representing the *Tg(mitfa:BRAF^{V600E})*. Fish 5 is a wild-type fish, as evidenced by only the 954 bp PCR product. Fish 6 is homozygous for *Tg(mitfa:BRAF^{V600E})* as determined by a bright 332 bp PCR product and the absence of the 954 bp band. The very bright primer bands are indicated, and fish 6 shows the presence of a nonspecific amplification (as discussed in **Note 8**)

Fig. 2 An example of the glass tanks used for temperature shift experiments. Separate glass tanks are filled with system water and contain filters and heaters to control the water quality and adjust the temperature. Additional thermometers are placed inside each tank to monitor the temperature. Each tank is also labeled with information regarding the fish and temperature settings

3.2 Inducing Melanoma Formation, Regression, and Recurrence

1. After selecting the $mitfa^{vc7}$ homozygotes and genotyping for $Tg(BRAF^{V600E})$, downshift the fish to <26 °C to allow the *mitfa* activity and induce melanoma (*see* **Note 9**). Use glass tanks with lids, filters, and heaters to adjust the temperature of water (*see* Fig. 2).
2. Once fish have tumors, they can be then upshifted to 32 °C to regress (*see an example* Fig. 3) and then back to <26 °C to recur (*see* **Note 10**).

3.3 Fixation

Depending on the purpose of study, fish can be culled and fixed at any point—after melanoma development, regression, or recurrence.

1. Sacrifice the fish using tricaine solution (*see* **Note 5**) and dissect it in half transversely (*see* **Note 11**). Incubate the tissues in 4% paraformaldehyde for 3 days at 4 °C with agitation.
2. Wash the samples with PBS and transfer to 0.5 M EDTA (pH 8) to decalcify for 5 days at 4 °C with agitation.
3. Wash the tissues once again with PBS and transfer to 70% ethanol.
4. Store at 4 °C until required.

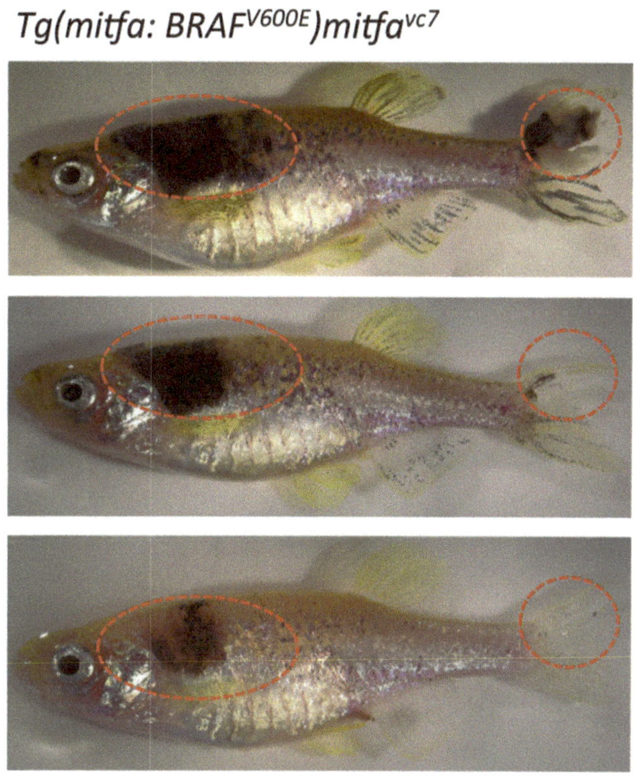

Fig. 3 An example of *Tg(mitfa:BRAF*^{V600E}*):mitfa*^{vc7} melanoma regression after a month at 32 °C. A *Tg(mitfa:BRAF*^{V600E}*):mitfa*^{vc7} zebrafish with melanoma was placed at 32 °C for regression. First signs of regression were observed within 10 days after the upshift, and a complete regression of the tail fin tumor as well as further regression of the back melanoma was observed after 1 month. Tumour sites are indicated by red ellipses

3.4 Embedding and Sectioning

1. Process the fixed tissue/adult/tumor as follows: 95 % ethanol (2 h), absolute alcohol (3 × 2 h, 2 × 3 h), xylene (2 × 2 h, 1 × 3 h), and molten paraffin wax (2 × 2 h, 1 × 3 h at 58 °C).
2. Embed processed tissue/adult/tumor into paraffin in the desired orientation (*see a schematic in* Fig. 4) into wax blocks to be sectioned.
3. Cut transverse sections of 5 μm thickness using microtome.
4. Float the wax sections in water bath at 40–45 °C, then place onto a glass slide.
5. Leave to dry for at least 2 h on a hot plate.

3.5 Hematoxylin and Eosin (H&E) Staining

1. De-wax the slides in xylene (2 × 5 min) (*see* **Note 12**).
2. Rehydrate the slides by washing in 99 %, 90 %, 70 %, 50 %, and 30 % alcohol solutions (5 min each), then wash in running water.

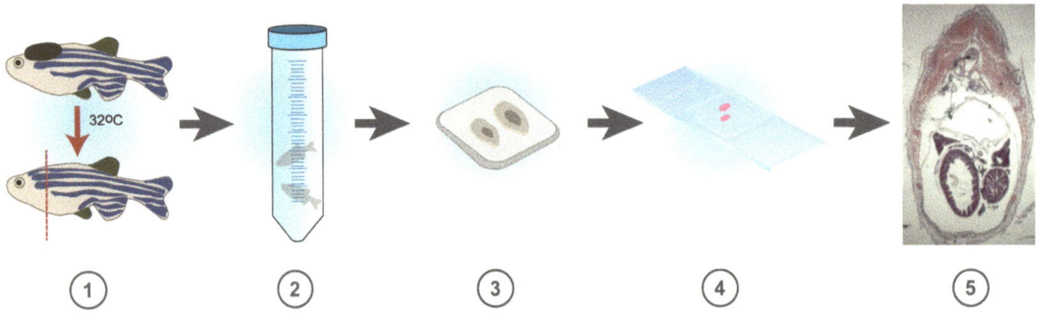

Fig. 4 A schematic of fish tissue processing for histology. After tumor regression at 32 °C, fish are cut in half along side of regression (1), fixed in 4 % PFA (2), processed and embedded into paraffin blocks (3). Blocks are then cut into thin sections placed on glass slides, H&E or IHC stained (4), and imaged under the microscope (5)

3. Stain for 4 min in Mayer's hematoxylin, and wash with running water.
4. Rinse with 95 % acid-alcohol solution for few seconds, then wash with water again.
5. Counterstain with eosin for 1–2 min.
6. Dehydrate the slides with 95 % alcohol (5 min) and 99 % alcohol (2 × 5 min), then clear with 2 × 5 min washes in xylene.
7. Use DPX to mount the slides with cover slips and leave to dry in the air.

3.6 Immunohistochemistry

1. Prepare the slides as in **steps 1** and **2** above (Subheading 3.5).
2. Bleach the slides for 15 min in 1 % KOH with 3 % H_2O_2 solution (*see* **Note 13**), then wash 3 × 5 min with water.
3. Perform antigen unmasking in 0.01 M citrate buffer or 1 mM EDTA pH 8 (depending on the antibody) using pressure cooker (*see* **Note 14**).
4. Wash with water 3 × 5 min, then incubate in 3 % hydrogen peroxide for 10 min and wash again with water (2 × 5 min).
5. Wash for 5 min in 1× TBS buffer.
6. Incubate for 30 min with serum-free protein block (e.g., DAKO) at room temperature.
7. Add the primary antibody diluted in antibody diluent (e.g., DAKO) overnight at 4 °C or for 1–2 h at room temperature in the dark (*see* **Note 15**).
8. Remove the primary antibody and wash the slides 3 × 5 min with 1× TBS buffer.
9. Overlay slides with HRP rabbit/mouse secondary antibody (e.g., DAKO EnVision kit) and incubate for 30 min at room temperature.
10. Wash the slides 2 × 5 min with 1× TBS buffer.

11. Visualize the staining by incubating in 1:50 of DAB chromogen/DAB substrate for 10 min at room temperature.
12. Wash 2 × 5 min with water, then stain with hematoxylin (4 min).
13. Rinse with running water and then acidic alcohol for few seconds and blue up with lithium chloride and wash with water again.
14. Dehydrate and clean the slides as in **steps 6** and 7 of Subheading 3.5.

4 Notes

1. The *mitfa*^{vc7} line was identified in an ENU-mutagenesis screen [9] (initially described as *mitfa*^{fh53}), and carries a mutation that enables the conditional temperature-dependent control of Mitfa activity in zebrafish [8]. The adult fish reared at normal system water temperature (28 °C) have almost no melanocytes/stripes and are "white".
2. The human *BRAF*^{V600E} in this construct is myc tagged and only expressed in melanocytes and can be detected in the fish using epitope-specific antibodies [4, 5].
3. Co-inject 25 ng/μl of plasmid with 35 ng/μl of transposase RNA [15] directly into the single-staged embryos.
4. Before moving into nursery tanks, it is a common practice to bleach the embryos at or around 24 hpf.
5. Make sure to use the most humane procedures and methods of killing and that they have all been approved by the local ethics committee.
6. Remember not to keep the fish under the anesthesia for too long. You can help the fish to recover by gently pumping the fresh system water into the gills using pastette.
7. It is always recommended to run positive and negative/blank control samples together with the genotyping samples.
8. The expected band for *Tg(mitfa:BRAF*^{V600E}*)* is at 332 bp and for the wild-type locus (no transgene insertion), a bright band is detected at 954 bp. In our experience, in some cases, in the PCR reaction for the wild-type locus, there can be a faint band visible at around 1200 bp; for example, *see* Fig. 1. The expected band for *mitfa*^{vc7} is 200 bp. The lower case "t" in the forward primer is a mismatch that creates Ddel site in the product amplified from the *vc7* mutant allele but not the wild-type allele, so digesting with Ddel will bring the *vc7* amplimer down in size to 175 bp but leave the wild-type amplimer alone.
9. If kept at <26 °C, fish usually start to get tumors at around 6 months of age. The tumor formation usually starts with pigmented lesions and fish nevi, which then progress to melanoma.

10. The first changes can be usually observed after 1–2 weeks after the upshift or downshift, and then depending on the tumor size, the full regression can be usually observed up to 2 months after the change of water temperature.
11. Depending on the aim of experiment, fish can be dissected in any desired way, but it is always good to cut it at least into half (e.g., across the tumor site or the site of regression) to increase the penetration of the fixative.
12. Remember to perform all of the paraformaldehyde, xylene, and DPX steps under the fume hood and wear protective clothing.
13. This step gets rid of the pigment on the sections and can be omitted if it is not necessary for particular staining.
14. Bring the buffer to the boiling point first (12 min in microwave on high power), then place the slides in the buffer, further heat for 7 min, and remove from the buffer to cool down on the bench (for 10–20 min). Remember to wear protective face shield and gloves when venting and opening the pressure cooker.
15. The concentration and incubation have to be optimized and depend on the primary antibody used. To confirm the melanoma status of the tumor, staining with antibodies against the myc tag (to verify the presence of mutant BRAF) and anti-melan-A is recommended.

Acknowledgments

We are grateful to the MRC Human Genetics Zebrafish Facility and to Amy Capper for setting up the tanks for temperature shift experiments, and Craig Nicol for the graphical support in Figure 4. The work was supported by the Medical Research Council, Cancer Research UK, and a European Research Council Consolidator Award (ZF-MEL-CHEMBIO 648489).

References

1. Davies H, Bignell GR, Cox C, Stephens P, Edkins E, Clegg S et al (2002) Mutations of the BRAF gene in human cancer. Nature 417:949–954
2. Bollag G, Tsai J, Zhang J, Zhang C, Ibrahim P, Nolop K, Hirth P (2012) Vemurafenib: the first drug approved for BRAF-mutant cancer. Nat Rev Drug Discov 11:873–886
3. Chapman PB, Hauschild A, Robert C, Haanen JB, Ascierto P, Larkin J et al (2011) Improved survival with vemurafenib in melanoma with BRAF V600E mutation. N Engl J Med 364:2507–2516
4. White R, Rose K, Zon LI (2013) Zebrafish Cancer: the state of the art and the path forward. Nat Rev Cancer 13:624–636
5. Patton EE, Zon LI (2004) Taking human cancer genes to the fish: a transgenic model of melanoma in zebrafish. Zebrafish 1:363–368
6. Patton EE, Widlund HR, Kutok JL, Kopani KR, Amatruda JF, Murphey RD et al (2005) BRAF mutations are sufficient to promote nevi formation and cooperate with p53 in the genesis of melanoma. Curr Biol 15:249–254
7. Howe K, Clark MD, Torroja CF, Torrance J, Berthelot C, Muffato M et al (2013) The

zebrafish reference genome sequence and its relationship to the human genome. Nature 496:498–503

8. Lister JA, Capper A, Zeng Z, Mathers ME, Richardson J, Paranthaman K et al (2014) A conditional zebrafish MITF mutation reveals MITF levels are critical for melanoma promotion vs. regression in vivo. J Invest Dermatol 134:133–140
9. Johnson SL, Nguyen AN, Lister JA (2011) Mitfa is required at multiple stages of melanocyte differentiation but not to establish the melanocyte stem cell. Dev Biol 350:405–413
10. Zeng Z, Johnson SL, Lister JA, Patton EE (2015) Temperature-sensitive splicing of mitfa by an intron mutation in zebrafish. Pigment Cell Melanoma Res 28:229–232
11. Cheli Y, Ohanna M, Ballotti R, Bertolotto C (2010) Fifteen-year quest for microphthalmia-associated transcription factor target genes. Pigment Cell Melanoma Res 23:27–40
12. Cronin JC, Wunderlich J, Loftus SK, Prickett TD, Wei X, Ridd K et al (2009) Frequent mutations in the MITF pathway in melanoma. Pigment Cell Melanoma Res 22:435–444
13. Konieczkowski DJ, Johannessen CM, Abudayyeh O, Kim JW, Cooper ZA, Piris A et al (2014) A melanoma cell state distinction influences sensitivity to MAPK pathway inhibitors. Cancer Discov 4:816–827
14. Van Allen EM, Wagle N, Sucker A, Treacy DJ, Johannessen CM, Goetz EM et al (2014) The genetic landscape of clinical resistance to RAF inhibition in metastatic melanoma. Cancer Discov 4:94–109
15. Kawakami K (2007) Tol2: a versatile gene transfer vector in vertebrates. Genome Biol 8(Suppl 1):S7

Chapter 11

Imaging of Human Cancer Cell Proliferation, Invasion, and Micrometastasis in a Zebrafish Xenogeneic Engraftment Model

Claudia Tulotta, Shuning He, Lanpeng Chen, Arwin Groenewoud, Wietske van der Ent, Annemarie H. Meijer, Herman P. Spaink, and B. Ewa Snaar-Jagalska

Abstract

The xenograft model, using the early life stages of the zebrafish, allows imaging of tumor cell behavior both on a single cell and whole organism level, over time, within a week. This robust and reproducible assay can be used as an intermediate step between in vitro techniques and the expensive, and time consuming, murine models of cancer invasion and metastasis.

In this chapter, a detailed protocol to inject human cancer cells into the blood circulation of a zebrafish embryo is described; the engraftment procedure is then followed by visualization and quantification methods of tumor cell proliferation, invasion, and micrometastasis formation during subsequent larval development. Interaction with the host microenvironment is also considered.

Key words Xenotransplantation, Tumor proliferation, Tumor invasion, Chemical treatment, Zebrafish embryo, Metastasis, Microenvironment

1 Introduction

Zebrafish (*Danio rerio*) has become an important animal model for cancer, immune, and stem cell research [1–3]. Many molecular and cellular components that operate during tumorigenesis are conserved between zebrafish and mammals [4]. Due to the transparency of zebrafish embryos and the availability of various tissue-specific fluorescent reporter transgenic families, high-resolution in vivo analysis of tumor progression as well as interaction between tumor cells and the host microenvironment is relatively easy [5, 6]. Furthermore, zebrafish develops histopathological characteristics that can be found in human tumors [7] and physiologically respond to a wide range of pharmacologically active compounds, which makes it a good model for drug screening purposes [8, 9].

Koichi Kawakami et al. (eds.), *Zebrafish: Methods and Protocols*, Methods in Molecular Biology, vol. 1451,
DOI 10.1007/978-1-4939-3771-4_11,

Xenogeneic transplantation is a widely used technique in cancer biology to study human tumors and zebrafish offers an innovative approach to investigate tumor behavior in parallel to the established chick embryo [10, 11] and mouse models [12, 13].

The engraftment of human tumors in the transparent zebrafish embryo bears the great advantage to follow the multistep process of cancer progression in a whole vertebrate, on a single-cell level, through high-resolution imaging. In the last decade, it has been shown that human tumors display the ability to proliferate at the implantation site, as in the case of yolk [14–17] and common cardinal vein [5], to migrate from the yolk sac toward other tissues [14, 16, 17], to induce angiogenesis once implanted in Duct of Cuvier [5], yolk [14], hindbrain [14], and perivitellin space [18–20], to extravasate [21], to induce micrometastasis with infiltration of blood vessels [22], and to interact with the host microenvironment [5]. In particular, the interaction with innate immune cells, often supportive of tumor progression, has been highlighted in a previous work from our group, where the neutrophils prepare the metastatic niche and enhance tumor invasion if VEGFR is chemically inhibited [5].

Importantly, the effect of drugs on transplanted tumors can be easily monitored in vivo, by adding the compounds directly into the water and the simultaneous treatment of multiple larvae guarantees large group numbers for statistical analysis. Recently, we demonstrated that the zebrafish xenograft assay is a robust model for examining the role of pharmacological modulators and genetic perturbation of TGF-β signaling in human breast tumor cells [23].

Here, we report a detailed method to study human cancer proliferation, invasion, and micrometastasis formation in a zebrafish xenogeneic model. We engraft human cancer cell lines, stably expressing fluorescent proteins, in a zebrafish embryo host, where blood vessels or immune cells are fluorescently traceable. Quantitative analysis of local tissue invasion and tumor proliferation on whole embryo and subsequent larval stages are performed, as well as qualitative validation based on whole-mount immunohistochemistry. Drug treatment setup to estimate inhibitory effects on tumor burden is described and finally the interactions between cancer cells and the host stroma are visualized.

2 Materials

2.1 Tumor Cell Suspension Preparation

Prewarm all media and reagents at 37 °C.

1. Cell culture flasks 25 cm² (Greiner bio-one).
2. Trypsin-EDTA: 0.25% Trypsin, 0.53 mM EDTA in HBSS (30-2101, ATCC®).
3. DMEM (Dulbecco's Modified Eagle Medium, Gibco® by life technologies).

4. Fetal Calf Serum (FCS).
5. DPBS (Dulbecco's Phosphate-Buffered Saline, GIBCO® by life technologies).
6. Human tumor cell lines stably expressing fluorescent proteins (here we use breast MDA-MB-231-B dsRed [24], prostate PC3-Pro4 mCherry [25], and A673 GFP Ewing sarcoma tumor cell lines).
7. Cell Tracker™ (CM-Dil, Molecular Probes®, Life technologies): resuspend the lyophilized pellet in DMSO at a concentration of 1 mg/ml and make aliquots for long term storage at −20 °C. Dilute the stock concentration to 5 μg/ml in DPBS; prepare it fresh on the day of cell inoculation (*see* **Note 1**).
8. 2 % PVP40 (Polyvinylpyrrolidone-40).

2.2 Preparation of Zebrafish Embryos and Microscopy for Phenotype Detection

1. 2-Day-old zebrafish larvae (reporter line with fluorescent blood vessels or fluorescent immune cells). Larvae must be without chorion.
2. Egg water: 60 μg/ml InstantOcean® salt in distilled water.
3. PTU (1-phenyl-2-thiourea, Sigma-Aldrich): prepare a 0.01 M stock solution (pH ~ 7) and store aliquots in 50 ml tubes, at −20 °C. Add 0.003 % PTU in each Petri dish.
4. Tricaine (MS-222): prepare a 0.4 % stock solution (pH ~ 7) and store aliquots at −20 °C. Use a 0.02 % working concentration.
5. Petri dish (92 × 16 mm).
6. Petri dish (92 × 16 mm) coated with 1.5 % (w/v) agarose, dissolved in egg water.
7. Pasteur plastic pipettes (3.5 ml).
8. Glass capillary needles: 1.0 OD × 0.78 ID × 100 L mm, 30-0038 Harvard Apparatus.
9. Flaming/Brown micropipette puller, model P-97, Sutter Instrument Co.
10. Microloader tips (20 μl Eppendorf ep T.I.P.S., 5242 956 003).
11. Forceps (Dumont#5) for manual chorion removal and needle cutting.
12. Pneumatic Pico Pump PV820, World Precision Instruments (WPI).
13. Micromanipulator (WPI).
14. Stereo microscope (Leica MS5).
15. 34 °C incubator.
16. Stereo fluorescence microscope (Leica MZ16FA).
17. Confocal microscope with 4× objective and motorized stage (Nikon A1R).
18. Glass-bottomed 96-well plate.

2.3 Immunohistochemistry

1. PBS-tween 0.1% (PBST).
2. Tween 20
3. PFA 4%: dilute 16% Formaldehyde solution (w/v) methanol-free (Thermo-Scientific) to 4% in PBST.
4. Proteinase K, recombinant PCR grade (03115879001, Roche): use 10 μg/ml.
5. Blocking solution: sheep serum (Sigma-Aldrich) or 10× blocking buffer (MitoSciences®). Prepare final dilutions in PBST (use 5% sheep serum).
6. p-Histone 3 (Ser10)-R, rabbit polyclonal IgG, Santa Cruz Biotechnology.
7. Goat anti-rabbit Alexa Fluor 405 or 633 (Molecular Probes®, Life Technologies).

3 Methods

3.1 Zebrafish Lines and Embryo Preparation

1. Choose a zebrafish line according to the readout of the assay. In order to assess tumor cell extravasation, invasion, and micrometastasis, a reporter line with fluorescent vasculature is recommended (generally we choose *Tg(kdrl:mCherry)* [26], *Tg(kdrl:EGFP)*^{s843} [27], or *Tg(fli1a:EGFP)* ^{y1} [28]). To visualize the interaction between malignant and immune cells, we use transgenic lines with GFP expressing neutrophils *Tg(mpx:GFP)*^{i114} [29] and macrophage reporter lines *Tg(mpeg1:EGFP)gl22* [30] and *Tg(mpeg1:mCherry)*^{UMSF001} [31] (*see* **Note 2**).
2. Keep eggs at 28–28.5 °C, in Petri dishes with egg water ($n \leq 100$). Add PTU between 8 hpf (hours-post-fertilization) and 1 dpf (day-post-fertilization) to avoid pigment formation (*see* **Note 3**).
3. Manually remove the chorion at 1 dpf and keep the embryos at 28–28.5 °C until cell injection.

3.2 Tumor Cell Suspension Preparation

On the day of the implantation, the adherent cell layer should reach confluence between 60 and 80%, when the growth is still in the Log Phase (*see* **Note 4**).

1. Remove old medium and perform a quick wash with Trypsin-EDTA (0.5 ml in a T25 flask).
2. Add 0.5 ml of Trypsin-EDTA and swirl the flask in order to homogeneously distribute the Trypsin all over the surface where cells have grown. Gently tap the flask on the side, whether cell detachment requires longer time.

3. When cells are detached, add 5 ml of complete medium and pipette up and down few times in order to make a single-cell suspension.
4. Transfer the single-cell suspension from the flask to a 15 ml tube and centrifuge for 5 min at 200 × *g* (Eppendorf 5702).
5. Remove the complete medium and resuspend the cell pellet in 5 ml of DPBS.
6. Centrifuge for 5 min at 200 × *g* and remove the supernatant. To continue, *see* **step** 7 if cells stably express a fluorescent protein or **steps 8** and **9** when labeling is required.
7. Resuspend the cell pellet in 1 ml of DPBS and transfer in a 1.5 ml Eppendorf tube. Centrifuge for 5 min at 135 × *g* (Eppendorf 5424), remove the supernatant and finally make a cell suspension in 5–10 μl 2 % PVP40 (*see* **Note 5**). Keep cells at room temperature during cell engraftment in embryos.
8. Resuspend the cell pellet in 1 ml CM-Dil solution and transfer in 1.5 ml Eppendorf tube. Incubate for 5 min at 37 °C and then for 15 min at 4 °C. Gently tap the tube, when cells have precipitated at the bottom.
9. Centrifuge for 5 min at 135 × *g*, remove the supernatant and add 1 ml of DPBS to wash off the residual CM-Dil. Perform last centrifugation step as previously described and resuspend the cells in 5–10 μl of 2 % PVP40 (*see* **Note 5**). The cell suspension is kept at room temperature until the end of the implantation procedure.

3.3 Engraftment of Human Tumor Cells in Zebrafish Embryos

1. Load the needle with the cell suspension and place it in the micromanipulator (*see* **Note 6**).
2. Add a few drops of egg water on the agarose-coated Petri dish.
3. Cut the end of the needle with a forceps, under the microscope (*see* **Note** 7).
4. Perform a test injection in the water drop and count number of cells.
5. Adjust injection pressure and injection time in order to engraft about 200–500 cells (*see* **Note 8**).
6. Anesthetize zebrafish larvae with Tricaine.
7. Place 10–20 anesthetized larvae on the agarose-coated Petri dish and remove the egg water with a transfer pipette (*see* **Note 9**).
8. Position the larvae under the objective (*see* **Note 10**).
9. In order to perform the implantation directly in the blood circulation, insert the needle through the yolk sac and reach the distal branch of the Duct of Cuvier (Fig. 1), close to the opening into the heart cavity (*see* **Notes 11** and **12**).

10. Transfer the implanted larvae in a new Petri dish, containing egg water and PTU.
11. Repeat until engraftment is performed in the desired number of larvae and then transfer to 34 °C.

3.4 Verification of Correct Tumor Cell Engraftment in Zebrafish Embryos

1. Anesthetize implanted larvae with Tricaine and remove excess egg water, in order to reduce sample movements during imaging, avoiding drying out.
2. Discard wrongly implanted larvae 3–5 h after injection, using a fluorescent microscope (*see* **Note 13**). When implantation in the blood circulation is examined, larvae with a copious tumor

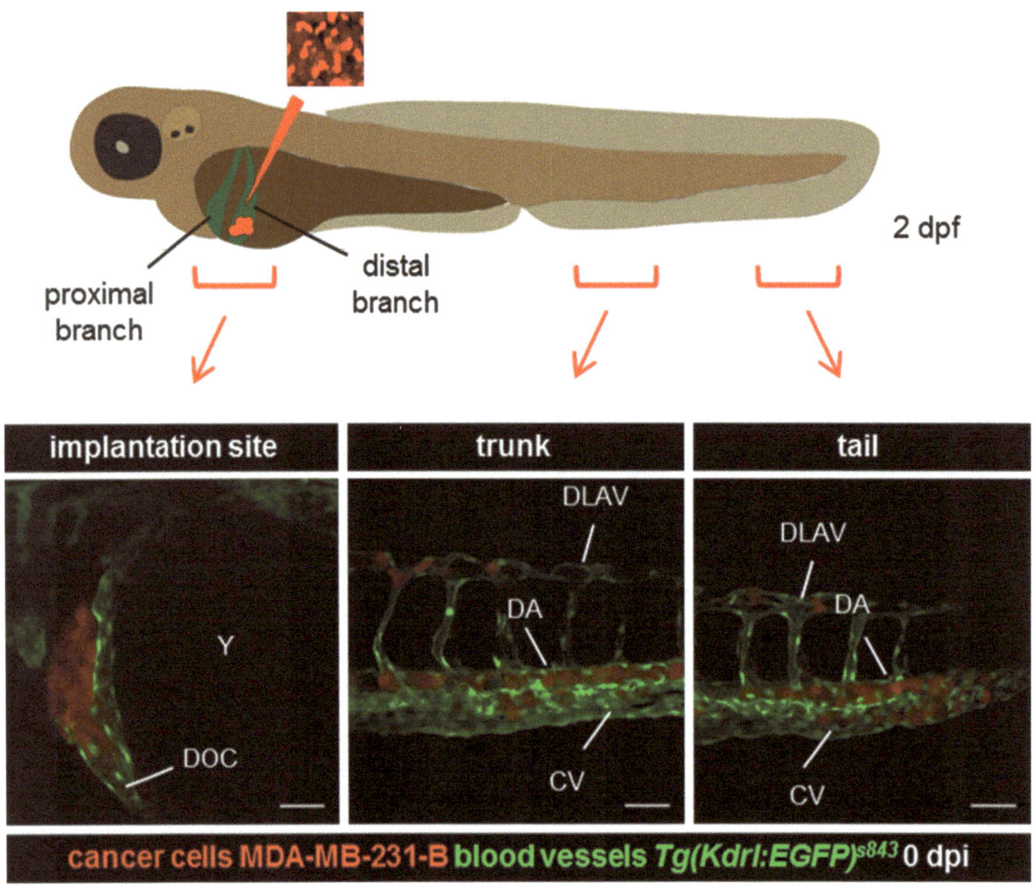

Fig. 1 Tumor cell engraftment in zebrafish embryo. Schematic representation of a 2-day-old zebrafish embryo and the implantation site; human cancer cells, stably expressing red fluorescent protein, are engrafted in the distal branch of the Duct of Cuvier. Tumor cell accumulation at the injection site (DOC) and circulation in the vasculature (DA, CV, DLAV), both in the trunk and tail regions, can be visualized at 0 dpi, in an embryo with green fluorescent blood vessels. *MDA-MB-231-B* triple negative breast cancer cell line, *Y* yolk, *DOC* duct of cuvier, *DA* dorsal aorta, *CV* caudal vein, *DLAV* dorsal longitudinal anastomotic vessels, *dpi* days-post-implantation. Images were acquired using a Leica TCS SPE (20× objective). Scale bar = 50 μm

cell leakage in the yolk and low number of cancer cells inside the blood vessels are excluded.

3. Take representative images of engrafted embryos (Fig. 1) and transfer afterwards in a new Petri dish, containing egg water and PTU. Establish the starting number for each group and place the sample back at 34 °C.

3.5 Phenotype Analysis of Tumor Cell Extravasation and Micrometastasis Formation in Zebrafish Larvae

The phenotype assessment can be done at any desired time points (choose at least two time points over the course of the experiment (2 and 4 dpi)) (*see* **Note 14**).

1. Cancer cells that survive in the blood circulation generally localize and accumulate in proximity of the circulatory loop between the dorsal aorta (DA) and the caudal vein (CV). Depending on tumor cell behavior, the phenotype quantification varies. Count the number of tumor cells that extravasate and invade the tail fin (Fig. 2a), below the CV, if single cells can be distinguished (*see* **Note 15**). When tumor masses are formed instead, quantification of proliferation should be assessed as described in Subheading 3.6.
2. Acquire representative images, using a fluorescent microscope. Invasive cancer cells can also be visualized in the tail fin, using the transmitted light (Fig. 2b).

3.6 Phenotype Analysis of Tumor Burden by Automated Imaging

1. The assessment of tumor cell proliferation can be performed at any desired time points, although required adaptation of the engrafted cells into the host should be taken into consideration and therefore analysis at early time points after implantation are not advised. We normally quantify cell proliferation at 4 or 6 dpi (*see* **Note 16**).
2. Screen larvae at the chosen time point, using a stereo fluorescent microscope (Fig. 2c), selecting the appropriate filters (generally dsRed and GFP filters). Discard malformed and dead larvae.
3. Fix embryos with 4% PFA overnight (ON) at 4 °C.
4. Perform PBS washes (3 × 10 min) in order to remove the fixative.
5. Use a glass-bottomed 96-well plate and place a larva in each well. Remove as much water as possible, without letting the sample dry out. Position the larva diagonally in each well, using a brush (*see* **Note 17**).
6. Position the 96-well plate under the confocal microscope with a motorized stage and use a 4× lens for the acquisition.
7. Set the position of each larva and define the Z-stacks. Perform automated image acquisitions, keeping the same parameters for each group.

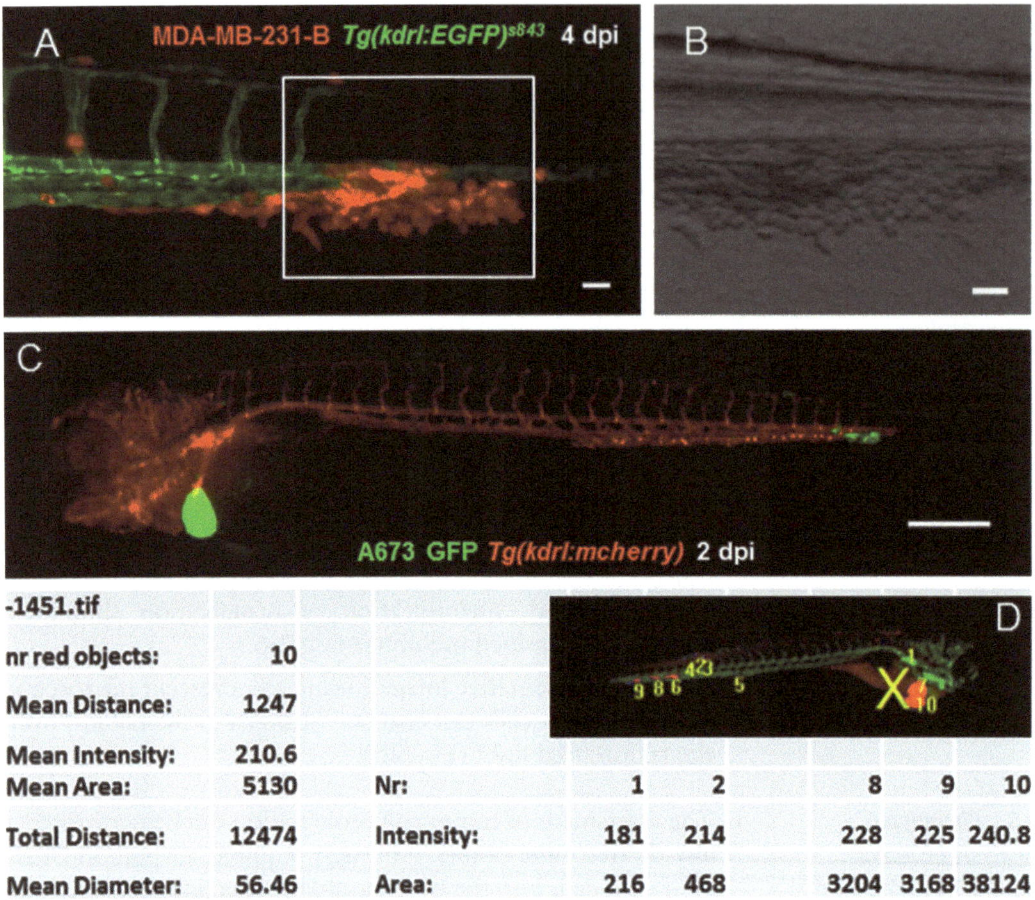

Fig. 2 Behavior of human cancer cells in zebrafish larvae. (**a**) Extravasation and invasion of the tail fin, by MDA-MB-231-B dsRed in *Tg(kdrl:EGFP)s843* at 4 dpi; the selected area correspond to bright field image in (**b**); scale bar = 50 μm. (**c**) A673 GFP behavior in whole *Tg(kdrl:mCherry)* larva: primary mass formation at the engraftment point and secondary mass localization in the tail at 2 dpi; scale bar = 500 μm. (**d**) Image-Pro Analyzer 7.0 output analysis is used to quantify tumor proliferation; tumor cell masses are identified as objects and the respective areas are calculated in a 4 dpi larva. Images were acquired with Leica MZ16FA stereomicroscope with Leica DFC420C camera in (**a**–**c**) and Nikon confocal microscope system A1R, with Plan Apo 4× objective in (**d**)

8. Import the TIFF files in ImageJ, in order to generate a folder containing a maximum projection for each embryo.
9. The folder created is opened in Image-Pro Analyzer 7.0 (Media Cybernetics). Signal thresholding outlines each larva and identifies cells and cell aggregates as objects, limiting the selection of cellular fragments; the threshold for the green and red signal is based on one larva and automatically applied to all images. Using a macro, values such as the number of objects, the area of each object and the intensity of the fluorescence are generated (Fig. 2d). To calculate tumor cell proliferation we use the formula mean area × number of objects.

10. It is possible to limit the analysis of tumor proliferation to a specific region. We restrict the analysis to the tail area, using a macro designed to exclude area of non-interest and therefore recalculating the number of objects and the mean area in the non-excluded segment (*see* **Note 18**).

3.7 Immuno-histochemistry to Verify Cancer Cell Proliferation at the Invasive Site

Whole-mount immunostaining with p-Histone H3 is an additional means to verify tumor proliferation activity at the site of invasion. High-resolution confocal imaging can be consequently performed (Fig. 3a, a′). Choose the stage of interest; here the protocol refers to 6 dpi larvae.

1. After fixation, wash sample with PBST (3 × 10 min).
2. Permeabilize the tissues with Proteinase K (10 μg/ml, diluted in PBST) for 3 h at RT, on a shaker.
3. Wash with PBST 4 × 5 min.

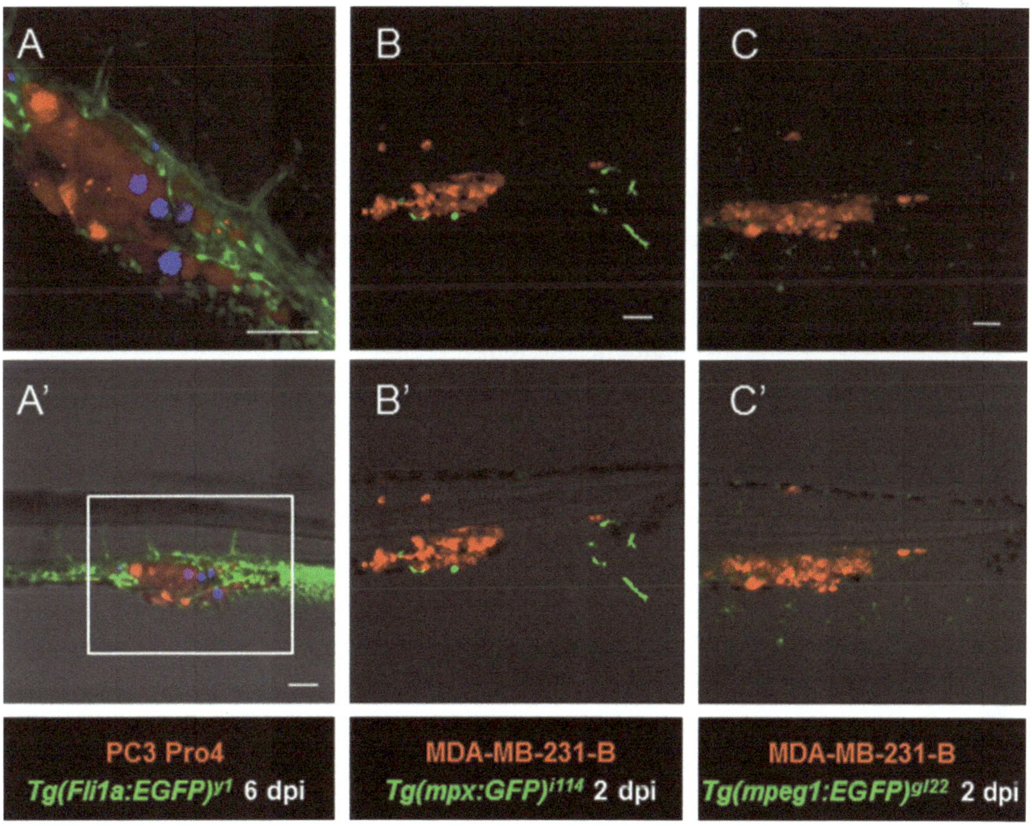

Fig. 3 Local proliferation of human tumor cells and their interaction with the zebrafish stroma. (**a** and **a**′) PC3 Pro4 mCherry tumor cells proliferate in the tail fin of 6 dpi *Tg(Fli1a:EGFP)***y1** zebrafish larva, as shown with p-H3 immunostaining in *blue*; in (**a**), higher magnification of *square* in (**a**′). (**b** and **b**′) Neutrophils infiltrate and surround the tumor site in the tail of 2 dpi *Tg(mpx:gfp)i114* zebrafish larva. (**c** and **c**′) Presence of macrophages is detected at the tumor invading edge and in the tail fin, in T*g(mpeg1:EGFP)gl22* larvae, at 2 dpi. Scale bars = 50 μm

4. Perform the blocking step with 5 % sheep serum or commercial blocking buffer, for 2 h at RT, on a shaker.
5. Apply the primary antibody rabbit p-Histone H3. Prepare 1:200 dilution in blocking solution. Incubate ON at 4 °C, using a shaker.
6. Remove the antibody and keep the solution; it can be reused multiple times.
7. Perform three quick washes in PBST and four washes for 10 min each time.
8. Repeat the blocking step as previously described.
9. Incubate with secondary antibody goat anti-rabbit conjugated with Alexa Fluor (we used 405 in this case, as human cells produce dsRed/mCherry proteins and zebrafish vessels are GFP positive . Dilution 1:200, incubation 3 for hours at RT).
10. Wash samples with PBST and store at 4 °C until imaging. Keep the secondary antibody for reuse for at least 3 other times.

3.8 Testing the Effect of Chemical Compounds on Cancer Proliferation and Micrometastasis Formation In Vivo

Drug treatment is performed on the same day of the implantation or the following day, after the verification of correct engraftment (*see* **Note 19**).

1. Place six larvae in a well of a 24-well plate. Generally, 5–6 wells are filled in per condition.
2. Prepare fresh working aliquots of the inhibitor prior to treatment.
3. Replace the egg water in each well with 1 ml of compound-containing solution.
4. Refresh the compound every day or every other day.
5. Keep larvae at 34 °C.
6. At 4 or 6 dpi, fix samples in 4 % PFA ON at 4 °C and proceed to image acquisition, data analysis, and quantification as described in previous sections.

3.9 Innate Immunity Interaction with Human Cancer Cells

The number of neutrophils and macrophages recruited to the tumor area is used as readout to study the interaction between the immune system and tumor cells (Fig. 3b, b′, c, c′).

1. Use reporter lines for neutrophils and macrophages or perform the L-plastin immunodetection, combined with chromogenic mpx or fluorescent TSA staining as previously described [32, 33] (*see* **Note 20**).
2. Choose the stage of interest. We perform the analysis at 2 and 4 dpi, to monitor changes in immune cell number infiltrating the tumor aggregate and localizing at the invasive front, over time.

3. Fix between 15 and 20 larvae with 4% PFA, ON, at 4 °C. Consider also not injected controls in the analysis.
4. Remove PFA, rinsing with PBST.
5. Place and align larvae on a confocal dish and remove as much PBST as possible.
6. Image the tail, using a 20× lens in a confocal microscope. Set 2 μm for each section and create a z-stack with a thickness of about 60 μm.
7. Analyze the image files in Image J. Create a maximum projection for each channel and make the overlay. Count the number of neutrophils and macrophages for each embryo of each group and evaluate statistical significance with a non-parametric test.

3.10 Statistics

Statistic tests are performed in GraphPad Prism 6. Unpaired Student's *t* test and one-way ANOVA with Bonferroni's *post hoc* test are used to compare means of two and more than two groups respectively to analyze tumor burden. Non parametric tests like Mann–Whitney and Kruskal–Wallis test with Dunn's *post hoc* test are used to assess significance after quantification of cancer cell invasion and immune cell recruitment.

4 Notes

1. The use of a transient dye allows phenotype verification when a tumor cell line is implanted for the first time in the embryo. However, due to the stability of the dye, fragmented dead cells will maintain fluorescence. Therefore, it is recommended to verify that the size of the fluorescent object is not smaller than the size of a cell. For further quantifications and analysis it is advised to use cell lines with stable expression of fluorescent proteins.
2. Choose the fish reporter, considering fluorescence combination with cell line. Either family or single crosses can be performed. When a family cross is setup, sort the embryos based on developmental stage, in order to further perform the cell implantation in homogeneously developed larvae. Embryos can be easily sorted between 4 and 8 hpf and maximum 100 individuals should be placed in a 92 × 16 mm Petri dish to ensure synchronized development.
3. Adding 0.003 % PTU prevents pigment formation and therefore facilitates image acquisition and analysis. Moreover, the presence of pigmented cells on the duct of Cuvier reduces the visibility of the injection site.

4. When the cell layer is too confluent and the cells have already entered the stationary phase, the reproducibility of the phenotype in vivo might not be reached.
5. The volume of PVP used for the final cell suspension, at the end of the preparation, is about 5–10 μl according to the starting cell density (5 μl if the cells are grown in a T25 flask or 10 μl whether a T75 is instead used). If required, a cell counting can be performed and the final volume of PVP should be determined considering a final cell density of 200–500 cells/nl.
6. Before loading, pipette up and down with a tip in order to have a homogeneous cell suspension and avoid possible cell aggregate formation.
7. Needle clogging might occur during implantation, therefore start with a small needle opening that could be cut further.
8. Start with low values for both parameters and increase if necessary. We normally use 20–30 psi injection pressure and 200–300 msec injection time. For a successful injection, balance injection pressure, injection time, and needle opening. Big needle opening and/or high injection pressure/time result in wounding, embryo damage, and possible edema formation.
9. Remove as much water as possible, avoiding larvae drying out.
10. Keep larvae separate from each other, using a brush if required.
11. Avoid performing cell implantation in the proximal branch of the Duct of Cuvier, as it results in cell leakage.
12. During the implantation procedure, needle clogging might occur. If cell aggregates form, first increase injection pressure and/or injection time to eject the clog. Restore the beginning parameters once cell injection runs normally. Secondly, cut further the needle and adjust injection parameters, to keep constant the number of cells injected and reduce variation. Replace the needle if cell aggregates prevent cell ejection. During implantation, systematically insert the needle in the agarose to eliminate yolk residues around the needle tip and perform random cell injection in the water to check for cell loading.
13. Screening immediately after the xenotransplantation should be avoided: an early assessment not always reveals tumor cell damage that is caused by high injection pressure or time and large needle opening; consequently, erroneous engraftments will be grouped in the sample.
14. If the verification of the phenotype is done for the first time, implanted larvae should be analyzed using a fluorescent microscope at least every other day (optimally every day).

15. Preferential localization and extravasation of engrafted tumor cells in the area where the circulatory loop is formed between DA and CV are non-tumor specific phenomena [5]; cancer cell lines that display low metastatic potential in other models, non-tumorigenic cell lines as well as polystyrene beads are able indeed to accumulate in the circulatory loop of the tail vascular system and to localize outside the blood vessels, in between the branches of the caudal vein plexus [23]. Although the extravasation is not tumor dependent, not all tumor cells adhere, with a mesenchymal cell-like morphology, to the external wall of the vascular endothelium. On the contrary, invasiveness and micrometastasis formation in the tail fin, at the end of the caudal haematopoietic tissue of a zebrafish larva (Fig. 2a, b) are tumor specific events and cannot not be observed in the engraftment of every tumor cell line [5, 23].
16. To assess tumor cell proliferation, the use of zebrafish transgenic lines with fluorescent vasculature is not strictly required, but it is advised to visualize larva outlines during image analysis.
17. Positioning the larvae diagonally in every well, it ensures the visualization from head to tail in a 4× objective.
18. All macros used in this analysis are written by H. de Bont (Department of Toxicology, LACDR, Leiden University). The automated image analysis was used to assess tumor cell dissemination in Ghotra et al. [34]. The method is adapted in van der Ent et al. [16] to study tumor cell proliferation upon yolk implantation and described in this chapter to quantify cancer cell burden when engraftment is performed in the Duct of Cuvier.
19. Carry out a toxicity test starting with the working concentration setup in vitro. Test different higher concentrations in the μM range in vivo. Establish the concentration to use and the compound refreshment frequency (generally every day or every other day), considering possible side effects on larval development and by assessing drug efficacy. Moreover the uptake of small molecule compounds by the zebrafish embryo through the skin allows the use of inducible systems in vivo.
20. Incubation time with proteinase K may vary according to developmental stage. In the referred protocol, Triton-X is used as surfactant, we use Tween20; instead of 1% BSA, sheep serum or commercial blocking buffer are used.

Acknowledgments

The cell lines reported in this chapter A673 GFP, MDA-MB-231-B DsRed, and PC3 Pro4 were kindly provided by C. Schleitoff and U. Dirksten (University of Münster), P. ten Dijke (LUMC), and G. van der Pluijm (LUMC), respectively. We thank H. de Bont for assistance with the Nikon A1R confocal microscope and Image-Pro analysis support.
The work was supported by the Netherlands Organization for Scientific Research (TOP GO Grant: 854.10.012).

References

1. White R, Rose K, Zon L (2013) Zebrafish cancer: the state of the art and the path forward. Nat Rev Cancer 9:624–636
2. Lieschke GJ, Trede NS (2009) Fish immunology. Curr Biol 19(16):R678–R682
3. Jagannathan-Bogdan M, Zon LI (2013) Hematopoiesis. Development 140(12):2463–2467
4. Amatruda JF, Patton EE (2008) Genetic models of cancer in zebrafish. In: Jeon KW (ed) International review of cell and molecular biology, vol 271. Elsevier, Amsterdam, pp 1–34
5. He S, Lamers GE, Beenakker JW et al (2012) Neutrophil-mediated experimental metastasis is enhanced by VEGFR inhibition in a zebrafish xenograft model. J Pathol 227(4):431–445
6. Feng Y, Santoriello C, Mione M et al (2010) Live imaging of innate immune cell sensing of transformed cells in zebrafish larvae: parallels between tumor initiation and wound inflammation. PLoS Biol. doi:10.1371/journal.pbio.1000562
7. Amatruda JF, Shepard JL, Stern HK et al (2002) Zebrafish as a cancer model system. Cancer Cell. doi:10.1016/S1535-6108(02)00052-1
8. Zon LI, Peterson RT (2005) In vivo drug discovery in the zebrafish. Nat Rev Drug Discov 4(1):35–44
9. Veinotte CJ, Dellaire G, Berman JN (2014) Hooking the big one: the potential of zebrafish xenotransplantation to reform cancer drug screening in the genomic area. Dis Model Mech 7(7):745–754
10. Murphy JB (1913) Transplantability of tissues to the embryo of foreign species. Its bearing on questions of tissue specificity and tumor immunity. J Exp Med 17:482–493
11. Ribatti D (2014) The chick embryo chorioallantoic membrane as a model for tumor biology. Exp Cell Res 328(2):314–324
12. Rygaard J, Povlsen CO (1969) Heterotransplantation of a human malignant tumour to "Nude" mice. Acta Pathol Microbiol Scand 77:758–760
13. Cekanova M, Rathore K (2014) Animal models and therapeutic molecular targets of cancer: utility and limitations. Drug Des Devel Ther 8:1911–1922
14. Haldi M, Ton C, Seng WL et al (2006) Human melanoma cells transplanted into zebrafish proliferate, migrate, produce melanin, form masses and stimulate angiogenesis in zebrafish. Angiogenesis 9(3):139–151
15. Corkey DP, Dellaire G, Berman JN (2011) Leukaemia xenotransplantation in zebrafish-chemotherapy response assay in vivo. Br J Haematol 153:786–789
16. van der Ent W, Jochemsen AG, Teunisse AF et al (2014) Ewing sarcoma inhibition by disruption of EWSR1-FLI1 transcriptional activity and reactivation of p53. J Pathol 233(4):415–424
17. van der Ent W, Burrello C, Teunisse AF et al (2014) Modeling of human uveal melanoma in zebrafish xenograft embryos. Invest Ophthalmol Vis Sci 55(10):6612–6622
18. Nicoli S, Ribatti D, Cotelli F et al (2007) Mammalian tumor xenografts induce neovascularization in zebrafish embryos. Cancer Res 67(7):2927–2931
19. Zhao C, Wang X, Zhao Y et al (2011) A novel xenograft model in zebrafish for high resolution investigating dynamics of neovascularization in tumors. PLoS One 6:e21768
20. Lee SL, Rouhi P, Dahl JL et al (2009) Hypoxia-induced pathological angiogenesis mediates tumor cell dissemination, invasion, and metas-

tasis in a zebrafish tumor model. Proc Natl Acad Sci U S A 106(46):19485–19490
21. Stoletov K, Kato H, Zardouzian E et al (2010) Visualizing extravasation dynamics of metastatic tumor cells. J Cell Sci 123(Pt 13):2332–2341
22. Tobia C, Gariano G, De Sena G et al (2013) Zebrafish embryo as a tool to study tumor/endothelial cell cross-talk. Biochim Biophys Acta 1832(9):1371–1377
23. Drabsch Y, He S, Zhang L et al (2013) Transforming growth factor-β signalling controls human breast cancer metastasis in a zebrafish xenograft model. Breast Cancer Res 15(6):R106
24. Wetterwald A, van der Pluijm G, Que I et al (2002) Optical imaging of cancer metastasis to bone marrow: a mouse model of minimal residual disease. Am J Pathol 160(3):1143–1153
25. Pettaway CA, Pathak S, Greene G et al (1996) Selection of highly metastatic variants of different human prostatic carcinomas using orthotopic implantationin nude mice. Clin Cancer Res 2(9):1627–1636
26. Wang Y, Kaiser MS, Larson JD et al (2010) Moesin1 and Ve-cadherin are required in endothelial cells during in vivo tubulogenesis. Development 137(18):3119–3128
27. Jin SW, Beis D, Mitchell T et al (2005) Cellular and molecular analyses of vascular tube and lumen formation in zebrafish. Development 132:5199–5209
28. Lawson N, Weinstein B (2002) *In vivo* imaging of embryonic vascular development using transgenic zebrafish. Dev Biol 248:307–318
29. Renshaw SA, Loynes CA, Trushell DMI et al (2006) A transgenic zebrafish model of neutrophilic inflammation. Blood 108:3976–3978
30. Ellett F, Pase L, Hayman JW et al (2011) Mpeg1 promoter transgenes direct macrophage-lineage expression in zebrafish. Blood 117(4):e49–e56
31. Bernut A, Herrmann JL, Kissa K et al (2014) Mycobacterium abscessus cording prevents phagocytosis and promotes abscess formation. Proc Natl Acad Sci U S A 111:E943–E952
32. Cui C, Benard EL, Kanwal Z et al (2011) Infectious disease modeling and innate immune function in zebrafish embryos. Methods Cell Biol 105:273–308
33. Loynes CA, Martin JS, Robertson A et al (2010) Pivotal advance: pharmacological manipulation of inflammation resolution during spontaneously resolving tissue neutrophilia in the zebrafish. J Leukoc Biol 87:203–212
34. Ghotra VP, He S, de Bont H et al (2012) Automated whole animal bio-imaging assay for human cancer dissemination. PLoS One 7(2):e31281

Chapter 12

Modeling Leukemogenesis in the Zebrafish Using Genetic and Xenograft Models

Vinothkumar Rajan, Graham Dellaire, and Jason N. Berman

Abstract

The zebrafish is a widely accepted model to study leukemia. The major advantage of studying leukemogenesis in zebrafish is attributed to its short life cycle and superior imaging capacity. This chapter highlights using transgenic- and xenograft-based models in zebrafish to study a specific leukemogenic mutation and analyze therapeutic responses in vivo.

Key words Transgenesis, Xenograft, Meganuclease, Patient-derived xenograft

1 Introduction

The zebrafish is an increasingly popular organism to study hematopoiesis and to model leukemia as well as various other human malignancies. There are a number of strategies described to model malignancies in the fish including: carcinogen exposure; transient approaches that modify gene expression, such as injections of mRNA and morpholinos; genetic approaches such as embryo mutagenesis or incorporation of oncogenic transgenes; and transplantation models, employing zebrafish allografts, or human xenograft models.

Transgenic models have been particularly useful in studying the oncogenic mechanisms underlying different mutations and the penetrance of the cancer phenotype arising from the genetic lesion. Melanoma, leukemia, liver, and pancreatic cancer are a few examples of malignancies that have been modeled using transgenic methodology in the zebrafish model (Table 1 outlines common cancer models available and their corresponding gene fusion/mutation used in the creation of the transgenic). Most of the early transgenic models were generated by injecting naked DNA into the single-cell stage of the zebrafish embryo, an approach that suffers from a low transgenesis rate (~1 %). Use of I-Sce

Koichi Kawakami et al. (eds.), *Zebrafish: Methods and Protocols*, Methods in Molecular Biology, vol. 1451, DOI 10.1007/978-1-4939-3771-4_12, © Springer Science+Business Media New York 2016

Table 1
Zebrafish transgenesis-based cancer models

Cancer	Gene insertion	Technology	Reference
Melanoma	*mitfa:BRAF-V600E*	Linearized DNA injection	[3]
Melanoma	*mitfa:EGFP:NRASQ61K*	Linearized DNA injection	[4]
T cell—acute lymphoblastic leukemia	*rag2:ICN1-eGFP*	Linearized DNA injection	[5]
T cell—acute lymphoblastic leukemia	*spi1:tel-jak2*	Linearized DNA injection	[6]
T cell—acute lymphoblastic leukemia	*rag2:c-myc*	Linearized DNA injection	[7]
T cell—acute lymphoblastic leukemia	*rag2:MYC-ER* *rag2:myr-mAkt2*	I-Sce meganuclease	[8]
B cell—acute lymphoblastic leukemia	*TEL-AML1*	Linearized DNA injection	[9]
Acute myeloid leukemia	*spi1:NUP98-HOXA9*	I-SceI meganuclease	[10]
Acute myeloid leukemia	AML-ETO1	Linearized DNA injection	[11]
Rhabdomyosarcoma	*rag2:KRASG12D*	Linearized DNA injection	[11]
Pancreatic cancer	*ptf1a:eGFP-KrasG12V*	BAC transgenesis	[12]
Liver cancer	*fabp10:krasV12*	Tol2 transposase	[13]

meganuclease significantly improves transgenesis to ~10% and efficiency can be further improved to ~50% through use of the tol2 transposase system [1, 2].

One of the advantages of transgenic zebrafish cancer models is that they can be made inducible. As such, the gene of interest (or the transgene) can be turned on at a specific developmental time point or restricted spatially to certain tissues or cells through the use of tissue specific promoters, providing an option to express mutated genes that are otherwise embryonically lethal. The CRE recombinase and the Tet-ON systems are two commonly used technologies to produce inducible transgenic constructs. The Cre system is based on the CRE recombinase of the P1 Bacteriophage, that is capable of catalyzing a site-specific recombination of two loxP sites [14]. In transgenic systems, Cre is used to excise a sequence that is sandwiched between two loxP sites. For example, this enzyme is used to excise a loxP-STOP-loxP cassette that has been cloned 5′ to the gene of interest. Cre recombinase is usually made inducible by an element responsive to an exogenous stimulus, such as a heat shock protein 70 promoter [15]. There is also a tamoxifen-inducible version of CRE that increases CRE specificity by retention of the recombinase in the cytoplasm until administration of tamoxifen, whereby it translocates to the nucleus in an active form to catalyze recombination between target LoxP sites (Fig. 1a) [16, 17]. In contrast, the Tet-ON tool kit permits only a temporal induction of gene expression, but provides other useful features like bidirectional transgenic activation, where two genes that are known to cooperatively contribute towards cancer development can be expressed under a single promoter (Fig. 1b) [18].

More recent tools to aid in developing transgenic zebrafish include the yeast galactose-inducible system in combination with the tol2 system. The system mediated by the transcription activator Gal4 and aided by consensus UAS binding sequence has been used

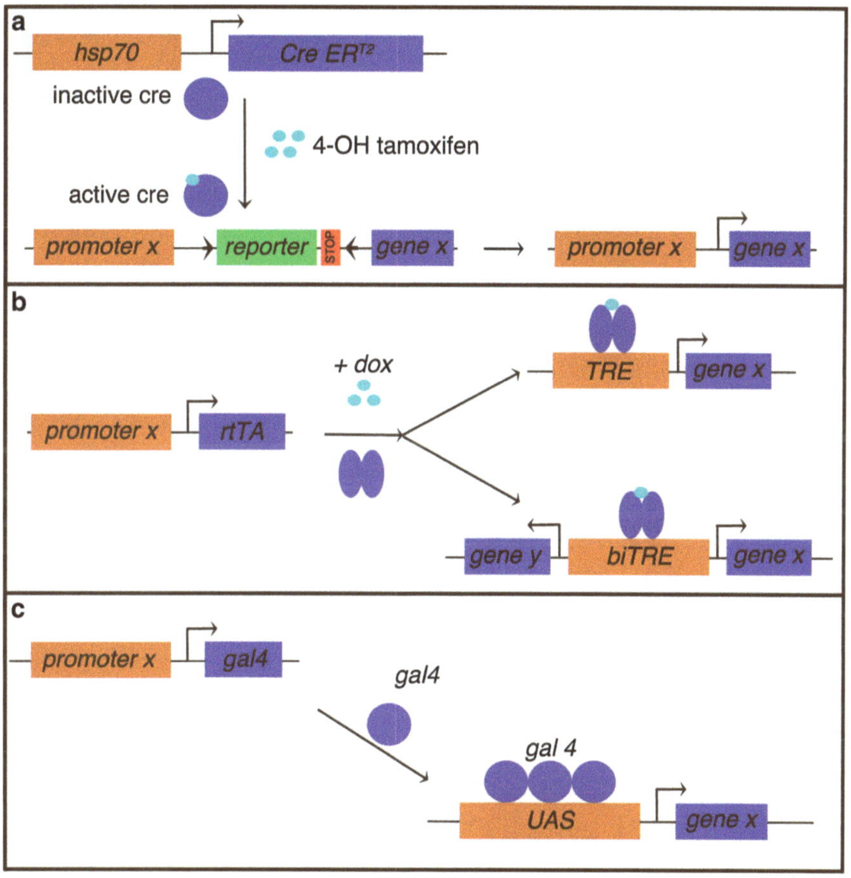

Fig. 1 (**a**) Tamoxifen-based CRE: The CRE ER^{T2} transcript is expressed under a heat shock promoter (hsp70) or tissue specific promoter dependent on the need of the transgenesis. The CRE protein even though secreted remains in an inactive form and needs the presence of four hydroxy tamoxifen to become active and cleave the loxP site (denoted by *arrows*). (**b**) Tet-ON system: The tetracycline response element is available as both unidirectional and bidirectional and includes a transactivator (rtTA) that can be expressed under a tissue specific promoter to make it tissue specific and inducible. (**c**) UAS-gal4 system: The gal4 gene is expressed under the promoter of interest (mostly in cases where the promoter doesn't have a strong expression pattern). gal4 binds and activates the UAS element tagged to a minimal promoter and the gene of interest is expressed under UAS. (*Rectangles* represent genes; *circle* or *ovals* represent proteins)

to promote the expression of a gene of interest, produce cell ablation, and label a subset of cells using various fluorescent proteins [19] (Fig. 1c). Typically, UAS binding sites followed by the gene of interest is injected into a fish already expressing Gal4 under either a constitutive promoter or a tissue/cell-type specific promoter. Another approach is to generate independent fish lines one expressing Gal4 and the other transgenic line carrying a UAS binding site linked to the gene of interest, such that crossing the two lines results in tissue/cell-specific expression of the gene of interest.

At the other end of the spectrum from these gain-of-function transgenic approaches is the generation of mutant zebrafish that develop cancer. One approach is to generate loss-of-function mutants of key tumor suppressor genes by creating insertions/deletions (indels) in the gene of interest using a number of genome engineering tools. These tools includes transcription activator-like effector nucleases (TALENs) and zinc finger endonucleases (ZFNs) that have been used to effectively knock out genes [20] through the use of the FokI nuclease, that is capable of creating a DNA double-stranded break. The efficiencies of these DNA endonuclease approaches have been variable, and they are generally laborious to clone. The latest technique to be added to this toolbox is the clustered regularly interspaced palindromic repeats (CRISPRs). CRISPR is distinct from the other genome engineering approaches, as it requires RNA to "guide" the site-specific activity of a bacterially derived DNA endonuclease called Cas9. CRISPR is immensely versatile and can be used to produce efficient "knock ins" and "knock outs" in the zebrafish genome. This technology is much more user-friendly in terms of design than its ancestors, various online tools . enable designing oligonucleotides to produce gene specific guide-RNA that can target a particular part of a gene and induce DNA breaks, including an online-tool pioneered in our laboratory called CRISPR-Multitargeter [21]. This technology has been reviewed in detail elsewhere [22].

Genetic manipulation of the zebrafish genome is complemented by evolving efforts in the transplantation of both transgenic fish-derived tumors into subsequent recipients and xenografting human cancer cell lines into zebrafish embryos. The latter has been successfully undertaken with a host of human cancers, such as melanoma [23], leukemia [24–26], pancreatic cancer [27], breast cancer [28], glioblastoma [29], neuroblastoma [unpublished observations] and colorectal cancer [24, 26], as well as primary patient-derived leukemia [25] and primary pancreatic cancer [30]. The development of the *roy*; *nacre* mutant zebrafish (*casper*) that lacks iridophores and melanocytes ensures persistent optical clarity to track cells in vivo, and as such has served as an invaluable tool, providing a major boost to the zebrafish as a xenograft platform [31]. The embryonic xenotransplantation method has enabled small molecule screens employing patient-derived xenografts (PDX) from biopsy material. PDX in zebrafish has several distinct benefits over rodent PDX models, including requiring very few cells from the biopsy, being relatively inexpensive, and is much more rapid than similar assays in mice (i.e., typically 3–5 days versus weeks or months in rodents) [32]. With these advantages, zebrafish PDX models could have great potential to be informative and perhaps influence therapeutic decisions, allowing personalized cancer therapy [25]. Another major advantage of the zebrafish xenograft model is that the ability to image a whole organism not only facilitates direct monitoring of tumor cell proliferation and migration in real time, but also permits the study of off-target toxicities of drugs, such as cardiotoxicity. In collaboration with Randall Peterson and colleagues, we recently conducted a phenotype-based screen using the xenograft model to demonstrate that administration of visnagin, a novel cardioprotectant, could prevent doxorubicin-induced myocardial damage without compromising leukemia cell cytotoxicity [33].

Leukemia therapy has progressed in recent years, in part due to the recognition of various candidate genes and related oncogenic pathways contributing to disease pathogenesis, and the ability to begin targeting these molecular circuits using small molecule inhibitors. Leukemia models based upon on key driver lesions identified in human leukemia have been created in zebrafish, including the c-Myc-based T-ALL model [7] and the TEL-AML1 (ETV6-RUNX1; t(8;21)) [9], AML1-ETO [34], NUP98-HOXA9 (t(7;11)) [10] fusion oncogene models of AML. Zebrafish xenograft models of leukemia are also at the forefront of this expanding field by virtue of the easy access to patient bone marrow and peripheral blood samples and the technical ease of working with the smaller size of these human cells compared to larger solid tumor cells characteristic of sarcomas and carcinomas.

In this chapter, we will outline how one can use either genetic or xenograft approaches to functionally study a specific gene found to be mutated in leukemia, including its potency and mechanisms through which it causes disease, as well as the opportunity to perform moderate throughput compound screens on genetically altered or transplanted zebrafish embryos.

By way of example, we will highlight the methodology that we followed in making and studying the *NUP98-HOXA9* transgenic fish [10] and our T-immunophenotype acute lymphoblastic leukemia (T-ALL) xenograft model [25]. However, other oncogenic transgenes or leukemia cell populations could easily be substituted.

2 Materials

2.1 Equipment

1. Microinjection station.
2. Needle puller [e.g., Sutter Instruments Co. Model P-97 settings: Pressure = 500, Heat = 600, Pull = 250, Time = 200 and velocity = 90].

3. 28 °C incubator.
4. 35 °C incubator.
5. Stereomicroscope.
6. Inverted microscope.
7. Needle puller.
8. All basic laboratory equipment.

2.2 Adult Zebrafish and Embryo Handling

1. Fish are raised in a specially designated room at a temperature of 27.5–28 °C, and a water conductivity of 800 μS, pH 7.5, with no ammonia, nitrites or hard metals in the fish water.
2. Mating tank with dividers.
3. Egg water.
4. Pronase (10 mg/mL).
5. Transfer pipettes.
6. Injection plate: 1.5 % agarose gel in Egg water poured in petri plate with mold forming lines to align embryos.
7. Thin-wall glass microinjection capillaries—1.0MM/4 In.
8. Proteinase K (10 mg/mL).
9. Tricaine-MS 222—A stock solution of 0.4 % is made and stored at −20 °C and a final concentration of 0.02 % is used.
10. Fish anesthetic solution: 0.02 % Tricaine in egg water.
11. Adult zebrafish Tg(*hsp70::Cre*).

2.3 Transgenic Model Development.

1. Human *NUP98-HOXA9* fusion gene containing plasmid.
2. *loxP-EGFP-STOP-loxP* cassette.
3. pI-Sce Mammalian expression vector [35].
4. I-Sce I meganuclease.
5. 0.1 % Phenol red.
6. Injection solution: 50 ng/μl of vector plus 0.5 μg/μl of I-SceI meganuclease with 0.1 % phenol red marker dye.
7. PCR-master mix.
8. 50 mM Sodium hydroxide solution.
9. 1 M Tris–HCl (pH ~ 8).
10. DNA midiprep kit.

2.4 Xenograft Model Development

1. Cell line carrying or engineered with the mutation of interest.
2. CellTracker CM-DiI dye (Invitrogen).
3. Cell staining solution: PBS containing 5 μg/mL CM-DiI.
4. Collagenase: Make stock 100 mg/mL with 1× Hanks Buffer Salt Solution.

5. DMSO.
6. Dilution buffer: PBS-5%FBS.

For Primary samples

7. PML antibody (Santa Cruz sc-5621).
8. DyLight Donkey Anti-Rabbit IgG 649 (Abcam).
9. Bone marrow medium: MarrowMAX.
10. Lymphoprep (Stemcell Technologies).
11. Shandon cytospin.

3 Methods

3.1 Transgenic Leukemia Model

3.1.1 Cloning of Transgenic Vector

1. Clone the *NUP98-HOXA9* gene into a vector containing the *spi1* promoter, downstream of *loxP-EGFP-STOP-loxP* site.
2. Assemble the construct into the pI-Sce vector [35] (Fig. 2).
3. Synthesize the *pI-Sce-spi1:loxP-EGFp-STOP-loxP:NUP98-HOXA9* vector using the DNA midi prep kit following the manufacturer's instructions.
4. Perform phenol-chloroform purification of the plasmid to achieve a high quality DNA content devoid of RNAses.

3.1.2 Breeding of Fish

1. Set up fish in a breeding chamber separated by a divider overnight with a male to female ratio of 1:2.
2. The next morning, pull the dividers and allow fish to breed.
3. Monitor fish tanks every 10 min; collect embryos in an embryo strainer and wash with RO water and egg water.
4. Allow embryos to grow or move to an injection station depending on the experimental needs.

3.1.3 Microinjection into Zebrafish Embryos

1. Orient the single-cell stage embryo in the injection plate in such a way that the needle can access the cell.
2. Prepare injection solution and inject into AB wild-type zebrafish embryos at the single-cell stage.
3. Inject the vector + meganuclease mixture into single-cell embryos.
4. Screen embryos 24 h after injection for the presence of EGFP (Fig. 3a). Separate out EGFP-positive embryos and raise these embryos. Discard embryos that lack fluorescence.
5. To evaluate a combinational mutagenesis phenotype with p53 deficient fish, inject the constructs into $p53^{M214K}$ fish (*see* **Note 1**).

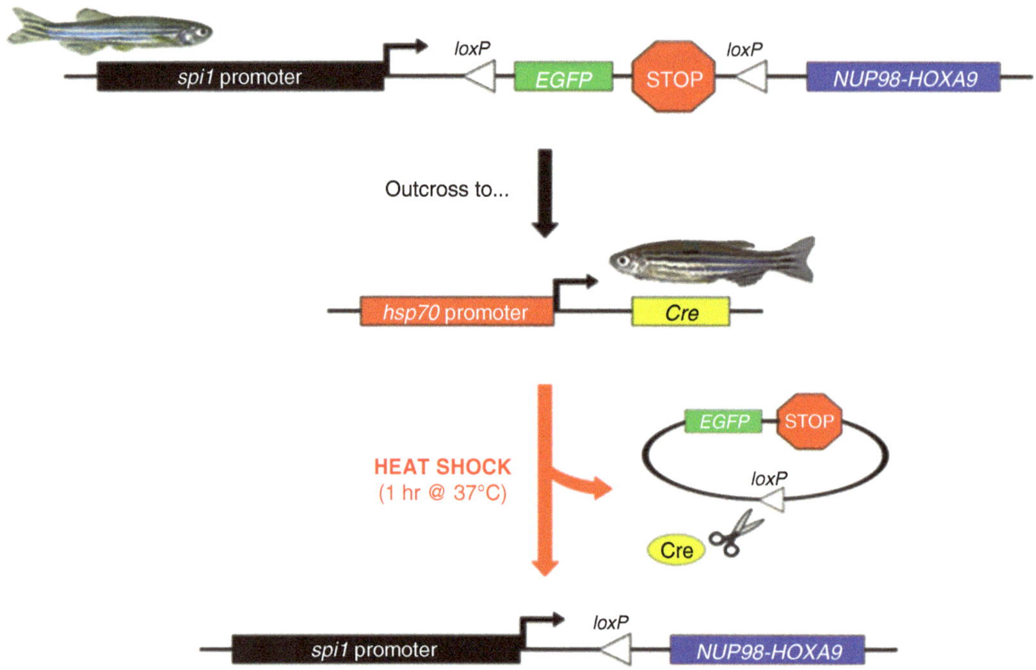

Fig. 2 Generating an inducible *NUP98-HOXA9* transgene: The human *NUP98-HOXA9* fusion gene is expressed under a 9 kilobase *spi1* promoter that is expressed in all myeloid lineages to produce a myeloid phenotype. The construct has a loxP-EGFP-Strong STOP-loxP cassette prior to the *NUP98-HOXA9* construct, such that the gene of interest can only be expressed when the loxP cassette is cleaved. Transgenic fish are outcrossed to fish having a heat shock inducible hsp70 promoter expressing CRE recombinase. Heat shocking for an hour at 37 °C activates CRE transcription. CRE protein subsequently excises the loxP cassette to activate *NUP98-HOXA9* expression. Reproduced with permission from "Forrester AM et al., NUP98-HOXA9-transgenic zebrafish develop a myeloproliferative neoplasm and provide new insight into mechanisms of myeloid leukaemogenesis. *British Journal of Haematology* 2011"

3.1.4 Genotyping of Injected Zebrafish

1. Fill a set of PCR tubes with 30 μl of 50 mM of sodium hydroxide solution and label accordingly to identify the fish from which the fin was collected.
2. Anesthetize the fish to be genotyped by exposure to 0.02% tricaine in the water of a special anesthesia tank.
3. After being anesthetized, transfer the fish to a Petri dish and clip a small piece of the fin (not more than 50% of the total fin area should be removed) using a scalpel and transfer the fin to the PCR reaction tube containing sodium hydroxide solution using forceps.
4. Once all the fins are collected in the PCR tubes, heat the samples to 95 °C for 10 min in a thermocycler.
5. Vortex the PCR tubes and incubate in ice until it becomes cool.
6. Add 10% of 1 M Tris–HCl (pH ~ 8) to the reaction to neutralize the reaction mixture.

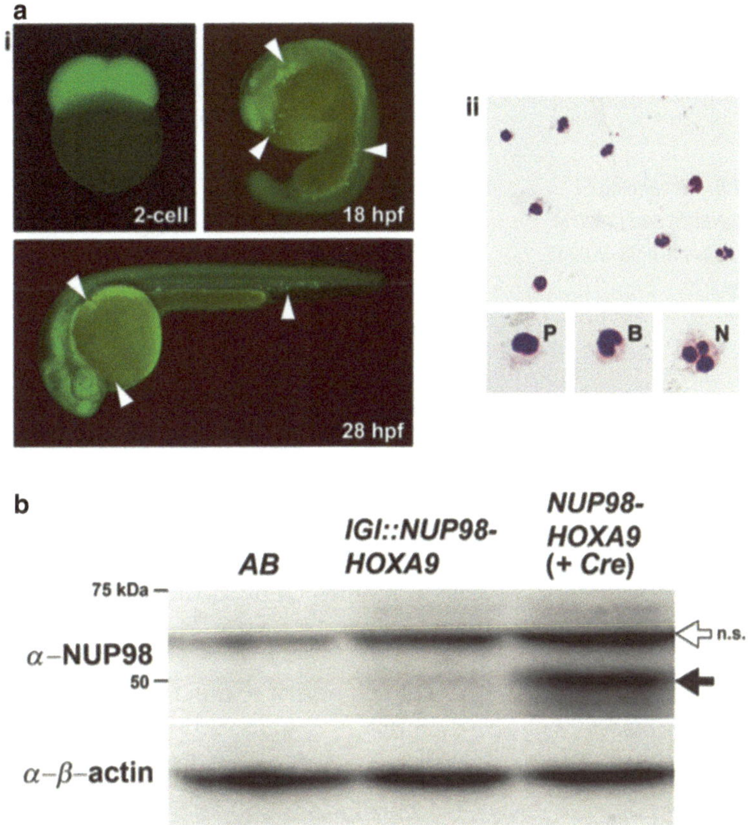

Fig. 3 *spi1* promoter results in myeloid-specific expression of EGFP and immunoblotting verifies the expression NUP98 upon heat shock: (**a**) (*i*) *Tg(spi1::loxp-EGFP-loxp-NUP98-HOXA9)* embryos demonstrate *EGFP* expression. At 2-cell stage the EGFP expression is ubiquitous in nature consistent with maternal expression. By 18 hpf, EGFP expression is restricted to blood cells, with off-target expression in the central nervous system and musculature that is still present at 28 hpf. At 28 hpf, discrete EGFP expression in punctate blood cells (*arrowheads*) at the ALPM in the head region spreading over the yolk sac, and at the PBI in the tail region. (*ii*) GFP+ cells from *Tg(spi1::loxp-EGFP-loxp-NUP98-HOXA9)* embryos were FACS-sorted and stained with Wright-Giemsa stain (*top*, 40× magnification). The cells demonstrate characteristic myeloid morphologies, such as (*bottom*, 100×, *l-r*) precursor (P), band form (B), and segmented neutrophils (N). (**b**) Immunoblotting using a monoclonal rat anti-Nup98 (clone 2H10) IgG2c-κ is capable of detecting the NUP98-HOXA9 fusion protein. NUP98-HOXA9 (*black arrow*) is expressed only in samples from CRE activated *Tg(spi1::loxp-EGFP-loxp-NUP98-HOXA9)* embryos at 28 hpf. An unknown non-specific band (*white arrow*, "n.s.") was also seen in all embryo lysates due to some off-target cross-reactivity of the antibody. *AB* wild-type, *Cre*, "*IGI::NUP98-HOXA9*," and rabbit anti-β-actin IgG serve as controls. *EGFP* enhanced green fluorescent protein, *ALPM* anterior lateral paraxial mesoderm, *PBI* posterior blood island. Reproduced with permission from "Forrester AM et al., NUP98-HOXA9-transgenic zebrafish develop a myeloproliferative neoplasm and provide new insight into mechanisms of myeloid leukaemogenesis. *British Journal of Haematology* 2011"

7. Briefly centrifuge at 200×*g* for 2 min and collect the supernatant.
8. The supernatant can be stored for up to a month in −20 °C and used for PCR reactions.

3.1.5 Generating F1 Generation and Creating an Active NUP98-HOXA9 Line

1. Outcross *NUP98-HOXA9* transgenic fish with *Tg(hsp70:Cre)* fish. Collect fertilized eggs and incubate embryos for an hour at 37 °C in order to activate the CRE recombinase.
2. CRE recombinase cleaves the *loxP* cassette out resulting in expression of the *NUP98-HOXA9* transgene downstream of the *spi1* promoter.
3. Perform a western blot to confirm the expression of NUP98-HOXA9 using a specific antibody against NUP98 (Fig. 3b).

At this point, several experiments can be carried out that include evaluating the changes in gene expression using quantitative PCR or RNAseq, evaluating the tumor forming ability of the cells as monitored by secondary transplantation and serial dilution-based studies (*see* **Note 2**), large scale drug library screening etc.

3.2 Xenograft: Leukemia Model

3.2.1 Preparing and Storing Primary Leukemia Cells

1. Dilute primary bone marrow samples 1:1 with dilution buffer.
2. Layer 2 volumes of the mixture on 1 volume of lymphoprep.
3. Centrifuge at 800×*g* for 20 min at room temperature without brakes (deceleration at "0") (*see* **Note 3**).
4. After centrifugation, collect the band of mononuclear cells located at the sample/medium interface with a Pasteur pipette without disturbing the lymphoprep mixture (*see* **Note 4**).
5. Wash cells again with dilution buffer twice followed by centrifugation at 200×*g* for 5 min.
6. Remove supernatant and resuspend cells in 900 μl of FBS.
7. Transfer cells to a cryovial and add 100 μl of DMSO and freeze cells at −80 °C for 1 day followed by transfer to liquid nitrogen.

3.2.2 Day 0: Breeding of Adult Casper Fish

1. Follow the method described in Subheading 3.1.2.

3.2.3 Day 1: Dechorionating

1. Raise fish to 24 h postfertilization (hpf). On the day of injection, treat embryos with pronase to enzymatically digest and weaken the chorion.
2. Monitor for dechorionation under stereomicroscope and allow the reaction to proceed until most of the fish are devoid of the chorion.
3. Following dechorionation, wash the embryos three times with egg water.

3.2.4 Day 2: Labeling the Leukemia Cell Lines and Primary Samples for Injection

1. Thaw primary cells 12 h before injection and allow them to grow in bone marrow medium.
2. OPTIONAL: Just prior to transplant, layer cells onto an equal volume of Lymphoprep and centrifuge for 30 min at 800 × *g*. Non-viable cells will pass through the gradient while the viable cell population will be retained within and can be harvested as previously described.
3. Spin leukemia cells (including primary leukemia cells or leukemia cell lines in culture) to be injected at 200 × *g* for 5 min.
4. Resuspend cells in cell staining solution and incubate at 37 °C for 5 min (*see* **Note 5**).
5. Transfer the tube to 4 °C and incubate for 15 min.
6. Wash the cells once with PBS and resuspend in the respective media
7. Adjust the cell concentration between 1 and 3×10^6 cells/mL by adding the appropriate amount of media. The cell concentration depends on the size of the cell and the aggressiveness of the cells.

3.2.5 Day 2: Injection

1. Anesthetize the fish with fish anesthetic solution.
2. Arrange embryos in the injection plate with the yolk sac facing the side of injection.
3. Pull capillary needles using a Micropipette needle puller.
4. Load cells from the top of the capillary tube and allow them to settle by gravity. The needle is manually cut such that each injection attempt releases about 50–100 cells into the yolk sac (*see* **Note 6**).
5. The pressure is set at ~5 psi at 0.5 s burst.
6. Be sure to wash embryos thoroughly after injection as residual tricaine in the egg water can damage embryos.
7. Following injection, allow embryos to recover for an hour at 28 °C and then shift to a 35 °C incubator (growing fish at 35 °C is crucial for the survival of the human cells [23]).

3.2.6 Day 4–8: Drug Screening

1. A drug toxicity curve is first performed on unengrafted fish in order to determine lethality of the drug. This can be done by placing the embryos in a 96-well plate and treating them with increasing concentrations of the drug and monitoring for survival. The starting dosage for toxicity curves is roughly around 10 times the maximum tolerated dose in cells. In the event there is no in vitro data available, begin with doses from 10 to 100 μM and then adjust the concentration depending on results.
2. Determine the maximum tolerated dose (MTD, drug dosage at which 80% of embryos survive) of the drug and derive the

MTD50, which is 50% of the maximum tolerated dose of the drug. Make sure the concentration of the drug that is used in the study is below the MTD50. For example, we found the MTD50 of Rapamycin and Tricirbine to be 1.25 μM and 100 μM, respectively [25].

3. Screen the fish injected with cells at 24 h post injection (hpi) (96 hpi for fish injected with primary samples) and select fish having a uniform bolus of cells (based on the fluorescent microscopy) (Fig. 4) (*see* **Notes 7** and **8**).
4. Separate positively screened embryos into groups of 15–20 and treat each group with a different drug (or combination of drugs) for the desired length of time (Fig. 5a).

3.2.7 Day 6–8: Dissociation Assay for Ex Vivo Quantification of Xenografted Leukemia Cells

Embryos are dissociated 24 h after transplantation of cells for baseline measurement and 48 h after administration of drug and vector to collect data on the effect of drug and vector control respectively.

1. Euthanize fish with tricaine overdose.
2. Add 54 μl of prewarmed collagenase P stock solution (100 mg/mL) to a group of 15–20 embryos suspended in 1 mL PBS.
3. Warm in a 37 °C incubator for approximately 20–30 min with intermittent pipetting using pipette/22 gauge needle and syringe.
4. Manually examine the reaction. Once you see a single-cell suspension, add 200 μl of fetal bovine serum (FBS) to each well/tube and mix with pipette.

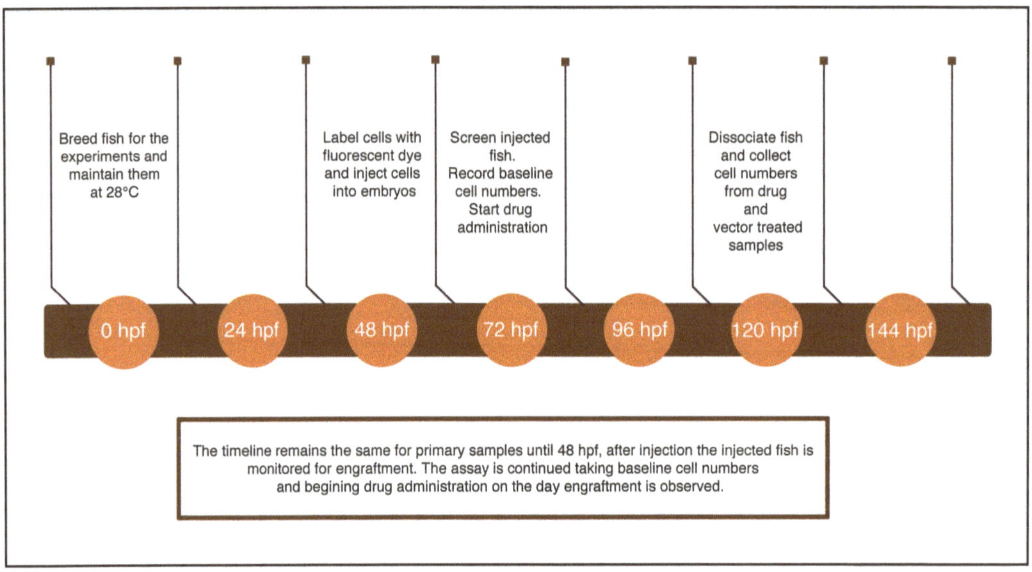

Fig. 4 Timeline for xenograft experiment with cell lines and primary samples: The timeline outlines the experimental steps to be undertaken on each day. Steps unique to primary patient-derived leukemia cells are indicated in the *box below*

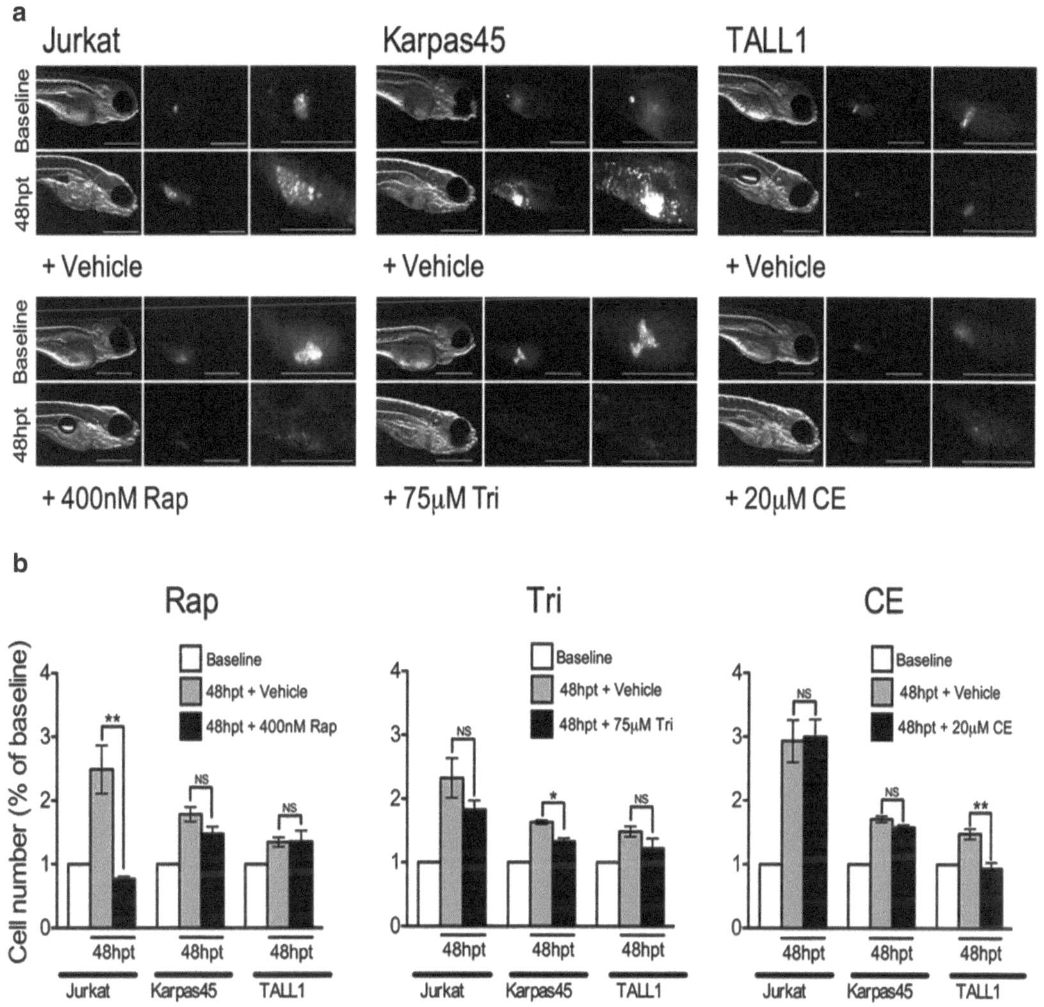

Fig. 5 In vivo inhibition of Notch, AKT, and mTOR in T-ALL cell lines with varied mutations show differential response in accordance to the mutations present in the cell lines. (**a**) Brightfield, fluorescent, and magnified fluorescent images (from *left* to *right*) of CM-Dil-labeled T-ALL cell lines transplanted into zebrafish embryos (baseline represents embryos 48 hpi, while 96 hpf and 48 hpt represents embryos that are 96 hpi and 144 hpf with or without drug). (*hpi* hours post injection, *hpf* hours post fertilization, *hpt* hours post treatment). (**b**) Ex vivo proliferation of T-ALL cell lines in the zebrafish xenograft model was performed as described in the text. Baseline number of cells was determined at 48 hpi and all drug treatments (at 48 hpt) was evaluated as a fold change of the baseline. Means ± SEM; $N=3$; $P^{*}<0.05$, $P^{**}<0.01$, $P^{***}<0.001$ for significant decrease in number of cells determined using 1-way ANOVA followed by Dunnett's multiple comparison test. N= the number of independent experiments, with 15–20 embryos per group per experiment. *hpt* hours posttreatment, *hpi* hours postinjection. Scale bars are 500 μM. Reproduced with permission from Bentley et al., "Focused chemical genomics using zebrafish xenotransplantation as a preclinical therapeutic platform for T-cell acute lymphoblastic leukemia. *Haematologica* 2014"

5. Centrifuge for 5 min at 200 × *g* at room temperature.
6. Add 10 µl/embryo of dilution solution and pass the cells through a 40 µM cell strainer.
7. Drop the cell suspension into a super frosted microscope slide as ten 10 µl boluses and wait for 5–10 min to allow the cells to settle such that all of them are in the same plane.
8. Collect images with fluorescent microscope for ten different boluses per sample using a 5× objective under a 5 × 4 mosaic using an Axiocam Rev 3.0 CCD camera and Axiovision Rel 4.0 software (Carl Zeiss Microimaging Inc.). As a measure of control for pipetting error, droplets that do not fit into the 5 × 4 mosaic are excluded from analysis.
9. The 3 × 2 mosaic that is internal to the 5 × 4 mosaic is counted for Cm-Dil labeled leukemia cells using a semi-automated cell quantification macro executed in ImageJ (NIH, Bethesda, Maryland) that is capable of converting the images to a single stack and adjust the fluorescence intensity threshold. The macro then counts the number of fluorescent cells in each image using the "Particle Picker" tool. The size of the particle is manually entered based on the size of the cells before injection (Fig. 5b).
10. For primary patient-derived samples, perform a cytospin at 27 × *g* for 10 min followed by immunofluorescence using promyelocytic leukemia (PML) directed antibody (PML protein is specific to humans and is absent in zebrafish) to quantify the number of cells (*see* **Notes 9** and **10**, Fig. 6).

3.2.8 Determination of Cell Count and Statistics

1. Collect data by dissociating injected fish at 24 hpi and consider this as the baseline reading.
2. Dissociate fish treated with vector control and drug 48 h post administration of the drug.
3. In the case of primary patient-derived samples, the number of cells inside the fish that express PML are quantified at various previously suggested time points.
4. The actual value for cell proliferation is calculated in terms of fold of increase in cell numbers using the baseline cell numbers.
5. The result should then be evaluated by unpaired student's *t*-test.

4 Notes

1. For generating mosaic transgenic animals containing a primary driver lesion and a secondary driver (or passenger mutation) to study the combinatorial effect in leukemogenesis, the current transgenic generated can be outcrossed to a second transgenic

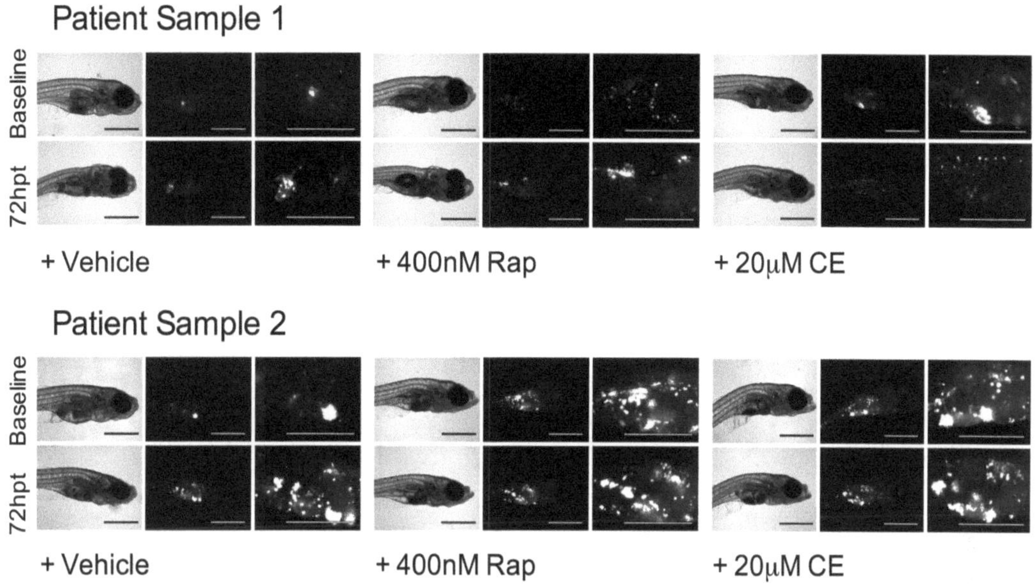

Fig. 6 T-ALL patient sample 1 responds in vivo to Notch pathway inhibition with compound E, but not to mTOR inhibition with rapamycin. Images of the zebrafish (brightfield, fluorescent, and magnified fluorescent: *left* to *right*) of transplanted CM-DiI labeled cells from T-ALL patient samples 1 and 2 at baseline (~48 hpi) and 72 hpt with or without drug. Approximately 500 cells were injected into each embryo. *hpi* hours postinjection; *hpt* hours posttreatment. Scale bars are 500 μM. Reproduced with permission from Bentley et al., "Focused chemical genomics using zebrafish xenotransplantation as a preclinical therapeutic platform for T-cell acute lymphoblastic leukemia. *Haematologica* 2014"

line or alternatively, a construct mixture containing one mutation can be injected into germline embryos carrying the other mutation to produce a compound embryo expressing both genetic lesions (e.g., mMyc/Bcl-2 [36], BRAFV600E + p53-/- [3], Fig. 7).

2. In order to check the penetrance of a cancer mutation and its ability to form tumor initiating cells, fish that develop tumors can be sacrificed, the tumor can be excised and then injected into a secondary recipient in various dilutions, with tumor development monitored with the amount of fluorescent cell proliferation as read out [7].
3. It is critical to have the brakes OFF to avoid disruption of gradient formation.
4. This is a technically crucial step; SepMate tubes (Stem cell technologies) can be used to reduce the complexity involved in this step.
5. CellTracker Cm-Dil can be replaced by using other fluorescent dyes; we have also using CellTracker Orange CMTMR dye in some of our studies. Also stable cell lines expressing fluorescent proteins can be used.

Fig. 7 Compound transgenic by crossing *Tg(spi1::loxp-EGFP-loxp-NUP98-HOXA9)* fish with p53^{M214K} mutant fish: *Tg(spi1::loxp-EGFP-loxp-NUP98-HOXA9)* fish with active CRE crossed with p53 loss of function mutation *M214K (p53M214K)* (*NHA9;tp53M214K*) show increased tumor mass relative to NUP98-HOXA9 inactivated transgenesis crossed with tp53^{M214K} transgenic (NHA9;tp53^{M214K}—Cre). The incidence of tumor was as high as 58 % in the compound transgenic compared to CRE activated *Tg(spi1::loxp-EGFP-loxp-NUP98-HOXA9)* (23 %)

6. Needles may need to be cut multiple times to optimize bore size.
7. Attempt to standardize positive embryo groups by selecting embryos with similar size boluses.
8. When using primary patient-derived samples, the number of days required for engraftment depends on the sample, so fish need to be monitored carefully every day for engraftment. The day at which engraftment is observed is treated as the baseline (day 0) and the drug administration process is started immediately (Fig. 4).
9. While using primary patient-derived samples, we perform an antibody-based quantification approach since rapid division of cells and greater number of days for engraftment collectively

impacts the fluorescence intensity of the dye, thus impacting the accuracy of fluorescence-based quantification [25].

10. Alternatively, a Ki67 antibody that does not react with zebrafish cells is can be used to differentiate human and zebrafish cells [26].

Acknowledgments

The authors would like to acknowledge members of Berman and Dellaire lab especially Dale Corkery, Chansey Veinotte, Sergey Prykhozhij, Nicole Melong, Babak Razaghi, Jaime Wertman, and Victoria Bentley for the critical review. Vinothkumar Rajan is funded through Cancer Research Trainee Program of the Beatrice Hunter Cancer Institute, Nova Scotia in partnership with the Canadian Imperial bank of Commerce and New Brunswick Health Research Foundation.

References

1. Kawakami K (2007) Tol2: a versatile gene transfer vector in vertebrates. Genome Biol 8(Suppl 1):S7
2. Thermes V, Grabher C, Ristoratore F, Bourrat F, Choulika A, Wittbrodt J, Joly J-S (2002) I-SceI meganuclease mediates highly efficient transgenesis in fish. Mech Dev 118(1–2):91–98, doi:http://dx.doi.org/10.1016/S0925-4773(02)00218-6
3. Patton EE, Widlund HR, Kutok JL, Kopani KR, Amatruda JF, Murphey RD, Berghmans S, Mayhall EA, Traver D, Fletcher CDM, Aster JC, Granter SR, Look AT, Lee C, Fisher DE, Zon LI (2005) BRAF mutations are sufficient to promote nevi formation and cooperate with p53 in the genesis of melanoma. Curr Biol 15(3):249–254, doi:http://dx.doi.org/10.1016/j.cub.2005.01.031
4. Dovey M, White RM, Zon LI (2009) Oncogenic NRAS cooperates with p53 loss to generate melanoma in zebrafish. Zebrafish 6(4):397–404. doi:10.1089/zeb.2009.0606
5. Chen J, Jette C, Kanki JP, Aster JC, Look AT, Griffin JD (2007) NOTCH1-induced T-cell leukemia in transgenic zebrafish. Leukemia 21(3):462–471, doi:http://www.nature.com/leu/journal/v21/n3/suppinfo/2404546s1.html
6. Onnebo SMN, Rasighaemi P, Kumar J, Liongue C, Ward AC (2012) Alternative TEL-JAK2 fusions associated with T-cell acute lymphoblastic leukemia and atypical chronic myelogenous leukemia dissected in zebrafish. Haematologica 97(12):1895–1903
7. Langenau DM, Traver D, Ferrando AA, Kutok JL, Aster JC, Kanki JP, Lin S, Prochownik E, Trede NS, Zon LI, Look AT (2003) Myc-induced T cell leukemia in transgenic zebrafish. Science 299(5608):887–890. doi:10.1126/science.1080280
8. Gutierrez A, Grebliunaite R, Feng H, Kozakewich E, Zhu S, Guo F, Payne E, Mansour M, Dahlberg SE, Neuberg DS, Jd H, Prochownik EV, Testa JR, Harris M, Kanki JP, Look AT (2011) Pten mediates Myc oncogene dependence in a conditional zebrafish model of T cell acute lymphoblastic leukemia. J Exp Med 208(8):1595–1603. doi:10.1084/jem.20101691
9. Sabaawy HE, Azuma M, Embree LJ, Tsai H-J, Starost MF, Hickstein DD (2006) TEL-AML1 transgenic zebrafish model of precursor B cell acute lymphoblastic leukemia. Proc Natl Acad Sci 103(41):15166–15171. doi:10.1073/pnas.0603349103
10. Forrester AM, Grabher C, McBride ER, Boyd ER, Vigerstad MH, Edgar A, Kai F-B, Da'as SI, Payne E, Look AT, Berman JN (2011) NUP98-HOXA9-transgenic zebrafish develop a myeloproliferative neoplasm and provide new insight into mechanisms of myeloid leukaemogenesis. Br J Haematol 155(2):167–181. doi:10.1111/j.1365-2141.2011.08810.x
11. Storer NY, White RM, Uong A, Price E, Nielsen GP, Langenau DM, Zon LI (2013) Zebrafish rhabdomyosarcoma reflects the developmental stage of oncogene expression during myogene-

sis. Development 140(14):3040–3050. doi:10.1242/dev.087858

12. Park SW, Davison JM, Rhee J, Hruban RH, Maitra A, Leach SD (2008) Oncogenic KRAS induces progenitor cell expansion and malignant transformation in zebrafish exocrine pancreas. Gastroenterology 134(7):2080–2090. doi:10.1053/j.gastro.2008.02.084
13. Nguyen AT, Emelyanov A, Koh CHV, Spitsbergen JM, Parinov S, Gong Z (2012) An inducible krasV12 transgenic zebrafish model for liver tumorigenesis and chemical drug screening. Dis Model Mech 5(1):63–72. doi:10.1242/dmm.008367
14. Nagy A (2000) Cre recombinase: the universal reagent for genome tailoring. Genesis 26(2):99–109. doi:10.1002/(SICI)1526-968X(200002)26:2<99::AID-GENE1>3.0.CO;2-B
15. Thummel R, Burket CT, Brewer JL, Sarras MP, Li L, Perry M, McDermott JP, Sauer B, Hyde DR, Godwin AR (2005) Cre-mediated site-specific recombination in zebrafish embryos. Dev Dyn 233(4):1366–1377. doi:10.1002/dvdy.20475
16. Hans S, Kaslin J, Freudenreich D, Brand M (2009) Temporally-controlled site-specific recombination in zebrafish. PLoS One 4(2):e4640. doi:10.1371/journal.pone.0004640
17. Hans S, Freudenreich D, Geffarth M, Kaslin J, Machate A, Brand M (2011) Generation of a non-leaky heat shock–inducible Cre line for conditional Cre/lox strategies in zebrafish. Dev Dyn 240(1):108–115. doi:10.1002/dvdy.22497
18. Campbell LJ, Willoughby JJ, Jensen AM (2012) Two types of tet-on transgenic lines for doxycycline-inducible gene expression in zebrafish rod photoreceptors and a gateway-based tet-on toolkit. PLoS One 7(12):e51270. doi:10.1371/journal.pone.0051270
19. Halpern ME, Rhee J, Goll MG, Akitake CM, Parsons M, Leach SD (2008) Gal4/UAS transgenic tools and their application to zebrafish. Zebrafish 5(2):97–110. doi:10.1089/zeb.2008.0530
20. Gjini E, Mansour MR, Sander JD, Moritz N, Nguyen AT, Kesarsing M, Gans E, He S, Chen S, Ko M, Kuang Y-Y, Yang S, Zhou Y, Rodig S, Zon LI, Joung JK, Rao A, Look AT (2014) A zebrafish model of myelodysplastic syndrome produced through tet2 genomic editing. Mol Cell Biol. doi:10.1128/mcb.00971-14
21. Prykhozhij S, Rajan V, Gaston D, Berman JN (2015) CRISPR MultiTargeter: a web tool to find common and unique CRISPR single guide RNA targets in a set of similar sequences. PLoS One 10(3):e0119372
22. Auer TO, Del Bene F (2014) CRISPR/Cas9 and TALEN-mediated knock-in approaches in zebrafish. Methods 69(2):142–150, doi:http://dx.doi.org/10.1016/j.ymeth.2014.03.027
23. Haldi M, Ton C, Seng W, McGrath P (2006) Human melanoma cells transplanted into zebrafish proliferate, migrate, produce melanin, form masses and stimulate angiogenesis in zebrafish. Angiogenesis 9(3):139–151. doi:10.1007/s10456-006-9040-2
24. Corkery DP, Dellaire G, Berman JN (2011) Leukaemia xenotransplantation in zebrafish – chemotherapy response assay in vivo. Br J Haematol 153(6):786–789. doi:10.1111/j.1365-2141.2011.08661.x
25. Bentley VL, Veinotte CJ, Corkery DP, Pinder JB, LeBlanc MA, Bedard K, Weng AP, Berman JN, Dellaire G (2014) Focused chemical genomics using zebrafish xenotransplantation as a preclinical therapeutic platform for T-cell acute lymphoblastic leukemia. Haematologica. doi:10.3324/haematol.2014.110742
26. Pruvot B, Jacquel A, Droin N, Auberger P, Bouscary D, Tamburini J, Muller M, Fontenay M, Chluba J, Solary E (2011) Leukemic cell xenograft in zebrafish embryo for investigating drug efficacy. Haematologica 96:612–616. doi:10.3324/haematol.2010.031401
27. Vlecken DH, Bagowski CP (2009) LIMK1 and LIMK2 are important for metastatic behavior and tumor cell-induced angiogenesis of pancreatic cancer cells. Zebrafish 6(4):433–439. doi:10.1089/zeb.2009.0602
28. Nicoli S, Ribatti D, Cotelli F, Presta M (2007) Mammalian tumor xenografts induce neovascularization in zebrafish embryos. Cancer Res 67(7):2927–2931. doi:10.1158/0008-5472.can-06-4268
29. Geiger GA, Fu W, Kao GD (2008) Temozolomide-mediated radiosensitization of human glioma cells in a zebrafish embryonic system. Cancer Res 68(9):3396–3404. doi:10.1158/0008-5472.can-07-6396
30. Marques I, Weiss F, Vlecken D, Nitsche C, Bakkers J, Lagendijk A, Partecke L, Heidecke C-D, Lerch M, Bagowski C (2009) Metastatic behaviour of primary human tumours in a zebrafish xenotransplantation model. BMC Cancer 9(1):128
31. White R, Sessa A, Burke C, Bowman T, LeBlanc J, Ceol C, Bourque C, Dovey M, Goessling W, Burns C (2008) Transparent adult zebrafish as a tool for in vivo transplantation analysis. Cell Stem Cell 2(2):183–189
32. Veinotte CJ, Dellaire G, Berman JN (2014) Hooking the big one: the potential of zebrafish xenotransplantation to reform cancer drug screening in the genomic era. Dis Model Mech 7(7):745–754. doi:10.1242/dmm.015784

33. Liu Y, Asnani A, Zou L, Bentley VL, Yu M, Wang Y, Dellaire G, Sarkar KS, Dai M, Chen HH, Sosnovik DE, Shin JT, Haber DA, Berman JN, Chao W, Peterson RT (2014) Visnagin protects against doxorubicin-induced cardiomyopathy through modulation of mitochondrial malate dehydrogenase. Sci Transl Med 6(266):266ra170. doi:10.1126/scitranslmed.3010189

34. Yeh J-RJ, Munson KM, Chao YL, Peterson QP, MacRae CA, Peterson RT (2008) AML1-ETO reprograms hematopoietic cell fate by downregulating scl expression. Development 135(2):401–410. doi:10.1242/dev.008904

35. Grabher C, Joly J-S, Wittbrodt J (2004) Highly efficient zebrafish transgenesis mediated by the meganuclease I-SceI. In: William Detrich H III, Westerfield M, Leonard IZ (eds) Methods in cell biology, vol 77. Academic, New York, pp 381–401, doi:http://dx.doi.org/10.1016/S0091-679X(04)77021-1

36. Langenau DM, Jette C, Berghmans S, Palomero T, Kanki JP, Kutok JL, Look AT (2005) Suppression of apoptosis by bcl-2 overexpression in lymphoid cells of transgenic zebrafish. Blood 105(8):3278–3285. doi:10.1182/blood-2004-08-3073

Chapter 13

Enumerating Hematopoietic Stem and Progenitor Cells in Zebrafish Embryos

Virginie Esain, Mauricio Cortes, and Trista E. North

Abstract

Over the past 20 years, zebrafish have proven to be a valuable model to dissect the signaling pathways involved in hematopoiesis, including Hematopoietic Stem and Progenitor Cell (HSPC) formation and homeostasis. Despite tremendous efforts to generate the tools necessary to characterize HSPCs in vitro and in vivo the zebrafish community still lacks standardized methods to quantify HSPCs across laboratories. Here, we describe three methods used routinely in our lab, and in others, to reliably enumerate HSPCs in zebrafish embryos: large-scale live imaging of transgenic reporter lines, Fluorescence-Activated Cell Sorting (FACS), and in vitro cell culture. While live imaging and FACS analysis allows enumeration of total or site-specific HSPCs, the cell culture assay provides the unique opportunity to test the functional potential of isolated HSPCs, similar to those employed in mammals.

Key words Hematopoietic stem cells, Live imaging, Flow cytometry, CFU-C assay, Zebrafish embryo

1 Introduction

Hematopoiesis is widely conserved across vertebrate species. Similar to mammals, zebrafish *H*ematopoietic *S*tem *C*ells (HSCs) first emerge in the ventral wall of the dorsal aorta, in a region named the Aorta-Gonad Mesonephros (AGM) [1, 2]; they later migrate to a transient secondary site of hematopoiesis, the Caudal Hematopoietic Tissue (CHT), before seeding the kidney marrow and the thymus, the sites of adult hematopoiesis [3, 4]. Many signaling pathways involved in HSC formation and maintenance are highly conserved across vertebrate species; recent reviews highlight this conservation of expression and function [5, 6]. In particular, the transcription factor Runx1 is required for HSC emergence in the AGM in all vertebrates thus far examined [7, 8], where it functions in endothelial-hematopoietic transition [9, 10]; Zebrafish have played a significant role in deciphering the function of Runx1, as well as identifying factors that alter its expression and the production of HSCs.

Koichi Kawakami et al. (eds.), *Zebrafish: Methods and Protocols*, Methods in Molecular Biology, vol. 1451,
DOI 10.1007/978-1-4939-3771-4_13, © Springer Science+Business Media New York 2016

With rapid and external development, an abundance of progeny, and transparency during embryonic stages, zebrafish is a well-suited model organism for the study of embryogenesis. Because zebrafish have all the same blood cell types as mammals and many mutant lines correlating human disease are available (reviewed in [6, 11]), they have proven particularly valuable for investigation of hematopoietic development. However, despite tremendous efforts to identify reliable HSC markers, given the paucity of cross reactivity for standard mammalian antibodies, the zebrafish community still lacks standardized methods to assess HSC number. In situ hybridization for the conserved HSC markers *runx1* and *cmyb* have been the most extensively used method to characterize alterations in HSC formation. While this assay reliably captures the general effect of genetic or chemical alteration on HSC development, it cannot be used to quantify those changes beyond illustrations of penetrance; similarly, due to contemporaneous expression in additional sites during embryogenesis, whole embryo qPCR analysis may not sufficiently capture real differences in local expression in the AGM region. Although not perfect, the use of transgenic reporter lines has proven valuable for identifying, isolating, and quantifying HSCs by FACS and/or live imaging; however as presented here, one needs to carefully select the appropriate lines to discriminate HSCs from endothelial precursors and differentiated blood populations, including that of other multipotent progenitors.

Here, we describe three protocols used routinely in our lab, and others, to enumerate HSCs in the zebrafish embryo: live imaging, Fluorescence-Activated Cell Sorting (FACS), and in vitro culture of embryonic HSCs. A detailed review of imaging live HSCs in zebrafish embryo has been published elsewhere [12], so here we focus on optimizing large-scale (up to 100 embryos in an hour) imaging and cell counting to efficiently enumerate HSCs in the AGM. Similarly, methods describing the in vitro culture assay for zebrafish Hematopoietic Stem and Progenitor Cells (HSPCs) have been published previously [13, 14], such that here we detail protocol modifications used for culturing embryonic HSPCs.

2 Materials

2.1 Transgenic Lines

1. *Tg(runx1P2 (runx1):eGFP)* [15].
2. *Tg(-6.0itga2b(CD41):eGFP)* [16].
3. *Tg(ptprc(CD45):dsRed)* [3].
4. *Tg(cmyb:eGFP)* [17].
5. *Tg(lmo2:dsRed)* [18].
6. *Tg(flk1:dsRed)* [19].
7. *Tg(gata1:dsRed)* [20].

2.2 Reagents

1. E3 medium:
   Utilize 14.6 g NaCl, 0.63 g KCl, 2.43 g $CaCl_2 \cdot 2H_2O$, and 1.99 g $MgSO_4$ to prepare 1 L of 50× E3 in H_2O. Autoclave to sterilize.
2. Tricaine solution (MS222) (Ethyl 3-aminobenzoate methane-sulfonate salt):
   Prepare a 4 g/L stock solution, pH ~7 (http://zfin.org/zf_info/zfbook/chapt10.html#wptohtml63). Dilute 1:25 to anesthetize the embryos. The stock solution can be kept for a month at room temperature or several months at −20 °C.
3. PBS (phosphate-buffered saline):
   0.9× solution diluted from 10× stock commercially available.
4. Liberase:
   Prepare stock solution (5 mg/mL), aliquot and store at −20 °C. Dilute to 75 μg/mL in 0.9× PBS.
5. Sytox Red (for cell viability):
   Dilute 1:1000 in 0.9× PBS (5 nM final).
6. Pronase:
   Prepare a stock solution at 50 mg/mL. Use at 1:50 (1 mg/mL) to remove the chorion.
7. ZKS culture medium [13, 14]:
   For 1 L of medium: 850 mL (50% L-15, 35% DMEM, 15% Ham's F-12), 150 mg Sodium Bicarbonate, 15 mL HEPES (1 M stock), 10 mL Penicillin (10000 U/mL)/Streptomycin (10 mg/mL), 10 mL L-glutamine (200 mM stock), 100 mL FBS (heat inactivated for 30 min at 56 °C), 2 mL Gentamicin sulfate (50 mg/mL stock).
8. Methylcellulose Media [13, 14]:
   For 20 mL of medium: 10 mL 2% methylcellulose stock (Sigma # 94378), 3.5 mL DMEM (High glucose), 1.5 mL Ham's F-12, 2 mL FBS (heat inactivated for 30 min at 56 °C), 2 mL 10% BSA, 300 μL HEPES (1 M stock), 200 μL Penicillin (10000 U/mL)/Streptomycin (10 mg/mL), 200 μL L-glutamine (200 mM stock), 40 μL Gentamicin sulfate (50 mg/mL stock).
9. Zebrafish Growth factors [13]:
   1% Carp serum, ZF Epo (0.1 μg/mL), ZF Gcsf (0.1 μg/mL), ZF TPO (0.03 μg/mL).

2.3 Instrumentation

1. Thermomixer or Water bath at 33 °C.
2. Centrifuge for 5 mL tubes.
3. Flow Cytometry Analyzer.
4. Fluorescence-Activated Cell Sorter.
5. Dissecting microscope with fluorescence source.

6. Tissue culture incubator (Temperature: 32 °C; Gas Exchange: 5% CO_2).
7. Inverted scope for cell enumeration with 10× and 20× objectives.

2.4 Software

1. Image analysis:
   ImageJ http://imagej.nih.gov/ij/ (free).
2. Flow Cytometry:
   FACSDiva (Becton Dickinson), Kaluza Flow Analysis Software (Beckman Coulter), FlowJo, (commercially available).
3. Statistical analysis:
   Excel (Microsoft Office), Prism 6 (GraphPad), both (commercially available).

2.5 Other Materials

1. Pasteur pipette, glass.
2. Double concavity microscope slides.
3. 1.5 mL tubes.
4. 5 mL FACS tubes.
5. 30 μm mesh filter (reusable).
6. 35 mm petri dish, 15 cm plates.
7. 3 mL syringes, 16 gauge needles.

3 Methods

3.1 Choosing Appropriate Transgenic Lines and Preparing for the Assay

HSPC emergence occurs from approximately 28 hpf to 52 hpf in the AGM of zebrafish embryos [3, 4]; HSPCs subsequently colonize the CHT from 36 hpf, the kidney glomerulus from 60 hpf and the thymus from 72 hpf [3, 4]; the kidney marrow and thymus remain active sites of hematopoiesis and lymphopoiesis in the adult, respectively. While many transgenic reporter lines are expressed in HSPCs, their expression pattern may not be specific to that of the HSCs and therefore cannot be used in isolation by fluorescence microscopy or FACS without caveats in data interpretation. Hemogenic endothelium is *runx1*$^{+}$ and nascent HSCs express *cmyb* and *itga2b* (*CD41*). However, Runx1:eGFP and cMyb:eGFP are also strongly expressed in neural progenitors and the developing eye; *cmyb:eGFP* is also expressed in Erythro-Myeloid Progenitors (EMPs), primitive myeloid cells and primordial germ cells. Likewise, CD41:eGFP exhibits high expression in thrombocytes and primordial germ cells, while also marking HSCs and EMPs (reviewed in [14]). Thus, as the number of available transgenic tools grows, one should consider a careful examination of the anatomic site(s) in vivo and/or utilize a combination of markers relevant to HSC biology such as Flk1:dsRed or CD45:dsRed to better enumerate HSCs in zebrafish embryo (*see* **Note 1**). Moreover, it should be mentioned that the folding kinetics (dsRed is up to 24 h delayed compared to GFP) [21] and the stability of fluorescent protein (no more than 24 h) is important to consider when evaluating cell-specific expression. For example, to analyze the expression of HSPCs at 72hpf+, it is recommended not to use Flk1:dsRed; cMyb:eGFP as the dsRed expression will likely have vanished in the majority of HSCs derived from AGM endothelium, leading to a over-interpretation of myeloid numbers.

1. At day 0, set up single pair matings. Use a divider to separate male and female fish and allow for a timely controlled experiment (*see* **Note 2**).
2. The morning of day 1, remove the divider and allow the fish to mate. Collect the embryos no more than 30 min after the female laid eggs (*see* **Note 3**) and use for microinjection if applicable. Place the embryos in an incubator at an appropriate temperature (25–28 °C) for later analysis.
3. In the afternoon of day 1 (late blastula—early gastrula stage), place about 10–20 embryos of the same clutch per condition in a 6-well plate (*see* **Note 4**). Do not combine clutches (*see* **Notes 2** and **4**) unless absolutely necessary (for example, chemical screening and/or cell sorting for genomic profiling). Expose to compound modifiers at the desired time point if applicable.
4. If the number of HSCs is assessed post 36 hpf but prior to 48 hpf, grow fish embryos at 25 °C for at least 48 h (embryos will be at 36 hpf at day 3) (*see* **Note 5**).
5. At day 3, assess embryo morphology and blood flow under a brightfield scope; blood flow, indicated by erythrocyte transit, should be rigorous in the tail by 36 hpf. Discard any "monsters" or embryos with significant delay compared to their control siblings. Select healthy-looking single or double positive embryos under a fluorescence scope for further analysis.

3.2 Site-Specific HSC Quantification by Fluorescence Microscopy

This section covers a method to manually count HSCs in the AGM and in the CHT after imaging of live individual embryos. Embryos may be taken out of tricaine after imaging and further developed to assess later developmental time points and hematopoietic sites in the same sample. In the AGM, Flk1:dsRed; cMyb:eGFP$^+$ HSCs can be counted after imaging under a fluorescence scope starting from 36 hpf when dsRed expression is very robust (*see* **Note 6**). In the CHT, HSCs can generally be enumerated from 38 hpf with *CD41:eGFP* transgenic line (*see* **Note 7**). Using a confocal microscope may improve the specificity of the method in regard to identification of double positive cells, however the cost and/or the duration of the experiment may make it less feasible for large-scale quantifications of several chemical and/or genetic variables to achieve statistical power. Combined with in situ hybridization data and compared to FACS analysis, enumeration by fluorescence microscopy proves highly reliable across the sample set and relatively efficient. The method below describes how to count Flk1:dsRed$^+$; cMyb:eGFP$^+$ HSPCs in the AGM. The same protocol is applicable to assess the number CD41:eGFP$^+$ HSCs in the CHT, using a single channel (eliminating thrombocytes and germ cells by size, brightness and location) or combined with other markers (e.g. Gata1:dsRed or CD45:dsRed).

1. Manually remove the chorion of embryos using forceps.
2. Place embryos in tricaine solution and let them stand for 5 min.
3. Place up to ten embryos in a double concave microscope slide.
4. Lay them on their side one-at-a-time in the center of the dish.
5. Focus on the cMyb:eGFP$^+$ cells in the dorsal aorta and image the embryo at 63–80× using a digital camera.
6. Repeat imaging for Flk1:dsRed$^+$ cells without changing the focus.
7. Repeat the same procedure for the other embryos (10–20 embryos per condition are usually necessary to reach statistical significance).
8. To number the HSCs in the AGM, open Flk1:dsRed and cMyb:eGFP images for each embryo, under ImageJ. When necessary, adjust the brightness/contrast to see Flk1:dsRed$^+$ and cMyb:eGFP$^+$ cells clearly (Image > Adjust > Brightness/Contrast). Keep track of the maximum and minimum value to adjust the parameters of all images equally (*see* **Notes 8** and **9**).
9. Open the "Channels Tool" dialog window (Image > Color > Channels Tool) and merge the red (Flk1:dsRed) and green (cMyb:eGFP) channels (Channels Tool > More > Merge Channels). A new dialog window opens, choose the appropriate channel for the Flk1:dsRed and cMyb:eGFP images and check "Create Composite" at the bottom of the window.
10. In the "Channels Tool" dialog window, click on any "Channel", then click "OK". You can now visualize each color independently by checking/unchecking the channel number.
11. Open the "Cell Counter" dialog window (K. De Vos; Plugins > Analyze > Cell Counter). Click on the "Initialize" button then check "Type 1" to start the manual cell count. A new window "Counter Window—Image" now opens.
12. In the Flk1:dsRed$^+$ domain (*see* **Note 10**), manually click in the vicinity of each Flk1:dsRed$^+$; cMyb:eGFP$^+$ cell to record the number of cells as illustrated in Fig. 1. The cell count is automatically recorded next to "Type 1" in the "Cell Counter" dialog window.

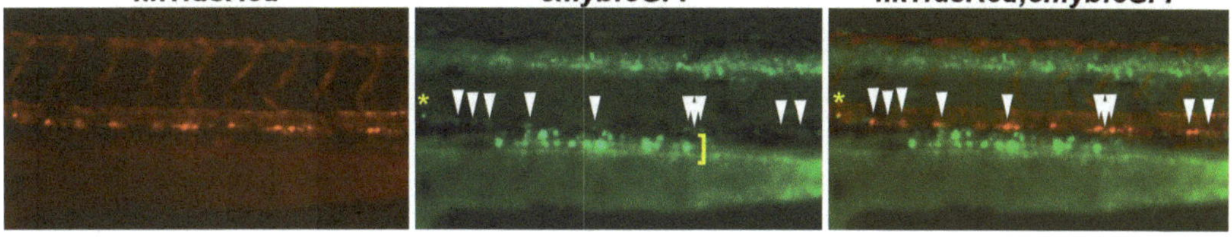

Fig. 1 Flk1:dsRed; cMyb:eGFP⁺ cells can be enumerated in the AGM by fluorescence microscopy. Flk1:dsRed⁺; cMyb:eGFP⁺ HSCs (*white arrowheads*) are distinguishable from germ cells (*yellow bracket*) and myeloid cells (*yellow star*) in circulation in the AGM region. While germ cells are identified based on their location and the reproducible lateral stripe pattern, myeloid cells are often larger and/or brighter cells that tend to be found in circulation rather than on the endothelial walls of the aorta

13. For each embryo, record the number of Flk1:dsRed; cMyb:eGFP⁺ cells on an Excel or Prism data sheet.
14. Tabulate the average and variance (e.g. mean and SD) for each cohort; run a statistical analysis (Student Test or Anova as appropriate) to evaluate the statistical significance of the differences examined across sample cohorts and/or compared to control(s).

3.3 Flow Cytometry

This section covers a method to dissociate embryos and analyze HSPCs by FACS in the developing embryo, up to 120 hpf taking advantage of available reporter lines. Here, we describe a protocol to quantify newly formed HSPCs marked by Flk1:dsRed⁺; cMyb:eGFP⁺ between 36 and 48 hpf. Please note that FACS analysis requires the appropriate negative and single positive controls to adjust the parameters (voltage and compensation); about five age-matched wild-type, *flk1:dsRed⁺* and *cmyb:eGFP⁺* embryos will be needed per sample. Due to the sensitivity and the features of each machine, we are unable to provide hard numbers for the settings or parameters of the FACS machine. **Steps 13–19** are usually done in parallel using control samples to adjust the voltages, the compensation and draw accurate gates. Additionally, gates are only required to obtain data of good quality but may be changed/updated to fit the population when all the samples are analyzed. To reach statistical significance, we recommend running at least five tubes with three to five embryos each (fixed number across the sample sets). Additional biologically independent experiments are encouraged to confirm the relative impact of a particular modulation (genetic and/or chemical) on HSCs.

1. To dissociate the embryos, place five control/*flk1:dsRed⁺*; *cmyb:eGFP⁺* embryos, presorted for coexpression using a fluorescence microscope, per 1.5 mL tube (*see* **Notes 11** and **12**).
2. Remove as much E3 medium as possible using a glass pipette.

3. Rinse with 0.9× PBS.
4. Remove PBS and add 500 μL of liberase solution.
5. Incubate for an hour at 33 °C in a Thermomixer (850 rpm) (*see* **Note 13**).
6. Dissociate the embryos by pipetting up and down with a P1000 pipette (*see* **Note 14**) (five to ten times). Only the spine should be left. If embryos are not fully dissociated, place them back at 33 °C and repeat the dissociation process every 30 min. Up to 72 hpf, fresh liberase should not require longer than 2 h to fully dissociate zebrafish embryos.
7. Filter the sample through a 30 μm mesh filter into a 5 mL FACS tube.
8. Immediately add 3 mL of 0.9× PBS through the filter to rinse the filter and dilute the liberase (*see* **Note 15**). For samples older than 48 hpf, supplement the 0.9× PBS with 1% Fetal Calf Serum and 1 mM EDTA.
9. Centrifuge all the tubes at 300 × *g* for 5 min at 4 °C.
10. Carefully drain off the liquid, wipe the tube opening with a tissue.
11. Resuspend in a solution of 0.9× PBS + Sytox Red (1:1000) to exclude dead cells from the analysis.
12. Cells can be left on ice prior to analysis for 1–3 h if necessary with no loss in number.
13. Using the flow cytometry software available for the users's FACS machine, create dot plots and draw gates for the following parameters (*y*-axis vs. *x*-axis) (*see* **Note 16**):
    (a) Fitness and gross selection of live single cells: SSC-A (log_{10}) vs. FSC-A (linear) (*see* **Note 17**). Gate the cells according to Fig. 2a, Box P1.
    (b) Excluding cell doublets: from the cells isolated in Box 1, SSC-W (linear) vs. SSC-H (log_{10}) (Fig. 2b, P2) then FSC-W (lin) vs. FSC-H (lin) (Fig. 2b, P3).
    (c) Selecting live cells: Using the cells isolated in Box 3, SSC-A vs. APC-A (if using Sytox Red) (Fig. 2c, P4).
    (d) Counting the population of interest (Flk1:dsRed$^+$; cMyb:eGFP$^+$): From the cells in P4, PE-A vs. FITC-A. If necessary, exclude autofluorescent cells (Fig. 2d, P5), then draw quadrant gates (Fig. 2e, Q1: dsRed$^+$, Q2: GFP$^+$; dsRed$^+$, Q3: GFP$^+$, Q4: GFP$^-$; dsRed$^-$).
14. Run the negative control on the FACS machine at a low or medium flow depending on the density of the cell suspension. It is not necessary to record at this step as the parameters are being set up and will likely be readjusted. DO NOT USE

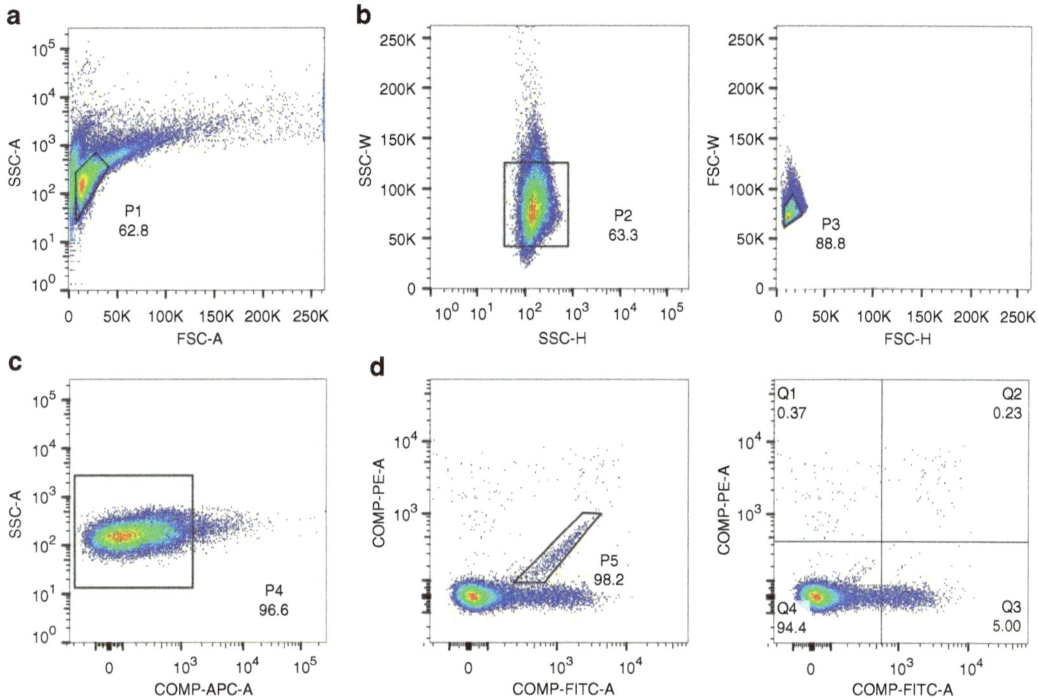

Fig. 2 Embryonic Flk1:dsRed⁺; cMyb:eGFP⁺ HSCs can be quantified by FACS. (**a**) P1 gating allows selection of intact cells compared to debris (on the *left* of the graph) and cell clumps (on the *right*). (**b**) P1 gating is further refined to isolate single cells using SSC-W/H and FSC-W/H (P2 and P3). (**c**) Live cells from P3 do not retain Sytox Red (P4) allowing exclusion of APC⁺ dead cells from the analysis. (**d**) Once single live cells are identified from the rest of the cell suspension, Flk1:dsRed⁺; cMyb:eGFP⁺ cells can be counted (Q3). FACS plots are representative of a single experiment

ENTIRE SAMPLE, as you will want it to verify the settings after all adjustments have been made.

15. Adjust the voltage for SSC-A and FSC-A so that the majority of the cells fall between 10^1 and 10^3 (SSC-A) and 0–100 (FSC-A) (Fig. 2a).
16. Adjust the voltage for APC-A. As most of the cells are alive, they do not retain the dye and are between 0 and 10^3 (Fig. 2c).
17. Adjust the voltage for PE-A and FITC-A so that the vast majority of cells are below 10^3. Some cells might appear positive for either FITC or PE due to autofluorescence (Fig. 2d, e).
18. Run the cMyb:eGFP⁺ control. GFP⁺ cells should be in Q3 only. If there are no cells in Q3, increase the voltage for FITC laser until cells are detected (about 4–5 % of cMyb:eGFP⁺ cells at 48 hpf). If necessary, adjust the compensation to have all the cells appear in Q3 with none in Q2 (PE-%FITC) (*see* **Note 18**). Again, DO NOT USE ENTIRE SAMPLE, as you will want it to verify the settings after all adjustments have been made.

19. Run the Flk1:dsRed$^+$ control. dsRed$^+$ cells should be in Q1 only. If there is no cell in Q1, increase the voltage for PE until cells are detected (about 0.3–0.4% of Flk1:dsRed$^+$ cells at 48 hpf). If necessary, increase the compensation to have all the cells in Q1 and none in Q2 (FITC-%PE).
20. Record (save/acquire file) the double negative and single positive controls as well as one tube of the sample set to confirm that the settings are optimized (no more than 5–10,000 events). If samples do not look as expected, repeat from **step 2** above, otherwise continue to run assay.
21. Record about 30,000 P4 events for the controls and up to 50,000 P4 events for the samples. To increase the accuracy of the cell count and reduce the abortion rate, we recommend not recording more than 2000 events/s.
22. Export the data in FCS format and analyze with your favorite flow cytometry software. We routinely use FlowJo but have also been successful using the flow cytometry manufacturer's software such as BD FACSDiva and Beckman Coulter Kaluza.

3.4 Cell Culture Assay

In this section, we describe a method to perform clonal analysis of isolated HSPCs from embryonic zebrafish between 36 hpf and 5 days. As an example, we will describe the cell sorting and culture of the CD41:eGFP$^+$; Gata1:dsRed$^-$ population, corresponding to HSCs at 5 days post-hybridization (dpf). Our protocol is an extension of the methylcellulose colony-forming assay protocol previously published by Stachura et al. [13] to study adult HSCs isolated from the zebrafish kidney marrow.

Note: Embryos should be bleached and sterile solutions utilized during incubation and isolation of cells in order to minimize bacterial and/or fungal contamination.

1. The number of embryos required will depend on the transgenic reporter line utilized and the age of the embryos. For *CD41:eGFP*$^+$; *gata1:dsRed*$^-$ (HSC fraction) embryos at 5 dpf, we recommend the use of 100 embryos per variable. Under our sorting conditions this will yield approximately 1500 cells for culture per condition.
2. Embryos are incubated in sterile E3 until the desired developmental endpoint for cell dissociation and cell sorting. (Optional: At this step embryos can be incubated with small molecules to test influence on HSC/HSPC functional potential in vivo prior to HSPC collection).
3. Add pronase for 10 min at room temperature to remove the chorion. Gently pipette the embryos up and down with a transfer pipette to dislodge (*see* **Note 19**).
4. Rinse three times in sterile E3 to remove any remaining pronase.

5. To reduce embryo aggregation, dechorionated embryos are incubated in 10 mM DTT in E3 for 30 min at room temperature.
6. Rinse three times in 0.9× PBS.
7. Place 50 embryos in 1.5 mL eppendorf tube and add 1 mL of liberase solution.
8. Incubate for 1 h at 33 °C in a Thermomixer (850 rpm) (*see* **Note 13**).
9. Gently dissociate the embryos by pipetting up and down with a P1000 pipette (20–30 times). Embryos should be fully dissociated with spinal cord remaining. If embryos are not dissociated placed back at 32 °C for another 30 min (*see* **Note 20**).
10. Filter the sample through a 30 μm mesh filter into a 5 mL FACS tube.
11. Wash filter with 3 mL of sterile 0.9× PBS.
12. Centrifuge tubes at 300 × *g* for 5 min at 4 °C.
13. Carefully remove supernatant and resuspend cells in 200 μL of 0.9× PBS 1% FCS per tube (*see* **Note 21**).
14. To exclude non-viable cells, Sytox Red is added followed by 10-min incubation of tubes on ice in the dark.
15. Prior to sorting, gently pipette the cell sample and filter a second time using a 30 μm mesh to avoid cell clumping that can clog the machine.
16. Sorter conditions will vary upon instrument, therefore include controls for each of the populations (i.e. negative controls: here, Gata1:dsRed$^+$ only, CD41:eGFP$^+$ only cells).
17. Collect cells of interest in a 5 mL FACS tube containing 500 μL of cold ZKS media.
18. To maximize cell viability, maintain a flow rate of under 2000 events/s.
19. Centrifuge sorted cells at 300 × *g* for 5 min and resuspend in 200 μL of ZKS media; place on ice for no longer than an hour until ready to plate for methylcellulose Colony-Forming assay.
20. Prior to sorting, prepare 1× methylcellulose medium with 1% carp serum, ZF Epo, ZF Gcsf, ZF Tpo (concentrations as indicated above). (Optional: small molecules can be added at this step to the medium to test the effect of specific compound on colony potential).
21. Aliquot 3.5 mL of methylcellulose per condition to be tested in 14 mL round bottom tubes.
22. Add 1×10^3 cells per mL into 1× complete methylcellulose, tightly cap tubes and gently vortex to mix cells.
23. Using a syringe, plate 1 mL of methylcellulose in 35 mm plates in triplicate (*see* **Note 22**).

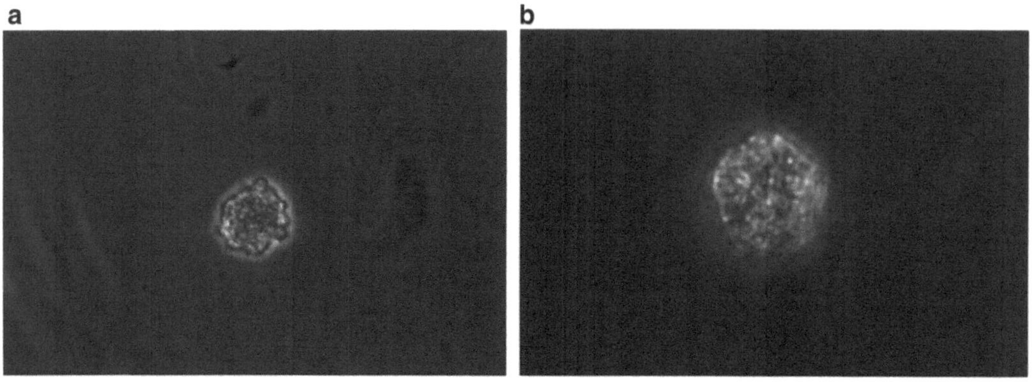

Fig. 3 Embryonic CD41:eGFP+; Gata1:dsRed− cells give rise to colony-forming units (CFUs) after 10 days of culture in methylcellulose. (**a**) Methylcellulose assay showing one embryonic zebrafish CFU at 10× magnification. (**b**) Higher magnification of an embryonic CFU (20×)

24. Place 35 mm plates in a humidified 15 cm plate and incubate at 32 °C; 5 % CO_2.
25. Colonies will become visible at 7–10 days post plating.
26. Colonies can be scored at 10 days post plating on an inverted microscope using the 10× objective (Fig. 3).
27. Count total CFU number per dish; statistics can be performed based on triplicate counts of replicate samples (*see* **Note 23**). Colonies can be further analyzed under a fluorescent microscope to quantify colony type. As an example sorted HSPCs cells from *CD41:eGFP*; *gata1:dsRed* embryos will give rise to Colony-Forming Units-Erythroid (CFU-E) colonies labeled by Gata1:dsRed, and CFU-Thrombocytes (CFU-T) labeled by CD41:eGFP; double positive colonies can be labeled as CFU-TE [22] (*see* **Note 24**).

4 Notes

1. Alternatively, *Tg(CD41:eGFP)* can be outcrossed to *Tg(gata1:dsRed)*. This allows discrimination of HSCs from CD41:eGFP; Gata1:dsRed+ thrombocytes and EMPs.
2. One downside of using transgenic reporter lines is the heterogeneity of the carriers; this has been assessed empirically in many laboratories and the one of the underlying mechanism has been published [23]. To limit the effect of heterogeneity on HSC enumeration, we use the offspring of an outcross between transgenic fish. We do not recommend using mass matings, but suggest pairwise matings be used instead so that all the embryos carry the same genetic material and express fluorescent protein at a similar level, in similar cells.

3. To decrease variability due to age differences, we recommend collecting the progeny within 30 min after the eggs were laid.
4. To reduce the cost associated with a large number of conditions, we recommend plating a maximum of 20 embryos per well in a 12-well plate. When using a drug, this allows incubation of embryos in 2.5 mL of E3 + drug (vs. 5 mL total volume in a 6-well plate).
5. Alternate incubation temperatures and associated growth rates. Embryos can be incubated at 25 °C and above from the moment the fish lay, however, we recommend not to place embryos at 22 °C prior to the mid-blastula transition at about 3 hpf.

Temperature (°C)	Incubation time (h)	Developmental stage (hpf)
22	24	12
25	24	18
28	24	24

6. Alternatively, *Tg(runx1:eGFP)* can used to count the cells budding. However, when imaged with a stereo microscope, the resolution of the resulting image may not be sufficient to distinguish budding cells from hemogenic endothelium. The expression pattern of GFP will be similar to the one obtained by *runx1/cmyb* in situ hybridization.
7. We do not recommend the use *Tg(flk1:dsRed; cmyb:eGFP)* embryos to evaluate the number of HSCs in the CHT, using a stereo microscope. As EMPs express *cmyb* transiently they are also GFP^+ at the time of analysis. Although EMPs are not Flk1:$dsRed^+$, one may not distinguish them from HSCs unless using a confocal microscope, which allows visualization of double positive vs. single positive cells.
8. Although they are siblings, embryos within a clutch may be variable in their level of transgene expression. For these embryos, it is possible to adjust the brightness and contrast so that it looks similar to the rest of the clutch. However, we recommend discarding these embryos from the analysis if they appear to be strong outliers.
9. This allows adjustment of fluorescence intensity without altering gamma correction.
10. Many cMyb:$eGFP^+$ cells are present in the AGM region, including germ cells in the pronephric ducts (bilateral stripes of large cMyb:$eGFP^+$ cells) and myeloid cells circulating or embedded in the aorta and the vein. Myeloid cells are larger and brighter than nascent HSCs and should be excluded from the count. If it appears that the myeloid cell population is significantly affected by the treatment, we suggest to repetition of

the experiment using *flk1:ntr-cerulean+*; *cmyb:eGFP+*; *lysd:dsRed+* embryos to distinguish HSCs from myeloid cells (*Tg(flk1:ntr-cerulean)* is an unpublished line from N. Chi, UCSD, *see* also ref. 6). Germ cell expression can be easily distinguished by adjusting the focal plane of the embryo (the AGM is more central).

11. Three to five embryos can be pooled per tube, five being ideal. However, we recommend using the same number of embryos in every tube to obtain more consistent data across the analysis.
12. To enumerate HSCs specifically in the AGM or in the CHT, the trunk or the tail of embryos may be clipped with forceps under anesthesia (do note this can alter total cell numbers due to erythrocyte loss); we recommend using an anatomical markers, like the yolk sac extension, to ensure consistent tissue harvest.
13. Instead of a Thermomixer, a water bath can be used to dissociate embryos. In this case, we recommend dissociating the embryos more often (every 20–30 min). For consistency, we suggest to use one or the other device within the same set of samples.
14. We do not recommend the use of a P200. Although the embryos may be dissociated faster, we have noticed cells are often damaged by the narrower tip.
15. We recommend performing this step immediately after filtering the sample as significant reductions in yield were noted if this step is performed after ALL the tubes were first filtered.
16. Other methods have been described to isolate single cells during flow cytometric analysis; to our knowledge, this one is the most sensitive, and is commonly used by flow cytometry experts.
17. Zebrafish cells are smaller than mammalian cells and require the FSC-A linear scale [20].
18. PE and FITC have partially overlapping emission spectra. When acquired at the same time, there may be a spillover from one channel to another one. Compensation allows one to mathematically adjust the spill over and have a specific stain (for a detailed method, *see* ref. 24). Setting up compensation varies by instrument and/or software and can be performed either during acquisition or later (depending on your machine); most analysis software offers compensation post-acquisition if the appropriate negative and single positive controls were acquired.
19. Pronase allows removal of the chorions of a large number of embryos in a relatively short time. Alternatively, chorions may be removed manually with forceps.

20. Dissociation time varies with embryo age and usually ranges from 30 min to 2 h. We recommend incubating embryos at 32 °C instead of 37 °C to increase cell viability.
21. Dissociated samples may be pooled at this step.
22. Methylcellulose is thick, therefore plates should be swirled to obtain even media distribution.
23. To facilitate colony counting, dishes can be placed under a grid petri dish (Stem cell Technologies).
24. Combinations of fluorescent reporter lines can be utilized to sort HSPCs and quantify multiple lineage commitment. For example, *CD41:Cerulean/gata1:dsRed/mpx:eGFP* triple labeled embryos can be used to identify CFU-Granulocyte Macrophage (CFU-GM, green), CFU-E (red), CFU-GEM (green/red), CFU-T (cerulean).

Acknowledgments

We thank David Traver (University of California, San Diego) and David Stachura (California State University, Chico) for sharing their protocols, assays, and reagents; and P. Crozier, R. Handin, and K. Poss for providing *runx1:eGFP*, *CD41:eGFP* and *flk1:dsRed* transgenic lines, respectively. We also thank David Stachura and Wolfram Goessling for critical reading of the manuscript. This work was supported by the Harvard Stem Cell Institute, the American Society of Hematology, and National Institutes of Health NIDDK 1R01DK098241-01A1 and 3R01DK098241-03S1.

References

1. Murayama E et al (2006) Tracing hematopoietic precursor migration to successive hematopoietic organs during zebrafish development. Immunity 25:963–975
2. Dzierzak E, Speck NA (2008) Of lineage and legacy: the development of mammalian hematopoietic stem cells. Nat Immunol 9:129–136
3. Bertrand JY, Kim AD, Teng S, Traver D (2008) CD41+ cmyb+precursors colonize the zebrafish pronephros by a novel migration route to initiate adult hematopoiesis. Development 135:1853–1862
4. Kissa K et al (2008) Live imaging of emerging hematopoietic stem cells and early thymus colonization. Blood 111:1147–1156
5. Ciau-Uitz A, Monteiro R, Kirmizitas A, Patient R (2014) Developmental hematopoiesis: ontogeny, genetic programming and conservation. Exp Hematol 42:669–683
6. Carroll KJ, North TE (2014) Oceans of opportunity: exploring vertebrate hematopoiesis in zebrafish. Exp Hematol 42:684–696
7. North TE et al (2002) Runx1 expression marks long-term repopulating hematopoietic stem cells in the midgestation mouse embryo. Immunity 16:661–672
8. Burns CE et al (2002) Isolation and characterization of runxa and runxb, zebrafish members of the runt family of transcriptional regulators. Exp Hematol 30:1381–1389
9. Kissa K, Herbomel P (2010) Blood stem cells emerge from aortic endothelium by a novel type of cell transition. Nature 464:112–115
10. Chen MJ, Yokomizo T, Zeigler BM, Dzierzak E, Speck NA (2009) Runx1 is required for the endothelial to haematopoietic cell transition but not thereafter. Nature 457:887–891

11. Iwanami N (2014) Zebrafish as a model for understanding the evolution of the vertebrate immune system and human primary immunodeficiency. Exp Hematol 42:697–706
12. Zhang P, Liu F (2011) In vivo imaging of hematopoietic stem cell development in the zebrafish. Front Med 5:239–247
13. Stachura DL et al (2011) Clonal analysis of hematopoietic progenitor cells in the zebrafish. Blood 118:1274–1282
14. Stachura DL, Traver D (2011) Cellular dissection of zebrafish hematopoiesis. Methods Cell Biol 101:75–110
15. Lam EYN et al (2009) Zebrafish runx1 promoter-EGFP transgenics mark discrete sites of definitive blood progenitors. Blood 113:1241–1249
16. Lin H-F et al (2005) Analysis of thrombocyte development in CD41-GFP transgenic zebrafish. Blood 106:3803–3810
17. North TE et al (2007) Prostaglandin E2 regulates vertebrate haematopoietic stem cell homeostasis. Nature 447:1007–1011
18. Zhu H et al (2005) Regulation of the lmo2 promoter during hematopoietic and vascular development in zebrafish. Dev Biol 281: 256–269
19. Kikuchi K et al (2011) Retinoic acid production by endocardium and epicardium is an injury response essential for zebrafish heart regeneration. Dev Cell 20:397–404
20. Traver D et al (2003) Transplantation and in vivo imaging of multilineage engraftment in zebrafish bloodless mutants. Nat Immunol 4:1238–1246
21. Lepilina A et al (2006) A dynamic epicardial injury response supports progenitor cell activity during zebrafish heart regeneration. Cell 127:607–619
22. Svoboda O et al (2014) Dissection of vertebrate hematopoiesis using zebrafish thrombopoietin. Blood 124:220–228
23. Akitake CM, Macurak M, Halpern ME, Goll MG (2011) Transgenerational analysis of transcriptional silencing in zebrafish. Dev Biol 352:191–201
24. Roederer M (2002) Compensation in flow cytometry. Curr Protoc Cytom Chapter 1, Unit 1.14–1.14.20

Chapter 14

Live Imaging of Host–Pathogen Interactions in Zebrafish Larvae

Molly A. Matty,* Stefan H. Oehlers,* and David M. Tobin

Abstract

Zebrafish larvae are a powerful platform for studying the innate immune response to infection. The small size and optical transparency of larval zebrafish allow for multiple subject, multidimensional, and longitudinal imaging experiments. This chapter describes protocols for infecting zebrafish larvae with their natural pathogen *Mycobacterium marinum*, rapid short-term imaging, long-term extended imaging, and drug treatment assays. These protocols can be easily adapted to image and manipulate host interactions with other pathogens.

Key words Zebrafish, Infection, Mycobacteria, Time-lapse imaging, Drug treatment, Innate immunity

1 Introduction

Exploring host–pathogen interactions often requires invasive techniques, endpoint analyses, or simplified models. The larval zebrafish model of mycobacterial infection allows for multiday, in vivo analyses in a live vertebrate. Using larval zebrafish, which contain an innate immune system that is relatively conserved with mammals [1], innate immunity can be examined in isolation, as zebrafish are thought to lack functional adaptive immunity until at least 3 weeks postfertilization [2]. The use of fluorescent proteins to label host cells and microbes provides a powerful in toto platform for visual analysis of interactions, not only between the host and the microbe, but also among host cells. Zebrafish are highly fecund, producing hundreds of larvae from a single spawning pair, small and genetically tractable, allowing for robust chemical and genetic interrogation of the host–pathogen interface [3, 4].

The zebrafish is a natural host for the pathogen *Mycobacterium marinum* (*Mm*). *Mm* is the closest genetic relative of the *M.*

*Author contributed equally with all other contributors.

Koichi Kawakami et al. (eds.), *Zebrafish: Methods and Protocols*, Methods in Molecular Biology, vol. 1451,
DOI 10.1007/978-1-4939-3771-4_14, © Springer Science+Business Media New York 2016

tuberculosis (*Mtb*) complex, sharing many of the same virulence mechanisms [5, 6]. *Mm* coevolved with ectothermic hosts, making zebrafish an excellent model for mycobacterial disease. The infection recapitulates many of the same complex interactions seen in humans, including granuloma formation and maturation [7, 8], and genetic determinants of pathogenesis in both host and bacteria [9–11]. Recent developments in genome editing, including CRISPR/Cas technologies, will facilitate future analyses of the host–pathogen interface [12–15].

2 Materials

2.1 Bacterial Preparation

1. 7H9 Liquid Culture Medium: For 1 L, dissolve 4.7 g Middlebrook 7H9 Broth Base (Difco) in water, add 0.2% (v/v) glycerol (we prepare a 10% (v/v) solution of glycerol for more accurate measurement) and bring to a final volume of 900 ml with water and autoclave.
2. 10× Oleic Albumin Dextrose Catalase (OADC Supplement—commercially available from Sigma-Aldrich, or can be made): 5% (w/v) BSA fraction V, 0.05% (w/v) oleic acid (we prepare a 1% w/v solution of oleic acid dissolved in 0.2 N NaOH and stored at −20 °C), 2% (w/v) dextrose, 0.85% (w/v) NaCl. Filter sterilize and store at 4 °C prior to addition to 7H9.
3. Polysorbate 80 (Tween 80): Use at 0.5% in 7H9.
4. Hygromycin B (Corning): Use at 50 μg/ml in 7H9 to select for plasmid-carrying bacteria.
5. *Mycobacterium marinum* M strain carrying pMSP12 (*see* **Notes 1** and **2**) [16].
6. Freezing media: 7H9 Liquid Culture Medium + 1X OADC
7. 27 G needles with syringes.
8. 5 μm filters with syringes.

2.2 Injection

1. Filtered fish water: Post UV-treatment system water is filtered using a 0.22 μm filter.
2. Phenylthiourea (PTU-Sigma-Aldrich): Add to filtered fish water at a final concentration of 45 μg/ml to inhibit pigmentation from 1 dpf.
3. Tricaine (Sigma-Aldrich): Use 1× concentration of 160 μg/ml in filtered fish water to anesthetize zebrafish or at lower concentrations where directed.
4. Low melting point agarose (Fisher Scientific): Prepare a 1% (w/v) solution in filtered fish water, melt by microwaving taking care that solution does not boil over and store at room temperature.

5. Bacterial injection solution: Mix the bacterial suspension aliquot (5 μl volume from Subheading 3.1) with phenol red (optional, aids visualization of injection success) and diluent (either 7H9 complete or PBS) to achieve a bacterial concentration of approximately 10^8 per ml. For aliquots prepared as per the instructions in Subheading 3.1 this would require the addition of 20 μl additional volume to the aliquot.
6. Borosilicate needle.
7. Microloader Eppendorf tips.
8. Microinjector.
9. Hemocytometer.
10. Mineral oil.
11. Glass depression slide.
12. Platinum wire manipulator.

2.3 Bacterial Preparation for Infection by Microinjection

Aliquots of *M. marinum* can be prepared for infections by microinjection and stored at −80 °C for at least a year [16]. Using aliquots of *M. marinum* rather than preparing inoculum for each infection experiment facilitates consistency between experiments.

1. Prepare 7H9 solutions.
   (a) 7H9 complete. Add 100 ml of 10× OADC supplement per liter of 7H9 liquid culture medium and 0.5 ml of Tween 80 per liter of 7H9 liquid culture medium. Store 7H9 complete at 4 °C.
   (b) Freezing media. Add 100 ml of OADC per liter of 7H9 liquid culture medium. Store freezing media at 4 °C.
2. Grow *M. marinum* with fluorescent marker in 7H9 complete at 33 °C until OD_{600} of culture is between 0.8 and 1 (*see* **Note 3**).
3. Pellet culture at 3000–4000 × *g* for 20 min.
4. Discard supernatant into appropriate biohazard waste disposal.
5. Resuspend pellet in 1 ml Freezing media and aliquot two 500 μl volumes into microfuge tubes.
6. Homogenize bacterial suspension by passage through 27 G needle ten times.
7. Quickly spin microfuge tubes at 800 × *g* for 30 s to 1 min. Any large cell clumps will form a soft pellet from this spin.
8. Passage suspension through 27 G needle a further ten times taking care to resuspend any soft pellet that may have formed in previous step.
9. Transfer suspension to syringe for filtering through 5 μm filter.
10. Repeat **item 9** for a total of three filtration steps.

11. Count fluorescent bacteria (FB) using serial dilutions to calculate FB/ml in suspension.
12. Adjust concentration of FB in suspension to around 5×10^8 FB/ml.
13. Aliquot 5 μl volumes of suspension into PCR tubes and freeze at −80 °C.

2.4 Assaying Bacterial Burden by Microscopy

1. Filtered fish water: Post UV-treatment system water is filtered using a 0.22 μm filter.
2. PTU (Sigma-Aldrich): Add to filtered fish water at a final concentration of 45 μg/ml to inhibit pigmentation from 1 dpf.
3. Tricaine (Sigma-Aldrich): Use 1× concentration of 160 μg/ml in filtered fish water to anesthetize zebrafish or at lower concentrations where directed.
4. ImageJ.
5. Platinum wire manipulator.

2.5 Initial Imaging

1. Filtered fish water: Post UV-treatment system water is filtered using a 0.22 μm filter.
2. PTU (Sigma-Aldrich): Add to filtered fish water at a final concentration of 45 μg/ml to inhibit pigmentation from 1 dpf.
3. Tricaine (Sigma-Aldrich): Use 1× concentration of 160 μg/ml in filtered fish water to anesthetize zebrafish or at lower concentrations where directed.
4. Low melting point agarose (Fisher Scientific): Prepare a 1% (w/v) solution in filtered fish water, melt by microwaving taking care that solution does not boil over and store at room temperature.
5. Glass bottom dish (Mattek, cat. no. P35G-1.5-20-C) or optically transparent bottom 96-well plates (Griener Bio-one, cat. no. 655090).
6. Platinum wire manipulator.

2.6 Extended Imaging

1. Filtered fish water: Post UV-treatment system water is filtered using a 0.22 μm filter.
2. PTU (Sigma-Aldrich): Add to filtered fish water at a final concentration of 45 μg/ml to inhibit pigmentation from 1 dpf.
3. Tricaine (Sigma-Aldrich): Use 1× concentration of 160 μg/ml in filtered fish water to anesthetize zebrafish or at lower concentrations where directed.
4. Optically transparent bottom 96-well plates (Griener Bio-one, cat. no. 655090).
5. Low melting point agarose (Fisher Scientific): Prepare a 1% (w/v) solution in filtered fish water, melt by microwaving taking care that solution does not boil over and store at room temperature.

6. Platinum wire manipulator.

2.7 Drug Treatment

1. Filtered fish water: Post UV-treatment system water is filtered using a 0.22 μm filter.
2. PTU (Sigma-Aldrich): Add to filtered fish water at a final concentration of 45 μg/ml to inhibit pigmentation from 1 dpf.
3. Tricaine (Sigma-Aldrich): Use 1× concentration of 160 μg/ml in filtered fish water to anesthetize zebrafish or at lower concentrations where directed.
4. Isoniazid (Sigma-Aldrich): Use at 200 μg/ml to inhibit *M. marinum* growth.
5. Sterile bacteriological Petri dishes.
6. Optically transparent bottom 96-well plates (Greiner Bio-one, cat. no. 655090).
7. Methylcellulose: Dissolve 4% (w/v) methylcellulose in filtered fish water at 4 °C, store at 4 °C.
8. Platinum wire manipulator.

3 Methods

3.1 Injection

We utilize multiple injection sites in the zebrafish to study different host–pathogen interactions (*see* Subheadings 3.1–3.4 for details).

1. Prepare bacterial solution immediately before injecting. We do not store mixed bacterial solution for longer than 2–3 h.
2. Prepare larvae. Dechorionate 1 dpf zebrafish manually or with pronase treatment and raise to desired age in fish water supplemented with PTU after 1 dpf. Anesthetize the zebrafish in fish water supplemented with tricaine. We anesthetize zebrafish immediately before injecting and do not leave zebrafish in tricaine for longer than 2–3 h.
3. Prepare to inject. Prepare borosilicate needle by backloading borosilicate needles with 5 μl of bacterial injection solution using microloader pipette tip. Attach borosilicate needle to microinjector. Break the tip off the needle. The optimal site of breakage must be established empirically but should result in a bore sized so that the bacteria can be injected, but not so wide that bacteria flow freely without injection. Adjust injection volume to deliver desired number of fluorescent bacteria by injecting into mineral oil over a hemocytometer (*see* **Note 4**). Transfer about 20 anesthetized zebrafish to a glass depression slide. Maneuver the fish into the proper orientation (Fig. 1) using a platinum wire pick. The figure provides suggested orientations that we have had success with (*see* **Note 5**).

4. Transfer infected fish to fresh fish water supplemented with PTU in bacteriological Petri dish and raise in 28 °C incubator.
5. Bacterial burden can be confirmed by imaging and manual counting immediately to a few hours after infection.

3.2 Caudal Vein Injection

We use this injection site to produce systemic *Mm* infections, image initial phagocytosis events and study granuloma formation. Bacteria are deposited directly into the blood stream and travel throughout the vasculature until they are phagocytosed by macrophages, which, at this stage, are particularly prevalent in the caudal hematopoietic tissue. This region of the zebrafish is also relatively thin and is thus amenable to single plane or short Z-stack plane imaging for rapid image acquisition. We most commonly perform caudal vein injection at 2 dpf, when the caudal vein is widest and the surrounding vasculature is not as prominent.

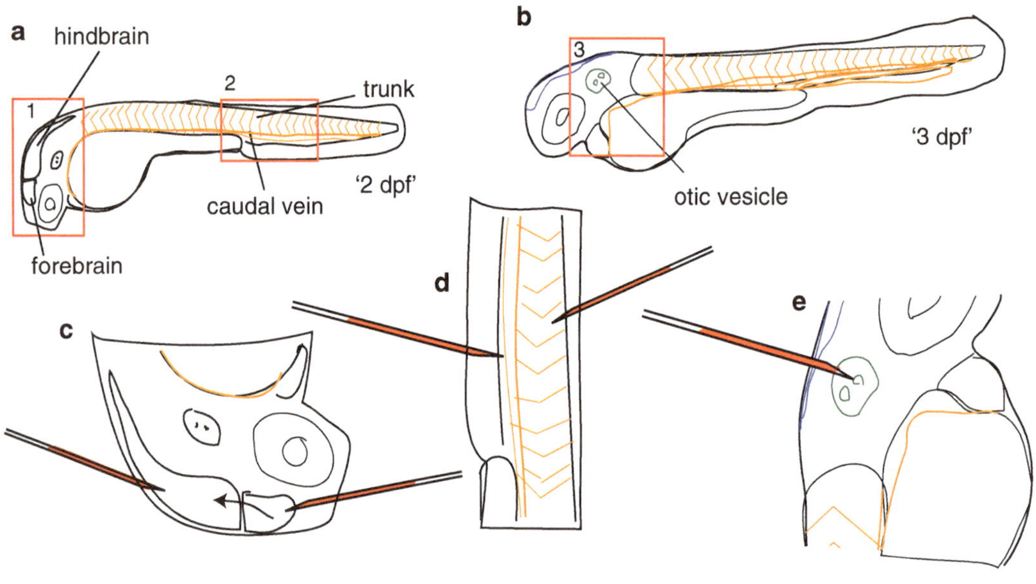

Fig. 1 Multiple injections sites of the zebrafish larva. (**a**) Schematic diagram of a 2 dpf zebrafish with *boxes* indicating locations of injection sites. *Box 1* indicates head ventricle sites shown at higher magnification in panel (**c**). *Box 2* indicates trunk and caudal vein injection sites shown at higher magnification in panel (**d**). Three day postfertilization larvae are used for alternate injection sites. (**b**) Schematic diagram of a 3 dpf zebrafish with a *box* indicating the site of the otic vesicle magnified in panel (**e**). Note the diminished size of the brain ventricles at this age. (**c**) Illustration of two approaches for performing hindbrain injections. Needle on the *left* illustrates a direct entry into the hindbrain ventricle. Needle on the *right* illustrates an injection through the anterior dorsal cavity with needle angle adjustment to enter the hindbrain ventricle. (**d**). *Left needle*—illustration of needle placement for a caudal vein injection, the needle is held at the ventral skin fold until the fish is moved onto the needle. *Right needle*—illustration of final needle location for a trunk injection, the needle is initially held at the dorsal skin fold. (**e**) Illustration of final needle location for direct injection into the right otic vesicle

1. Orient the zebrafish and rest the needle at the juncture of the ventral skin fold and body wall, about three to four somites posterior of the cloaca (Fig. 1).
2. Using the platinum wire manipulator, push the zebrafish toward the needle. The needle should slip into the caudal vein or dorsal aorta. Position the tip of the needle in the center of either vessel.
3. Inject using the predetermined volume required to deliver the desired dose of bacteria. Successful needle placement will result in phenol red indicator solution flowing through the dorsal aorta and caudal vein.

3.3 Hindbrain Ventricle and Otic Vesicle Injection

These injections should be carried out on 1–2 dpf and 3 dpf zebrafish, respectively. At 1–2 dpf, the hindbrain ventricle is large, soft, and more easily penetrated than later stages of development. At 3 dpf, the otic vesicle is much larger and this region of the head is flatter allowing easier penetration. Both structures can be used as a defined point in the zebrafish to examine leukocyte migration to a distal stimulus. We perform infections of these structures to measure leukocyte migration (speed, pattern, and total number) to an infection site and the contribution of specific chemokine/cytokine signaling to these parameters. These structures are relatively thick, and imaging requires significant Z-stack depth, which can limit the speed of image acquisition.

1. Hindbrain Ventricle: Orient the zebrafish by forming a 45° angle between the needle and the top of the head (Fig. 1). Hold the larvae still on the opposite side of the fish from the needle with the platinum wire manipulator, somewhere near the heart or mouth. An alternative strategy is to inject through the forebrain, moving the trajectory of the needle point after entry into the forebrain, as shown in Fig. 1. Using the platinum wire manipulator, push the zebrafish toward the needle. Do not over-push! The tip of the needle must not break internal membranes, or bacteria will readily diffuse into adjacent tissues. Inject the bacterial solution. Successful needle placement will result in only the hindbrain filling with phenol red indicator.
2. Otic Vesicle: Orient the zebrafish to rest the needle against the dorsal edge of the right otic vesicle (Fig. 1). We inject into the right otic vesicle because we primarily utilize an inverted microscope. Use the platinum wire manipulator to hold the fish still, near the heart, and push the zebrafish toward the needle. Do not over-push! The tip of the needle must not break the capsule of the otic vesicle internally (other than the initial external puncture wound), as this causes a wound response and excess recruitment. Inject the bacterial solution.

Successful needle placement will result in only the otic vesicle filling with phenol red indicator solution.

3.4 Trunk Injection

This infection site is dorsal to the caudal hematopoietic tissue, but not as populated with immune cells. This injection site can be utilized between 1 and 5 dpf, but 2 dpf is preferred because interstitial pressure appears to increase drastically after 2 dpf, making the delivery of consistent injection volumes more difficult. We utilize the trunk injection site for examining the role of vasculature in the infection and pathogenesis, as the intersomitic vasculature is highly stereotyped and ectopic vessel sprouting can be identified easily, or as a nonencapsulated site to observe leukocyte migration to a site where granulomas form [8].

1. Orient the zebrafish and rest the needle at the juncture of the dorsal skin fold and body wall (Fig. 1, *see* **Note 6**).
2. Using the platinum wire manipulator, push the zebrafish toward the needle. The needle should remain medial and stay in line with the skin fold as much as possible as the zebrafish is pushed. Position the tip of the needle roughly halfway ventrally towards the midline of the somites.
3. Inject using the predetermined volume required to deliver the desired dose of bacteria. Successful needle placement will result in phenol red indicator solution spreading anteriorly and posteriorly from the site of injection.

3.5 Assaying Infection Burden by Microscopy

There are a variety of image analysis tools available to examine the infection outcome in the zebrafish larvae. The descriptions provided here are for imaging on an inverted scope with an automated stage and filter-switching controls. Image quantification is optimized for using ImageJ, a public domain image processing program [17]. We assess bacterial burden using fluorescent strains of bacteria by measuring the fluorescent area covered by the bacteria per zebrafish as an estimation of bacterial number. Bacterial load can thus be rapidly assessed from images, without the need to lyse, and thus kill, zebrafish for colony forming unit plating.

1. Prepare fish for imaging.
   (a) Transfer infected zebrafish into fish water supplemented with tricaine.
   (b) Transfer individual fish onto imaging container (i.e., Optically transparent bottom 96-well plate).
      - Optional: embed the fish in agarose or methylcellulose for precise control of orientation (*see* Subheadings 3.6–3.8).
2. Image larvae
   (a) Center the zebrafish in the image and focus on the center of the zebrafish. We normally utilize a 2.5× objective lens that allows the capture of the entire length of the zebrafish larva.

(b) Set light source intensity and exposure time to ensure maximum signal is captured with minimum capture of background fluorescence from sources such as the yolk sac.

(c) Capture all images with the same exposure settings.

3. Image quantification.

(a) Export images in an ImageJ compatible format such as TIFF, uncompressed, 8-bit files or JPEG. The proprietary Zeiss camera software exports each channel from the merged proprietary format as an individual file when TIFF or JPEG files are created.

(b) Open bacteria-fluorescence channel file in ImageJ.

(c) If your file contains multiple fluorescent channels, extract only information from one fluorescence color.

- Image > color > split channels
  - Close all but the channel color of interest (e.g., red).

(d) Threshold image to remove background signal (Fig. 2).

- Image > adjust > threshold
  - We routinely threshold our signal with minimum limits of 30 to maximum limits of 255 for an 8-bit image. The exact threshold limits will vary based on the saturation of your image but ideally will completely cut out yolk autofluorescence.

(e) Use the selector tool to outline the zebrafish, as this will help exclude fluorescence from any surrounding debris and also allows multiple zebrafish to be imaged in a single image to speed up image acquisition.

(f) Quantify area of infected region.

- Analyze > Analyze particles.

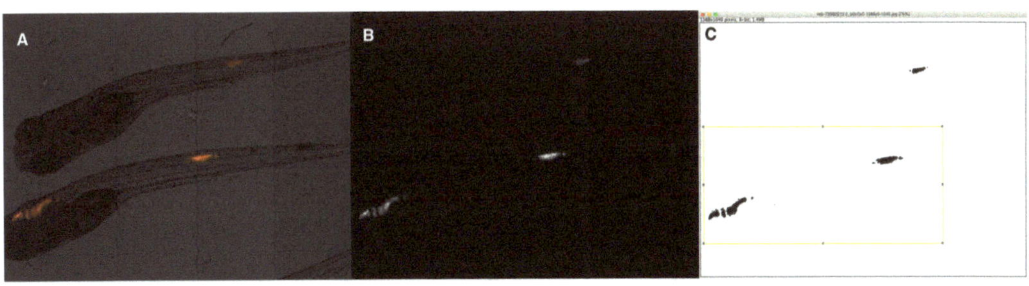

Fig. 2 Quantitation of burden: thresholding an image and removing background. (**a**) 6 dpi/8 dpf zebrafish infected with red fluorescent *M. marinum*. (**b**) After splitting the channels, only the red fluorescence signal remains. (**c**) Use the threshold command to choose the best threshold limits for your set of images. Keep these numbers constant throughout your dataset (e.g., 30–255 for this representative 8-bit image). Select the region(s) of interest to analyze only one fish and to exclude debris (*yellow box*)

- A table will pop up with the number of fluorescent foci, area of fluorescence, and other configurable information.
- Copy this data into spreadsheet or statistical analysis software to compare infection burden between groups (e.g., drug treatments, genotypes, and infection time points).

3.6 Initial Imaging

Initial time course imaging experiments encompass events such as leukocyte migration to the site of infection and phagocytosis of injected bacteria (Fig. 3). Because these processes are rapid, we typically image either one fish at a time in a single high magnification field of view or multiple zebrafish fish in a single field of view at lower magnification. This imaging can also be carried for extended periods of time with larger numbers at lower temporal resolution using the methods presented in Subheading 3.7. Using fluorescently labeled cells, interactions with the bacteria are easily followed.

1. Prepare solutions:
   (a) 1% low melting point agarose.
   - Melt 1% low melting point agarose solution by microwaving.
   - Aliquot 1 ml of 1% low melting point agarose solution into a microcentrifuge tube.
   - Store aliquot in a 42 °C water bath next to dissecting microscope. We store aliquots containing tricaine for up to 48 h.
   - Optional: supplement aliquot with 80 μg/ml tricaine. This step may not be required for zebrafish less than 5 dpf.
2. Mount infected fish in glass bottom dish or 96-well imaging plate.
   (a) Anesthetize larva with 160 μg/ml tricaine.
   (b) Transfer fish to imaging container, removing as much water as possible.
   (c) Pipette 50 μl of 1% low melt agarose on top of the fish.
   (d) Manipulate each zebrafish into desired position taking care to push each zebrafish to the bottom of the agarose for a consistent Z-axis coordinate.
   (e) After the agarose has solidified, add fish water supplemented with tricaine to cover the agarose to prevent desiccation and keep zebrafish fully anesthetized (Fig. 4).
3. Setup the microscope to encompass region(s) of interest.
   (a) Image in as many channels as necessary, ensuring that there is limited bleed-through in fluorophores. For instance we often utilize a yellow fluorescent protein filter to image

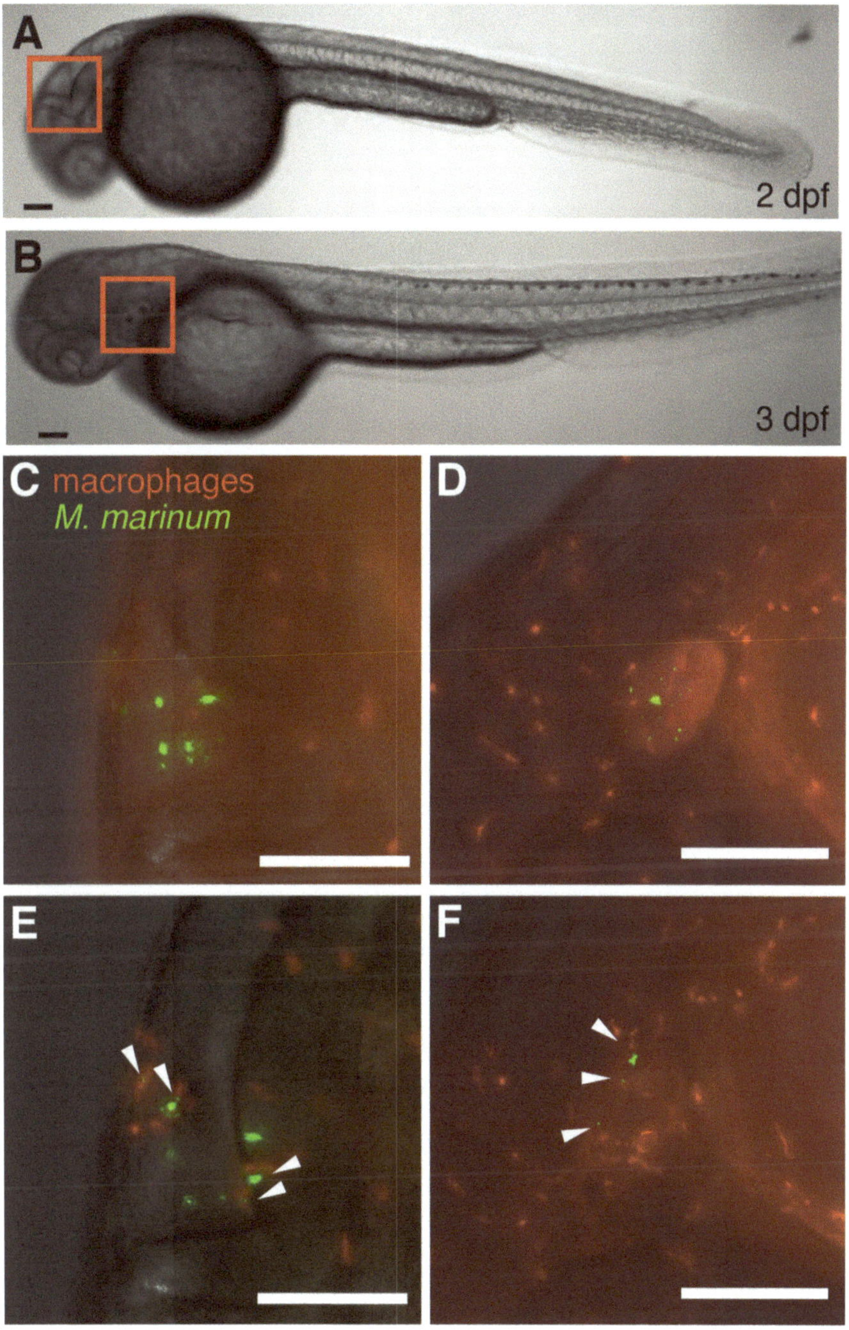

Fig. 3 Initial imaging of infected zebrafish larvae. (**a**) 2 dpf zebrafish imaged at 5×. Hindbrain ventricle (**c** and **e**) is highlighted in the *red square*. (**b**) 3 dpf zebrafish larva imaged at 5×. Otic vesicle (**d** and **f**) is highlighted in the *red square*. (**c** and **d**) Ten minutes post infection (10 mpi) larvae infected with green fluorescent *M. marinum*. Few red macrophages have recruited to the otic vesicle (**c**) and hindbrain (**d**). (**e** and **f**) Two hours post infection (hpi), numerous *red* macrophages have migrated to the infection sites and some green *M. marinum* are intracellular. C-F are 20× images. All scale bars are 100 μm

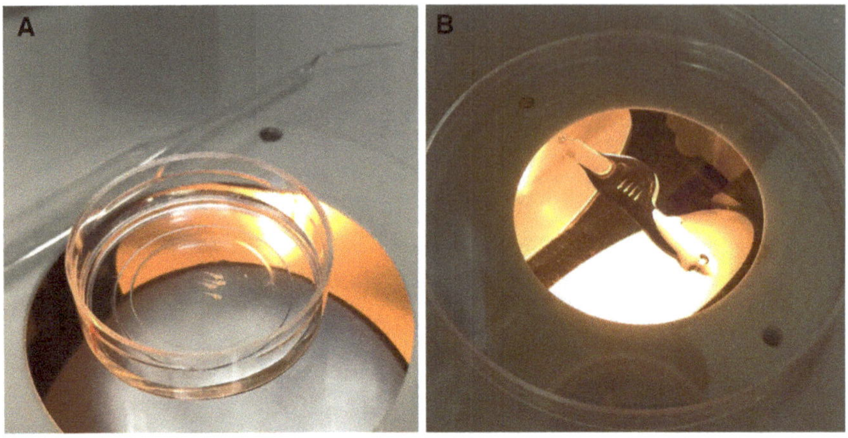

Fig. 4 Imaging setup in dishes. (**a**) Four zebrafish arrayed and embedded in agarose for rapid timelapse imaging in a glass bottom dish. In this arrangement, multiple larvae can be imaged at low magnification or single specimens can be assessed at high magnification. (**b**) Zebrafish arrayed in methylcellulose for high-content imaging. When immobilized this way, fish are stable for single timepoint imaging and can be easily rearranged

green fluorophores when simultaneously imaging a cyan wavelength fluorophore.

(b) Ration the use Z-stacks through thick regions like the ear and hindbrain to balance the need for depth resolution with the speed of image acquisition.

(c) We have effectively imaged zebrafish continuously in this configuration for up to 4 h without excessive photobleaching or light-induced damage.

3.7 Extended Time-Lapse Imaging of Infection

In addition to imaging host–pathogen interactions early in infection that lead to granuloma formation, our laboratory has utilized extended time lapse imaging to analyze host–pathogen interactions around these nascent granulomas [8]. This protocol and experimental setup is optimized for imaging on an inverted microscope with automated stage and filter-switching controls. We are thus able to perform our extended time lapse imaging in 96-well format and observe multiple zebrafish mounted at different depths in individual wells.

1. Prepare 0.75 % low melting point agarose with 0.5× tricaine.
   (a) Melt 1 % low melting point agarose solution by microwaving.
   (b) Aliquot 750 μl of 1 % low melting point agarose solution into a microcentrifuge tube.
   (c) Supplement aliquot with 250 μl filtered fish water and 80 μg/ml tricaine.

(d) Store aliquot in a 42 °C water bath next to dissecting microscope. We store aliquots containing tricaine for up to 48 h.

2. Mount larvae in 96-well imaging plate.
   (a) Anesthetize larvae with 160 μg/ml tricaine.
   (b) Transfer larvae to 96-well imaging plate (1 per well) and remove most of the transferred liquid with a 200 μl pipette.
   (c) Pipette 40 μl 0.75 % low melting point agarose with 0.5× tricaine solution into each well (*see* **Note 7**).
   (d) Manipulate each zebrafish into desired position taking care to push each zebrafish to the bottom of the well for a consistent *Z*-axis coordinate.
   (e) When agarose is set, pipette 100 μl fish water supplemented with PTU into the well to prevent zebrafish from desiccating and developing pigment.
   (f) Fill edges of 96-well imaging plate with fish water to maintain humidity inside the plate and seal edges of plate by wrapping with parafilm.
3. Image larvae on microscope with automated stage and filter control in warm room or chamber (*see* **Note 8**).

3.8 Drug Treatment of Infected Zebrafish

Zebrafish are often utilized for in vivo chemical screening because of their low cost, optical transparency, small size, and permeability to small molecules [18, 19]. The zebrafish-*M. marinum* infection model is susceptible to interrogation with small molecules by immersion exposure. Current and preclinical TB drug efficacy has recently been explored in the zebrafish-*Mm* model, confirming the use of this system for investigating future therapies [20]. Small molecule screening in zebrafish provides an alternative to the limitations of simplified cell culture models and time-consuming small mammal models.

1. Separate infected embryos into desired groups by pipetting into fish water supplemented with PTU in sterile bacteriological Petri dishes.
2. Add drugs and vehicle controls at desired doses directly into fish water (*see* **Note 9**).
3. Grow fish in 28 °C incubator until desired timepoint, changing drugs as necessary (*see* **Note 10**).
4. If multiple timepoints are required we generally anesthetize zebrafish in 96-well plates to allow tracking of individual zebrafish throughout the timecourse of infection (*see* **Note 11**). The method presented in **step 5** is also applicable to multiple

timepoint imaging but is more time consuming and requires more individual manual manipulation of zebrafish.

(a) Cryo-anesthetize zebrafish by placing 96-well plate at 4 °C for 5–10 min.

(b) Wipe the bottom of the 96-well plate with 70% ethanol to delay condensation and remove debris.

(c) Image zebrafish and perform desired image analyses (*see* Subheading 3.3).

(d) Return 96-well plate to 28 °C incubator until next timepoint.

5. If desired timepoint is terminal we anesthetize zebrafish with tricaine and mount in methylcellulose for high-content imaging. It is also possible to mount the zebrafish in agarose but we find that methycellulose is an easier mounting media, as there is no need to wait for the media to solidify.

(a) Anesthetize zebrafish by adding tricaine to a final concentration of 160 μg/ml.

(b) Apply methylcellulose to glass slide or plastic Petri dish (Fig. 4).

(c) Transfer anesthetized zebrafish to methylcellulose with a minimum of fish water.

(d) Manipulate zebrafish into desired position and depress to the bottom of the methylcellulose.

(e) Image zebrafish and perform desired image analyses (*see* Subheading 3.5).

(f) Immobilized zebrafish can be recovered from methylcellulose for further incubation or other end point analysis by flooding methylcellulose with filtered fish water to dilute the methylcellulose.

4 Notes

1. *Mm* is a BSL2 pathogen, which greatly simplifies work with the model compared to *Mtb* but may require additional training from your institution to mitigate risk to personnel and to deal with bioharzardous waste generated from these experiments.
2. This preparation technique only works for strains cultured in hygromycin. We have been unable to reliably prepare consistent aliquots of bacteria from strains cultured in kanamycin because the solution forms a precipitate that clogs the needle and yields fewer bacteria.

3. We generally grow 10–20 ml cultures of bacteria in 50 ml centrifuge tubes for shaking cultures or untreated tissue culture flasks for static cultures.
4. We typically inject between 10 and 20 nl. We initially setup a hemocytometer with mineral oil and inject into the mineral oil to measure the volume for a given setting. Injection volume can be adjusted by modulating either the duration or strength of the air pulse, or by rebreaking the needle to increase the bore width. It is recommended that new users start the needle calibration process with a narrow bore break.
5. We typically setup our injecting stations with the needle holding micromanipulator to the left of the microscope leaving a right-handed user's dominant hand free to operate the platinum wire manipulator as if it was a pencil. This setup is easily reversed for a left-handed user.
6. Caudal vein, hindbrain, and trunk injections can be imaged from either side of the fish, while the otic vesicle injections require imaging from the same side as injection for optical clarity. The easiest way to prevent confusion is to always inject the otic vesicle on same side.
7. The length of time before the agarose sets is determined by molten agarose and ambient temperatures, concentration of agarose (0.75 % solutions set noticeably slower than 1 % solutions), and the volume of agarose used (larger volumes take longer to set).
8. To prevent photobleaching damage and reduce stress to larvae being imaged, we typically use fluorescent light intensity of about 50% from a controllable LED or mercury lamp light source. Zebrafish tissues suffer more damage, and subsequently zebrafish do not survive as long, when imaged for fluorophores excited at lower wavelengths, such as Cerulean and Turquoise.
9. To ensure evenness of drug distribution and sufficient space for normal development we prefer to maintain fish in 20 ml volumes in 100 × 15 mm (diameter × depth) Petri dishes. We have carried out screening experiments where up to four zebrafish are housed in individual wells of a 96-well plate but have found that these wells are more susceptible to being overgrown by microbes than the 100 × 15 Petri dishes.
10. Drug stability in water has to be worked out empirically. A recent paper by Ordas et al. [20] describes methods to analyze the bioavailability of drugs in *M. marinum* infected zebrafish. We routinely change drug containing fish water daily or every second day for maximum drug effectiveness from the start of infection. To model the application of a drug after the establishment of mycobacterial granulomas, we generally start drug treatment at 3 dpi once granulomas have started to form.

11. Some suggested timepoints for analysis of different stages of *M. marinum* pathogenesis are 2 dpi: prior to granuloma formation; 3–4 dpi: nascent granulomas have formed.

Acknowledgments

We thank Dana Sisk for expert management of our zebrafish facility and Rebecca Beerman, Mark Cronan, and Kevin Takaki for technical advice. This work was supported by funding from a National Science Foundation Graduate Research Fellowship (M.A.M.), an Australian National Health and Medical Research Council CJ Martin Early Career Fellowship (S.H.O.), and a Mallinckrodt Scholar Award, a Searle Scholar Award, a Vallee Foundation Young Investigator Award, an NIH Director's New Innovator Award (1DP2-OD008614), and the Duke University Center for AIDS Research, a National Institutes of Health (NIH)-funded program (5P30 AI064518) (D.M.T.). The authors declare no conflict of interest.

References

1. Rauta PR, Nayak B, Das S (2012) Immune system and immune responses in fish and their role in comparative immunity study: a model for higher organisms. Immunol Lett 148(1):23–33. doi:10.1016/j.imlet.2012.08.003
2. Lam SH, Chua HL, Gong Z, Lam TJ, Sin YM (2004) Development and maturation of the immune system in zebrafish, Danio rerio: a gene expression profiling, in situ hybridization and immunological study. Dev Comp Immunol 28(1):9–28
3. Patton EE, Zon LI (2001) The art and design of genetic screens: zebrafish. Nat Rev Genet 2(12):956–966. doi:10.1038/35103567
4. Peterson RT, Fishman MC (2011) Designing zebrafish chemical screens. Methods Cell Biol 105:525–541. doi:10.1016/B978-0-12-381320-6.00023-0
5. Stinear TP, Seemann T, Harrison PF, Jenkin GA, Davies JK, Johnson PD, Abdellah Z, Arrowsmith C, Chillingworth T, Churcher C, Clarke K, Cronin A, Davis P, Goodhead I, Holroyd N, Jagels K, Lord A, Moule S, Mungall K, Norbertczak H, Quail MA, Rabbinowitsch E, Walker D, White B, Whitehead S, Small PL, Brosch R, Ramakrishnan L, Fischbach MA, Parkhill J, Cole ST (2008) Insights from the complete genome sequence of Mycobacterium marinum on the evolution of Mycobacterium tuberculosis. Genome Res 18(5):729–741. doi:10.1101/gr.075069.107
6. Tobin DM, Ramakrishnan L (2008) Comparative pathogenesis of Mycobacterium marinum and Mycobacterium tuberculosis. Cell Microbiol 10(5):1027–1039. doi:10.1111/j.1462-5822.2008.01133.x
7. Cronan MR, Tobin DM (2014) Fit for consumption: zebrafish as a model for tuberculosis. Dis Model Mech 7(7):777–784. doi:10.1242/dmm.016089
8. Oehlers SH, Cronan MR, Scott NR, Thomas MI, Okuda KS, Walton EM, Beerman RW, Crosier PS, Tobin DM (2015) Interception of host angiogenic signalling limits mycobacterial growth. Nature 517(7536):612–615. doi:10.1038/nature13967
9. Cosma CL, Klein K, Kim R, Beery D, Ramakrishnan L (2006) Mycobacterium marinum Erp is a virulence determinant required for cell wall integrity and intracellular survival. Infect Immun 74(6):3125–3133. doi:10.1128/IAI.02061-05
10. Volkman HE, Clay H, Beery D, Chang JC, Sherman DR, Ramakrishnan L (2004) Tuberculous granuloma formation is enhanced by a mycobacterium virulence determinant. PLoS Biol 2(11):e367. doi:10.1371/journal.pbio.0020367

11. Tobin DM, Vary JC Jr, Ray JP, Walsh GS, Dunstan SJ, Bang ND, Hagge DA, Khadge S, King MC, Hawn TR, Moens CB, Ramakrishnan L (2010) The lta4h locus modulates susceptibility to mycobacterial infection in zebrafish and humans. Cell 140(5):717–730. doi:10.1016/j.cell.2010.02.013
12. Auer TO, Duroure K, De Cian A, Concordet JP, Del Bene F (2014) Highly efficient CRISPR/Cas9-mediated knock-in in zebrafish by homology-independent DNA repair. Genome Res 24(1):142–153. doi:10.1101/gr.161638.113
13. Irion U, Krauss J, Nusslein-Volhard C (2014) Precise and efficient genome editing in zebrafish using the CRISPR/Cas9 system. Development. doi:10.1242/dev.115584
14. Jao LE, Wente SR, Chen W (2013) Efficient multiplex biallelic zebrafish genome editing using a CRISPR nuclease system. Proc Natl Acad Sci U S A 110(34):13904–13909. doi:10.1073/pnas.1308335110
15. Shin J, Chen J, Solnica-Krezel L (2014) Efficient homologous recombination-mediated genome engineering in zebrafish using TALE nucleases. Development 141(19):3807–3818. doi:10.1242/dev.108019
16. Takaki K, Davis JM, Winglee K, Ramakrishnan L (2013) Evaluation of the pathogenesis and treatment of Mycobacterium marinum infection in zebrafish. Nat Protoc 8(6):1114–1124. doi:10.1038/nprot.2013.068
17. Schneider CA, Rasband WS, Eliceiri KW (2012) NIH Image to ImageJ: 25 years of image analysis. Nat Methods 9:671–675. doi:10.1038/nmeth.2089
18. Astin JW, Jamieson SM, Eng TC, Flores MV, Misa JP, Chien A, Crosier KE, Crosier PS (2014) An in vivo antilymphatic screen in zebrafish identifies novel inhibitors of mammalian lymphangiogenesis and lymphatic-mediated metastasis. Mol Cancer Ther 13(10):2450–2462. doi:10.1158/1535-7163.MCT-14-0469-T
19. Yeh JR, Munson KM (2010) Zebrafish small molecule screen in reprogramming/cell fate modulation. Methods Mol Biol 636:317–327. doi:10.1007/978-1-60761-691-7_20
20. Ordas A, Raterink RJ, Jansen HJ, Cunningham F, Wiweger MI, Jong-Raadsen S, Bates RH, Barros D, Meijer AH, Vreeken RJ, Ballell-Pages L, Dirks RP, Hankemeier T, Spaink HP (2014) Testing tuberculosis drug efficacy in a zebrafish high-throughput translational medicine screen. Antimicrob Agents Chemother. doi:10.1128/AAC.03588-14

Chapter 15

Methodologies for Inducing Cardiac Injury and Assaying Regeneration in Adult Zebrafish

Jinhu Wang and Kenneth D. Poss

Abstract

The zebrafish has emerged as an important model organism for understanding the cellular and molecular mechanisms of tissue regeneration. Adult zebrafish efficiently replace cardiac muscle after partial resection of their ventricle, or after transgenic ablation of cardiomyocytes. Here, we describe methodology for inducing these injuries and assaying indicators of regeneration.

Key words Zebrafish, Heart, Surgery, Ablation, Cardiomyocyte, Regeneration

1 Introduction

The adult mammalian heart has an extremely limited ability to repair itself through regeneration of new cardiomyocytes (CMs) and instead heals tissue damage by scarring. Nearly four decades ago, it was reported that newts survive removal of up to 25 % of the cardiac ventricle, and CMs near the amputation plane initiate a proliferative response that involves DNA synthesis and mitosis [1, 2]. The lack of molecular genetic tools for salamanders in the following years severely limited investigation of the underlying mechanisms. In 2002, Poss and colleagues reported that zebrafish fully regenerate cardiac muscle after a similar partial ventricular resection, through a mechanism in which new muscle creation occurs more efficiently than scarring. More recently it has been demonstrated that zebrafish efficiently regenerate muscle removed by severe cryoinjury [3–5], or even after extreme injuries that genetically ablate 60 % or more of the animal's total CMs [6].

The zebrafish is a central vertebrate model system used by several hundred laboratories and there is a growing molecular toolbox. Studies from multiple groups over the past decade have revealed that zebrafish heart regeneration involves two key components: (1) proliferation of existing CMs as the primary muscle source [7, 8]; and (2)

Koichi Kawakami et al. (eds.), *Zebrafish: Methods and Protocols*, Methods in Molecular Biology, vol. 1451, DOI 10.1007/978-1-4939-3771-4_15, © Springer Science+Business Media New York 2016

an environment aided by nonmuscle epicardial and endocardial cells that stimulates regeneration by these CMs [9–12].

This work has altered the way mammalian cardiac repair is considered. Most notably, increased attention has been paid to CM division in adult mice and humans, and multiple studies have now observed that adult mammalian CMs undergo low but measurable turnover. Additionally, recent evidence indicates that CM proliferation is stimulated at low levels after injury to the adult mouse heart [13–15]. Studies of the epicardium in adult mice have similarly paired with findings in zebrafish, and epicardial cells are now being implicated in mammalian cardiac repair mechanisms. In vertebrates, epicardial cells can act as modulators of vascularization, muscle survival or regeneration, inflammation, and extracellular matrix regulation [10, 12, 16–22]. Thus, evidence indicates that the analogous regenerative machinery present in zebrafish also exists in the mammalian heart, but is not activated to the same extent for significant regeneration. This fundamental concept supports the continued use of zebrafish as a platform to reveal potential methods to gauge and stimulate human heart regeneration.

Here, we describe procedures for cardiac surgery and genetic CM ablation in adult zebrafish. Additionally, we describe key assays for CM proliferation, a hallmark of heart regeneration, using nuclear markers of CMs and proliferation indicators.

2 Materials

1. Zebrafish mating tank (1 l)
2. Stainless steel straight microscissors (Fig. 1a)
3. Stainless steel curved microscissors (Fig. 1a)
4. Stainless steel microforceps (Fig. 1a)
5. Sponge (1.5 × 5 × 3 cm) with groove cut in center using scissors (0.5 × 2.5 cm; Fig. 1a)
6. Plastic spoon
7. Plastic transfer pipette
8. Laboratory tissue wipers
9. Dissection microscope (Fig. 1a)
10. Glass bowl (90 × 50 mm; Fig. 1a)
11. Razor blade or scalpel
12. SuperFrost Plus Microscope Slides
13. Cover glasses
14. Hydrophobic pen
15. Coplin jar for slide staining

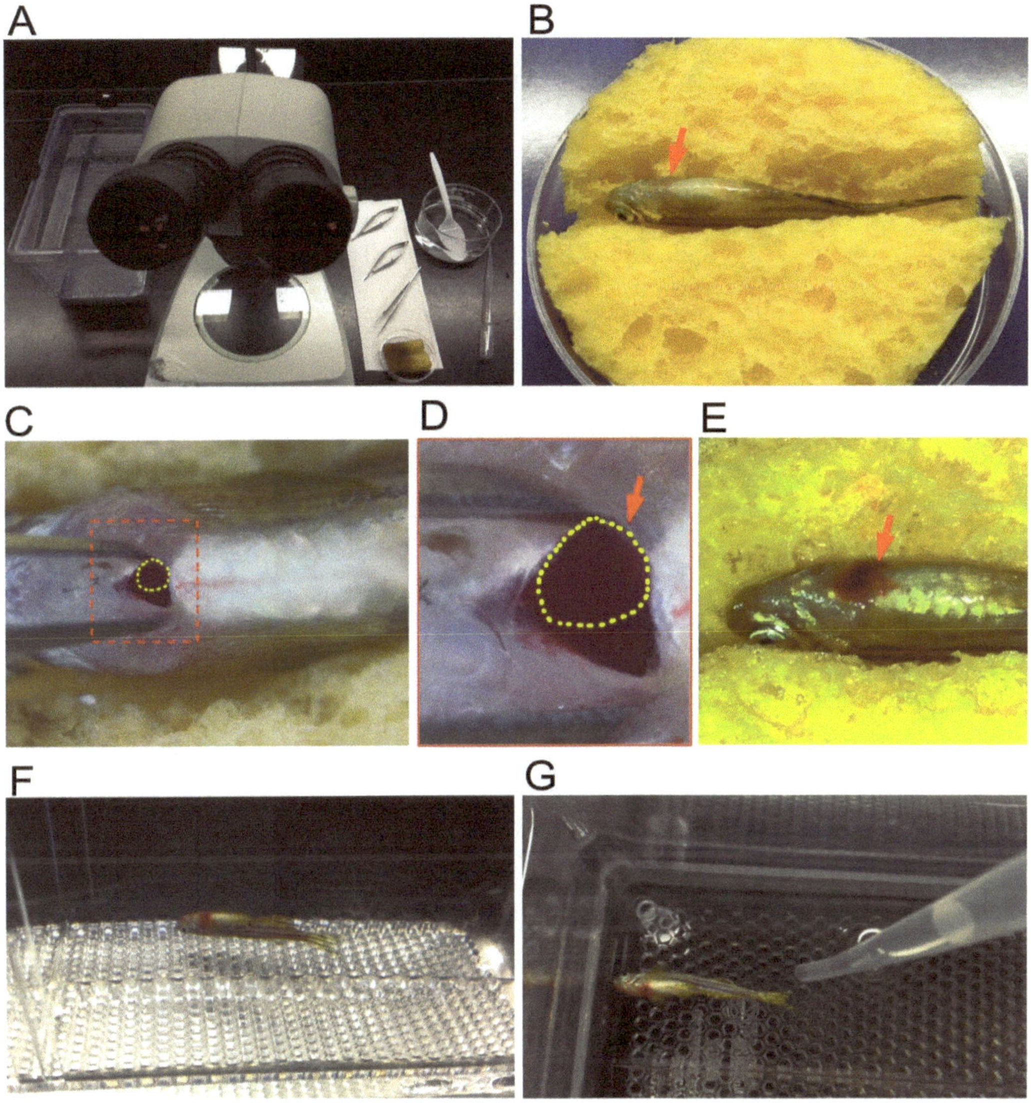

Fig. 1 Resection of the ventricular apex in adult zebrafish. (**a**) Required equipment, including mating tank with aquarium water, dissecting microscope, microscissors, microforceps, pipette, sponge, and glass bowl. (**b**) An anesthetized zebrafish placed ventral side up into a moist, slotted sponge. *Arrow* indicates the heart area. (**c**) View of the heart after puncturing the skin and pericardial sac. *Yellow dashed line* indicates the exposed ventricular apex. (**d**) High magnification of *boxed area* in (**c**). *Yellow dashed line* indicates the exposed ventricular apex (*red arrow*). (**e**) View of the blood clot near the injured heart after surgery. (**f**) A freshly injured fish remains on the bottom of the tank after it is returned to fish water. (**g**) Stimulate fish to gill by vigorously squirting water with a pipette

16. White tape
17. Thermal Cycler
18. Taq and 10× Buffer
19. Deoxynucleotide (dNTP) Solution Mix

20. 4-Hydroxytamoxifen (4-HT). 4-HT powder is dissolved in 100 % ethanol to a 1 mM stock solution that can be stored at 20 °C for 2 weeks. The stock solution can be diluted in aquarium water for different working concentrations. Handle 4-HT with care.
21. Tricaine solution. For tricaine stock solution (4 mg/ml = 0.4 %), dissolve Ethyl 3-aminobenzoate powder in 900 ml of water, then adjust pH to 7.4 with 1 Tris-HCL, pH 9.5 solution. Add water to 1000 ml. Aliquot in 50 ml tubes, label, and store frozen (keep working stock at 4 °C).
22. Fish Fix. Combine all the following reagents except paraformaldehyde. Move solution to hood and warm, slowly adding the PFA; stir until dissolved completely. Adjust to pH 7.4 and filter sterilize. Will keep at 4 °C for 1 week or freeze in 10 ml aliquots.

Paraformaldehyde	8 g
Sucrose	8 g
1 M $CaCl_2$	24 μl
0.2 M Na_2HPO_4	77 ml
0.2 M NaH_2PO_4	23 ml
dH_2O	up to 200 ml

23. Fish Fix Buffer. Combine the following reagents with stir bar, adjust pH to 7.4 if necessary (work without heat). Filter sterilize and freeze in 50 ml tubes.

Sucrose	8 g
1 M $CaCl_2$	24 μl
0.2 M Na_2HPO_4	77 ml
0.2 M NaH_2PO_4	23 ml
dH_2O	up to 200 ml

24. Fin digestion solution: 10 mM Tris–Cl pH 7.5; 50 mM KCl; 0.3 % Tween-20; 1 mM EDTA pH 8.0; 200 μg/ml Proteinase K.
25. Citrate Buffer (10 mM Citric Acid, 0.05 % Tween 20, pH 6.0). Mix 1.92 g Anhydrous citric acid in 1 l dH_2O. Adjust pH to 6.0 with 10 N NaOH. Add 0.5 ml of Tween-20 and mix well. Store this solution at room temperature for 3 months or at 4 °C for longer storage.
26. 30 % sucrose solution
27. TBS Tissue Freezing Medium
28. PBS solution

29. PBST: 1× PBS + 0.1 % Tween-20.
30. Horse Serum
31. NCS-PBT: 10 % NCS (heat inactivated newborn calf serum) + 89 % PBST + 1 % DMSO.
32. Vectashield HardSet Mounting Medium with DAPI
33. F59 anti-myosin antibody
34. Mouse anti-PCNA antibody
35. Rabbit anti-Mef2 antibody
36. Alexa Fluor 594 anti-mouse antibody
37. Alexa Fluor 594 anti-rabbit antibody
38. Alexa Fluor 488 anti-mouse antibody
39. Zebrafish care and use. Outbred EK, AB, or EK/AB mixed background zebrafish (4–12 months of age) are used for ventricular resection surgeries or for CM ablation, whereas transgenic animals are used for CM ablation [6]. Animal density is maintained at approximately 4–5 fish per liter (*see* **Note 1**).

3 Methods

Carry out cardiac injury procedures at either room temperature or 25–29 °C (the temperature of aquarium water).

3.1 Zebrafish Heart Surgery

1. Anesthetize zebrafish (4 months to 1 year old) in 0.02 % Tricaine in a glass bowl for 1–2 min. Anesthesia is monitored by watching for slowing of gill movements and by checking for a response to tail pinch; wait for fish to stop moving, but be careful not to anesthetize any longer than necessary (*see* **Notes 2** and **5**).
2. Use a plastic spoon to transfer anesthetized zebrafish; put fish ventral side up on a moist, slotted sponge (Fig. 1b). Visually locate the beating heart (Fig. 1b).
3. While using microforceps pressed to the left and right of the heart to pull the skin taut, use the straight microscissors to puncture the skin just posterior to the heart, being careful not to puncture the underlying atrium or ventricle (*see* **Note 3**). Cut carefully towards the anterior to make a small incision (~1 mm) that penetrates the skin and muscles above the posterior medial margin of the heart (Fig. 1c, d). The ventricle is exposed by opening the incision with microforceps and applying gentle abdominal pressure. Take care to expose the ventricle (pink muscular tissue) and not the atrium (dark red tissue).

4. Use microforceps inserted carefully into the incision, to the left and right of the heart, to open the incision while applying gentle pressure on the abdomen with the blunt edge of the scissors. Use the curved microscissors (tips pointing up) to quickly remove 10–20 % of the ventricular apex (Fig. 1e), and then allow the heart to slip back inside the body. Ensure the resected heart tissue is completely and visibly removed (*see* **Note 2**).
5. Use a tissue to blot the incision and to prevent blood from entering the gills. When bleeding has stopped, return the fish to fresh fish water (Fig. 1f). Stimulate the fish to move its gills by vigorously squirting water over the gills with a pipette (Fig. 1g). The fish will begin gill movements after several seconds and will then try to swim; continue squirting water over the gills until regular swimming movements are observed (*see* **Notes 4** and **6**).

3.2 Zebrafish CM Ablation

Inducible Cre recombinase-based approaches have been established in adult zebrafish, typically employing cell type-specific promoters to facilitate lineage tracing [7]. To ablate CMs with temporal precision, a dual transgene system was developed in which the first transgene has a CM-specific *cmlc2* promoter driving a 4-hydroxytamoxifen (4-HT)-inducible Cre recombinase (CreER). The second transgene has a constitutive *β-actin2* promoter driving a *loxp*-flanked transcriptional stop sequence followed by partially disabled Diphtheria toxin A-chain (DTA) [6]. In adult *cmlc2:CreER; bactin2:loxp-mCherry-STOP-loxp-DTA* (*bact2:RSD*) double transgenic fish, activation of the Cre recombinase by 4-HT treatment allows recombination in CMs to release expression of the toxin DTA, thereby killing CMs.

1. Heterozygous *cmlc2:CreER* fish are mated with heterozygous *bact2:RSD* fish. The embryos are screened for mCherry red fluorescence with strong and ubiquitous expression throughout whole body by 4 days post fertilization.
2. Screen for *cmlc2:CreER* positive fish at 2 months of age by PCR amplification. Use 0.02 % tricaine to anesthetize fish as described above, and then use a plastic spoon to place it on a piece of tape adhered to the lab bench. Using a razor blade or scalpel, cut the caudal fin tip (~2 mm) and put the fin tissue into a tube with 50 μl Fin digestion solution using tweezers. Fin digest protocol: incubate fins at 55 °C for 60 min, then incubate fins at 95 °C for 15 min. Take 2 μl of each fin digest solution in a PCR tube or plate for PCR (total volume: 25 μl per reaction). Run the PCR product on a 1.5–2 % agarose gel. Cre-positive fish will show two bands: the *cre* PCR product (size is 200 bp) and the *βactin2* PCR product (size is 400 bp).

PCR primers:	PCR program:
Cre forward: CGTACTGACGGTGGGAGAAT	(1) 94 °C 2 min
Cre reverse: GTGGCAGATGGCGCGGCAACA	(2) 94 °C 30 s
β-Actin forward: TGATGAGGCTCAGAGCAAGA	(3) 61.5 °C 30 s
β-Actin reverse: CACAATCCACACTTCCATGC	(4) 72 °C 30 s
Mix these primers together to make a primer mix (2–4) × 45 containing 10 μM of each primer.	(5) 72 °C 5 min
	(6) 4 °C hold

3. Myocyte ablation. Four month-old *cmlc2:CreER*; *bact2:RSD* animals are treated for 12–16 h in 0.1 μM 4-HT in fish water in a mating tank (*see* **Note** 7). Use 20–50 ml of diluted 4-HT per fish. 0.1 % ethanol can be used as a vehicle control. The fish should be kept in a dark environment during incubation.

3.3 Histological Analysis for Cardiac Muscle Ablation and Regeneration

To assess CM ablation and recovery, myofibers can be stained with an antibody directed against myosin heavy chain (MHC) (Fig. 2).

1. Adult zebrafish hearts are extracted and fixed in fish fix solution for 1 h at room temperature (RT). After fixation, hearts should be washed in fish fix buffer 3 × 5 min, then in 30 % sucrose overnight at 4 °C.
2. Hearts are embedded in TBS Tissue Freezing Medium, frozen on dry ice, cryosectioned at 10 μm thickness, and dried for 1 h at RT or 30 min on a 37 °C slide warmer.
3. Circle sections with a hydrophobic pen (e.g., Vector ImmEdge pen) and allow to dry. Then, wash slides 4 × 5 min in PBST in

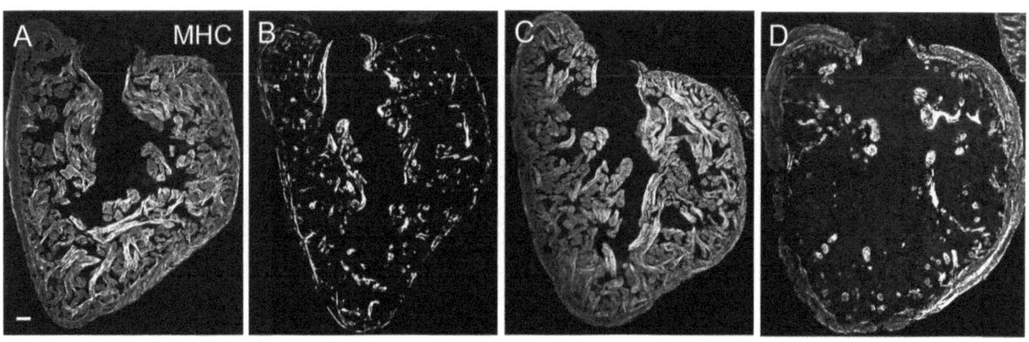

Fig. 2 Regeneration of ventricular cardiomyocytes after ablation-induced injury. (**a–d**) Myosin heavy chain (MHC) staining of ventricular sections from *cmlc2:CreER; bact2:RSD* fish treated with vehicle (**a**), or 0.1 μM 4-HT, at 7 (**b**) or 30 (**c**) days post treatment. (**c**) One example of an ablated heart with successful recovery. (**d**) One example of an ablated heart with greater than 60 % cardiomyocyte loss by 7 dpi that displays partial regeneration at 30 days post treatment. Scale bars, 50 μm

a Coplin jar. All of the following incubations are done in NCS-PBT (with 300 μl of each solution added on top of sections).

4. Block with 2% horse serum in NCS-PBT, in a humidified chamber at 37 °C for 30 min. Remove block (do not wash), then incubate with mouse anti-myosin heavy chain (F59, stored at 4 °C) diluted 1:150 in NCS-PBT, in a humidified chamber at 37 °C for 3 h.
5. Wash 4 × 5 min in PBST in a Coplin jar. Incubate with secondary (fluorescent) antibody in a humidified chamber at 37 °C for 1 h (e.g., Alexa Fluor 594 goat anti-mouse diluted 1:200 in NCS-PBT).
6. Wash 3 × 5 min in PBST. Mount slides using one drop of Vectashield with DAPI, or stain with DAPI and cover slip slides with appropriate mounting medium. Store in the dark at 4 °C, until ready for analysis by fluorescence microscopy. The ablation and regeneration of ventricular muscle are shown in representative sections in Fig. 2.

3.4 Assays for CM Proliferation During Heart Regeneration

Zebrafish fully regenerate cardiac muscle after heart surgery and ablation within 1 month, by CM hyperplasia [6, 7, 23]. CM proliferation can be evaluated by co-staining for Mef2, a nuclear marker of CMs, and the marker Proliferating cell nuclear antigen (PCNA) (Fig. 3).

1. Adult zebrafish hearts should be extracted, fixed, embedded, and cryostat sectioned as described above. Circle sections with a hydrophobic pen and allow to dry. Then, perform treatment with citrate buffer to remove native fluorescence and denature proteins. (Perform these steps in a chemical fume hood.) If not

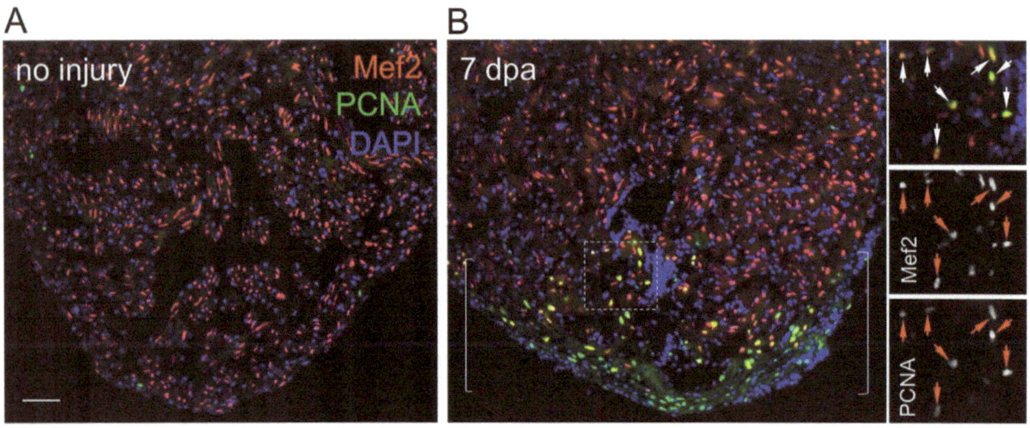

Fig. 3 Ventricular cardiomyocyte proliferation after cardiac surgery, assessed by Mef2 and PCNA staining. (**a**) Uninjured hearts show rare Mef2$^+$PCNA$^+$ cardiomyocytes. (**b**) Injured ventricular apices at 7 days after resection injury (dpa) display many Mef2$^+$PCNA$^+$ cardiomyocytes in wound area. *Bracket*, injury site. *Inset* in (**b**), high magnification of *boxed area*. *Arrows*, proliferating cardiomyocytes. Scale bars, 50 μm

using Citrate buffer, incubate slides in 10% SDS for 5–10 min as a pretreatment.

2. Place citrate buffer in a Coplin jar. Place Coplin jar in a beaker with water reaching midway to the height of the Coplin jar. Place the Coplin jar lid on loosely, and cover the beaker with an ice bucket lid. Heat to 98 °C (boiling—check with thermometer). Add slides and boil at 98 °C for 20 min. Remove the Coplin jar from the beaker and allow to cool in the hood for 20 min.
3. Recircle sections with a hydrophobic pen if necessary. Wash slides 4×5 min in PBST.
4. Block with 2% horse serum in NCS-PBT, in a humidified chamber at 37 °C for 30 min. Remove block (do not wash) and then incubate with 1:75 rabbit anti-Mef2 (stored at 4 °C) diluted 1:75, and mouse anti-PCNA (stored at −20 °C) diluted 1:200 in NCS-PBT, in a humidified chamber at 37 °C for 3 h.
5. Wash 4×5 min in PBST. Incubate with secondary (fluorescent) antibodies in a humidified chamber at 37 °C for 1 h (e.g., Alexa Fluor 594 goat anti-rabbit and Alexa Fluor 488 goat anti-mouse, both diluted 1:200 in NCS-PBT).
6. Wash 3×5 min in PBST. Mount slides using Vectashield with DAPI, or stain with DAPI and cover slip slides with appropriate mounting medium. Store in the dark at 4 °C. Representative sections showing CM proliferation in normal and injured ventricles are shown in Fig. 3.

4 Notes

1. All handling of zebrafish pre- and postcardiac injury should follow institutional guidelines and should be approved by institutional animal care and use committees.
2. The age of zebrafish should be greater than 3 months old. Ninety percent of the fish should survive surgery, with all deaths occurring on the day of surgery. Removal of greater than 25% of the ventricle drastically reduces survival.
3. Do not puncture the heart with scissors during the ventral incision into the pericardial cavity, as extensive bleeding will occur.
4. Revival times following anesthetization may vary. Keep stimulating the amputated fish for at least 5 min by squirting water over the gills until regular gill movement is observed.
5. Tricaine concentration is critical for animal revival and survival after heart surgery. Ensure the fish are anesthetized such that they are unresponsive to a tail pinch, typically within 1 and 3 min. If animals go under too quickly, adjust tricaine concentration by adding aquarium water.

6. The wounds bleed profusely for 5–30 s before clotting. Body wall incisions are not sutured, and heal within 1–2 days.
7. CM ablation may be variable among different clutches. To assess ablation within a given clutch, test the efficiency of myocyte ablation using different doses of 4-HT ranging from 0.1 to 5 μM, treated for 12–16 h.

Acknowledgments

This work was funded by postdoctoral fellowships from the American Heart Association to J.W. and grants from NIH (HL081674, HL131319) and American Federation for Aging Research to K.D.P. We thank Amy Dickson and Anne Knecht for reviewing and contributing to this manuscript.

References

1. Bader D, Oberpriller JO (1978) Repair and reorganization of minced cardiac muscle in the adult newt. J Morphol 155:349–357
2. Oberpriller JO, Oberpriller JC (1974) Response of the adult newt ventricle to injury. J Exp Zool 187:249–253
3. Schnabel K, Wu CC, Kurth T, Weidinger G (2011) Regeneration of cryoinjury induced necrotic heart lesions in zebrafish is associated with epicardial activation and cardiomyocyte proliferation. PLoS One 6:e18503
4. Gonzalez-Rosa JM, Martin V, Peralta M, Torres M, Mercader N (2011) Extensive scar formation and regression during heart regeneration after cryoinjury in zebrafish. Development 138:1663–1674
5. Chablais F, Veit J, Rainer G, Jazwinska A (2011) The zebrafish heart regenerates after cryoinjury-induced myocardial infarction. BMC Dev Biol 11:21
6. Wang J, Panakova D, Kikuchi K, Holdway JE, Gemberling M, Burris JS, Singh SP, Dickson AL, Lin YF, Sabeh MK, Werdich AA, Yelon D, Macrae CA, Poss KD (2011) The regenerative capacity of zebrafish reverses cardiac failure caused by genetic cardiomyocyte depletion. Development 138:3421–3430
7. Kikuchi K, Holdway JE, Werdich AA, Anderson RM, Fang Y, Egnaczyk GF, Evans T, Macrae CA, Stainier DY, Poss KD (2010) Primary contribution to zebrafish heart regeneration by gata4(+) cardiomyocytes. Nature 464:601–605
8. Jopling C, Sleep E, Raya M, Marti M, Raya A, Belmonte JC (2010) Zebrafish heart regeneration occurs by cardiomyocyte dedifferentiation and proliferation. Nature 464:606–609
9. Kikuchi K, Holdway JE, Major RJ, Blum N, Dahn RD, Begemann G, Poss KD (2011) Retinoic acid production by endocardium and epicardium is an injury response essential for zebrafish heart regeneration. Dev Cell 20:397–404
10. Kikuchi K, Gupta V, Wang J, Holdway JE, Wills AA, Fang Y, Poss KD (2011) tcf21+ epicardial cells adopt non-myocardial fates during zebrafish heart development and regeneration. Development 138:2895–2902
11. Kim J, Wu Q, Zhang Y, Wiens KM, Huang Y, Rubin N, Shimada H, Handin RI, Chao MY, Tuan TL, Starnes VA, Lien CL (2010) PDGF signaling is required for epicardial function and blood vessel formation in regenerating zebrafish hearts. Proc Natl Acad Sci U S A 107:17206–17210
12. Lepilina A, Coon AN, Kikuchi K, Holdway JE, Roberts RW, Burns CG, Poss KD (2006) A dynamic epicardial injury response supports progenitor cell activity during zebrafish heart regeneration. Cell 127:607–619
13. Senyo SE, Steinhauser ML, Pizzimenti CL, Yang VK, Cai L, Wang M, Wu TD, Guerquin-Kern JL, Lechene CP, Lee RT (2013) Mammalian heart renewal by pre-existing cardiomyocytes. Nature 493:433–436
14. Bersell K, Arab S, Haring B, Kuhn B (2009) Neuregulin1/ErbB4 signaling induces cardio-

myocyte proliferation and repair of heart injury. Cell 138:257–270

15. Bergmann O, Bhardwaj RD, Bernard S, Zdunek S, Barnabe-Heider F, Walsh S, Zupicich J, Alkass K, Buchholz BA, Druid H, Jovinge S, Frisen J (2009) Evidence for cardiomyocyte renewal in humans. Science 324:98–102
16. Huang GN, Thatcher JE, McAnally J, Kong Y, Qi X, Tan W, DiMaio JM, Amatruda JF, Gerard RD, Hill JA, Bassel-Duby R, Olson EN (2012) C/EBP transcription factors mediate epicardial activation during heart development and injury. Science 338:1599–1603
17. Smart N, Bollini S, Dube KN, Vieira JM, Zhou B, Davidson S, Yellon D, Riegler J, Price AN, Lythgoe MF, Pu WT, Riley PR (2011) De novo cardiomyocytes from within the activated adult heart after injury. Nature 474:640–644
18. Zhou B, Honor LB, He H, Ma Q, Oh JH, Butterfield C, Lin RZ, Melero-Martin JM, Dolmatova E, Duffy HS, Gise A, Zhou P, Hu YW, Wang G, Zhang B, Wang L, Hall JL, Moses MA, McGowan FX, Pu WT (2011) Adult mouse epicardium modulates myocardial injury by secreting paracrine factors. J Clin Invest 121:1894–1904
19. Wang J, Karra R, Dickson AL, Poss KD (2013) Fibronectin is deposited by injury-activated epicardial cells and is necessary for zebrafish heart regeneration. Dev Biol 382:427–435
20. Smart N, Risebro CA, Melville AA, Moses K, Schwartz RJ, Chien KR, Riley PR (2007) Thymosin beta4 induces adult epicardial progenitor mobilization and neovascularization. Nature 445:177–182
21. Song K, Nam YJ, Luo X, Qi X, Tan W, Huang GN, Acharya A, Smith CL, Tallquist MD, Neilson EG, Hill JA, Bassel-Duby R, Olson EN (2012) Heart repair by reprogramming non-myocytes with cardiac transcription factors. Nature 485:599–604
22. Qian L, Huang Y, Spencer CI, Foley A, Vedantham V, Liu L, Conway SJ, Fu JD, Srivastava D (2012) In vivo reprogramming of murine cardiac fibroblasts into induced cardiomyocytes. Nature 485:593–598
23. Poss KD, Wilson LG, Keating MT (2002) Heart regeneration in zebrafish. Science 298:2188–2190

Chapter 16

Studying Lipid Metabolism and Transport During Zebrafish Development

Erin M. Zeituni and Steven A. Farber

Abstract

The zebrafish model facilitates the study of lipid metabolism and transport during development. Here, we outline methods to introduce traceable fluorescent or radiolabeled fatty acids into zebrafish embryos and larvae at various developmental stages. Labeled fatty acids can be injected into the large yolk cell prior to the development of digestive organs when the larvae is entirely dependent on the yolk for its nutrition (lecithotrophic state). Once zebrafish are able to consume exogenous food, labeled fatty acids can be incorporated into their food. Our group and others have demonstrated that the transport and processing of these injected or ingested fatty acid analogs can be followed through microscopy and/or biochemical analysis. These techniques can be easily combined with targeted antisense approaches, transgenics, or drug treatments (*see* **Note 1**), allowing studies of lipid cell biology and metabolism that are exceedingly difficult or impossible in mammals.

Key words Zebrafish, Lipid, Transport, Metabolism, Microscopy, Thin layer chromatography

1 Introduction

In all cells, lipids are essential molecules that support cellular membrane structure, promote inter- and intracellular signaling, and serve as fundamental sources of fuel. In metazoans, lipids play additional essential roles in metabolic functions and physiology. Chronic diseases (affecting more than ½ of adults in the US) cause poor health, disability, and death, and represent the bulk of the US health-care expenditures [1]. The misregulation of lipids observed in a variety of human dyslipidemias is a major factor in these very prevalent chronic diseases [2–6]. A better grasp of how altered lipid metabolism contributes to specific disease pathologies requires extensive studies of the fundamentals of lipid metabolism, storage, and transport in experimentally tractable whole-animal models. Zebrafish are emerging as an ideal model for examining not only the fundamentals of lipid

Koichi Kawakami et al. (eds.), *Zebrafish: Methods and Protocols*, Methods in Molecular Biology, vol. 1451,
DOI 10.1007/978-1-4939-3771-4_16, © Springer Science+Business Media New York 2016

biology, such as dietary lipid absorption and lipoprotein biology, but also aspects of lipid biology related to obesity, metabolic disorders, and cardiovascular disease [7–11].

Zebrafish provide a unique model where in just 6 days, one can study lipid metabolism in live metabolic organs such as the liver and intestine. Several groups, including our own, have developed methods to image fatty acid metabolism using fluorescently labeled fatty acids coupled with a variety of fluorescence microscopy techniques [9, 12–16]. Further, we have pioneered the use of both fluorescently and radiolabeled lipids for biochemical studies of metabolism and transport at multiple stages of zebrafish embryo and larval development [12, 14, 17]. These methods are outlined in this chapter.

Genes involved in lipid and lipoprotein metabolism are frequently expressed early in embryonic development [18–21] and are occasionally required for proper embryo development and/or viability. These essential genes are poor candidates for genetic screens involving the uptake of exogenous food after 5 dpf because the animals are either too deformed or dead. However, these genes can be examined successfully using pharmacological approaches. Because of this issue, it was imperative to design a method to study lipid metabolism at earlier developmental stages. Herein, we describe an isotopic labeling technique using either fluorescent or radioactive fatty acids that are injected directly into the yolk of lecithotrophic embryos and larvae (24 hpf to 4 dfp). We have shown that these fatty acids then undergo the expected metabolic processing and are incorporated into more complex lipid species that can then be subject to transport via lipoproteins [21]. These processes can be examined in detail through microscopy or thin layer chromatography (TLC).

At 6 dpf, zebrafish larvae have fully absorbed their yolk and begin ingesting exogenous food. This allows researcher to use a variety of foods that can contain tractable labeled nutrients, together with ingestible drugs to study the metabolism and transport of dietary lipids [12, 13]. In 2011, we introduced a method, outlined in this chapter, in which fluorescently labeled fatty acids could be incorporated into liposomes that are then fed to zebrafish [12]. We demonstrated that this technique could be used to study the metabolism of different fatty acid chains, the requirement of an established microbiota for proper lipid absorption, and the influence of drug treatments on lipid absorption [22]. The fluorescent liposome ingestion technique has been enthusiastically taken up by the zebrafish community and has accelerated the use of zebrafish as a model to study lipid metabolism and disease (cited in over 30 publications since 2011).

Both the yolk injection and the fluorescent liposome ingestion techniques described herein could be readily combined with the wealth of morpholinos, genetic screens, transgenic lines, disease models, and drug treatments available for studying lipid metabolism and transport in zebrafish. With the need for new models to study lipid biology and disease, zebrafish will be leading the pack.

2 Materials

2.1 Embryo/Larvae Collection and Cleaning

1. Crossing cages for single pairs or in-tank crosses.
2. 2.5 in. nylon mesh strainer (e.g., Progressus Brand; 970775).
3. 100×20 mm sterile plastic Petri dishes (Becton Dickinson).
4. Embryo medium (EM) (Zebrafish Book, General Methods, Water).
5. Wide bore pipettes (e.g., Kimble Chase, 63A53WT).
6. Pipet pump (e.g., VWR Pipette Pump, 10 mL, 53502-233).
7. Incubator set to 29 °C with light cycle capabilities (14 h light; 10 h dark).
8. Stereomicroscope

2.2 Storage of Chicken Egg Yolk

1. Chicken eggs bought from the grocery store. Eggs with vitamin fortification or omega fatty acid supplementation should be avoided.
2. 500 mL glass beaker.
3. 1.5 mL Eppendorf tubes.
4. Freezer box.
5. –80 °C freezer.

2.3 Labeling Oil with Fluorescently Tagged Fatty Acids

1. Fluorescently tagged fatty acids (i.e., BODIPY®FL C_{16} (D-3821), BODIPY®FL C_{12} (D-3822), BODIPY®FL C_5 (D-3834); Invitrogen) stored in chloroform at 0.5 μg/mL in brown glass vials at –20 °C in chloroform.
2. 1.5 mL Eppendorf tubes.
3. N_2 gas for drying down lipid.
4. Canola Oil (from grocery store).

2.4 Labeling Oil with Radiolabeled Fatty Acids

1. Tritiated fatty acids (i.e., lignoceric acid [C24:0-ART 0865], oleic acid [C18:1-ART 0198], or palmitic acid [C16:0-ART 0129]; American Radiolabeled Chemicals) stored at –20 °C in glass vials (2 mL Amber Vial with Teflon Cap; Thermo Fisher Scientific).
2. 1.5 mL Eppendorf tubes.
3. Speed vacuum.
4. Canola Oil (from grocery store).

2.5 Injecting Labeled Oil Droplet into Yolk of Embryos or Larvae

1. Incubator set to 28.5 °C.
2. EM warmed to 28.5 °C.
3. Hollow glass capillary (e.g., Glass Capillary with Filament; Narishige; GD-1).
4. Micropipette puller.
5. Fine forceps.
6. Lab Tape.
7. Stereomicroscope.
8. Labeled Canola Oil (*see* Subheading 3.3 or 3.4).
9. N_2 gas.
10. N_2 gas pressure injector (e.g., PLI 100, Harvard Apparatus).
11. 50 cc syringe.
12. Embryos/larvae of appropriate age for experiment.
13. Tricaine solution (0.03 % tricaine in EM).

2.6 Preparing BODIPY® Fatty Acid Analogs

1. Fluorescently tagged fatty acids (i.e., BODIPY®FL C_{16} (D-3821), BODIPY®FL C_{12} (D-3822), BODIPY®FL C_5 (D-3834); Invitrogen) stored in brown glass vials at −20 °C in chloroform at 0.5 μg/mL.
2. N_2 gas for drying down lipid.
3. EM.
4. Ethanol.
5. Aluminum foil.

2.7 Preparing Chicken egg Yolk Emulsion and Labeling Liposomes with BODIPY® Fatty Acid Analogs

1. Frozen chicken egg yolk aliquot (*see* Subheading 3.2).
2. Spatula to remove egg yolk from Eppendorf tube.
3. Two 50 mL plastic conical tubes.
4. 2.5 in. nylon mesh strainer (e.g., Progressus Brand; 970775).
5. 15 mL plastic conical tube (one per feeding condition).
6. Sonicator with 1/4th inch tapered microtip (e.g., Sonicator Ultrasonic Processor 6000, Misonix Inc., Farmingdale, New York, USA). Set a program for 5 s total processing time, 1 s on, 1 s off, output intensity: 3 W.
7. Vortex.

2.8 Feeding Fluorescently Labeled Liposome Solution to Larvae 6 dpf and Older

1. Zebrafish larvae 6 dpf and older.
2. 35 × 10 mm plastic Petri dish or 6-well plastic culture dish.
3. Three 100 × 20 mm sterile plastic Petri dishes filled with EM.
4. One 100 × 20 mm sterile plastic Petri dish filled with Tricaine solution.
5. Poker to position larvae. (Pokers are made by super gluing fishing line (10 lb, 0.012 in. diameter) into the end of a glass

capillary tube with approximately 1 cm of overhang. The glass capillary tube is then wrapped in lab tape.)

6. Stereomicroscope.
7. Aluminum foil or freezer box lid to cover dishes/plates and protect BODIPY® from light.
8. Incubated orbital shaker set to 29 °C and 30 RPM.

2.9 Long-Term Live Imaging by Upright Microscopy Using an Immersion Objective

1. 70 °C heat block.
2. 42 °C heat block.
3. 1.2% low melt agar in EM. To make aliquots ahead of time, melt the 1.2% agar in EM and aliquot 1 mL into 1.5 mL Eppendorf tubes. Aliquots can be stored at 4 °C for up to 1 year.
4. Tricaine solution.
5. Zebrafish embryos/larvae treated with fluorescent fatty acid analogs (*see* Subheading 3.5, or Subheading 3.8).
6. Wide bore pipettes (e.g., Kimble Chase, 63A53WT).
7. Pipet pump.
8. 35 × 10 mm sterile plastic Petri dish.
9. Poker to position larvae. (Pokers are made by super gluing fishing line (10 lb, 0.012 in. diameter) into the end of a glass capillary tube with approximately 1 cm of overhang. The glass capillary tube is then wrapped in lab tape.)
10. EM.
11. Microscope with fluorescence capabilities (e.g., LeicaSP5 Confocal on a DM6000 microscope, PMT detectors; Leica Microsystems, Germany). Excitation of BODIPY® can be accomplished with a 488 nm laser. The stage must be able to accommodate a 60 × 15 mm plastic Petri dish.
12. Immersion objective (e.g., HCX IR APO L 25× (2.5 mm WD, 0.95NA) objective, Leica, Germany).

2.10 Long-Term Live Imaging by Inverted Microscopy Using a Standard Objective

1. 70 °C heat block.
2. 42 °C heat block.
3. 1.2% low melt agar in EM. To make aliquots ahead of time, melt the 1.2% agar in EM and aliquot 1 mL into 1.5 mL Eppendorf tubes. Aliquots can be stored at 4 °C for up to 1 year.
4. Tricaine solution.
5. Zebrafish embryos/larvae treated with fluorescent fatty acid analogs (*see* Subheading 3.5, or Subheading 3.8).
6. Wide bore pipettes (e.g., Kimble Chase, 63A53WT).
7. Pipet pump.

8. EM.
9. Glass bottom Petri dish
10. Metal block stored in freezer.
11. Kimwipes.
12. Inverted microscope with fluorescence capabilities (e.g., LeicaSP5 on a DMI6000 microscope, PMT detectors; Leica Microsystems, Germany). Excitation of BODIPY® can be accomplished with a 488 nm laser.
13. Standard objectives (e.g., HCX PL APO NA 1.10 W CORR CS objective; Leica, Germany).

2.11 Short-Term Live Imaging by Upright or Inverted Microscopy Using a Standard Objective

1. Prepared slides: a 22 × 22 mm glass coverslip is glued to one end of a 25 × 75 × 1 mm glass slides using QuickTite® Instant Adhesive Gel (LocTite® Item#39202, Henkel Corporation, USA). This provides a ledge that will partially protect the zebrafish larvae from compression.
2. 3 % methylcellulose.
3. Poker to position larvae. (Pokers are made by super gluing fishing line (10 lb, 0.012 in. diameter) into the end of a glass capillary tube with approximately 1 cm of overhang. The glass capillary tube is then wrapped in lab tape.)
4. Tricaine solution
5. Zebrafish embryos/larvae treated with fluorescent fatty acid analogs (*see* Subheading 3.5 or Subheading 3.8).
6. Wide bore pipettes (e.g., Kimble Chase, 63A53WT).
7. Pipet pump.
8. 22 × 30 mm glass coverslip.
9. QuickTite® Instant Adhesive Gel (LocTite® Item#39202, Henkel Corporation, USA).
10. Kimwipes.
11. Inverted or upright microscope with fluorescence capabilities (*see* microscopes used in Subheadings 2.9, **steps 11** and **12** or 2.10, **step 12**). Excitation of BODIPY® can be accomplished with a 488 nm laser.
12. Standard objectives (e.g., HC PlanApo CS, NA 1.4 objective; Leica, Germany).

2.12 Isolation of Total Lipids Following the Bligh and Dyer Method [23]

1. Zebrafish embryos/larvae.
2. Tricaine solution.
3. 1.5 mL Eppendorf tubes.
4. Glass pasteur pipets.
5. 2 mL pipet bulb.

6. Dry Ice.
7. –80 °C Freezer.
8. Aluminum foil or freezer box lid to protect BODIPY®-labeled samples from light.
9. Ice.
10. DI water.
11. Sonicator with 1/4th inch tapered microtip (e.g., Sonicator Ultrasonic Processor 6000, Misonix Inc., Farmingdale, New York, USA). Set output intensity to 3 W.
12. Chloroform:methanol (1:2).
13. Vortex.
14. Chloroform.
15. 200 mM Tris-HCL, pH 7.
16. Microcentrifuge.

2.13 Running TLC Plate Using a Two-Solvent System

1. Two TLC solvent tanks.
2. Polar solvents: ethanol, triethylamine, water.
3. Nonpolar solvents: petroleum ether, ethyl ether, acetic acid.
4. Lipid extract samples (*see* Subheading 3.12, **step 1**).
5. Lipid standards.
6. Speed vacuum.
7. Chloroform:methanol (2:1).
8. Silica gel chromatography plates (LK5D, Whatman).
9. Aluminum foil or dark box to protect BODIPY® label from light.

2.14 Analyzing TLC of Fluorescent Lipids

1. Fluorescence scanner (e.g., Typhoon Scanner, GE Healthcare, Pittsburgh, PA, USA).
2. ImageQuant software (e.g., GE Healthcare, Pittsburg, PA, USA) or Image J software (NIH, USA).

2.15 Analyzing TLC of Radiolabeled Lipids

1. Bioscan radio-TLC Imaging Scanner (Bioscan, AR-2000).
2. Peak Analyzer Pro software package (8.6, OriginLab).

2.16 Quantifying Ingestion of BODIPY®-Labeled Liposomes

1. Lipid extraction samples (*see* Subheading 3.12).
2. Speed vacuum.
3. Chloroform.
4. Single channel silica gel chromatography plate
5. Aluminum foil or dark box to protect BODIPY® label from light.
6. Fluorescence scanner.
7. ImageQuant software or Image J software.

3 Methods

Zebrafish are emerging as an ideal model for examining the fundamentals of lipid biology, such as dietary lipid absorption and lipoprotein biology, as well as aspects related to human diseases. Our lab and others have established tractable methods for studying lipid transport, metabolism, and signaling in zebrafish larvae. Here, we outline methods to introduce fluorescent or radiolabeled fatty acids into zebrafish embryos and larvae at various stages of development. After the injection or ingestion of these labeled fatty acid analogs, their transport and metabolism can be followed through microscopy or biochemical analysis. The methods outlined herein can easily be combined with morpholinos, transgenics, or drug treatments, allowing flexibility and adaptability in using zebrafish as a model of lipid biology.

3.1 Embryo/Larvae Collection and Cleaning

1. Prepare single pair crosses or in-tank crosses of zebrafish.
2. The next day collect zebrafish embryos in a nylon mesh strainer. Run system water through the nylon mesh strainer to clean embryos. Transfer embryos to sterile plastic 100 × 20 mm Petri dishes containing system water.
3. Using a low-magnification dissecting scope, select embryos of the same developmental stage (*see* **Note 2**). Use wide bore pipets and a pipet filler to sort embryos into 100 × 20 mm Petri dishes containing EM (80 larvae per dish).
4. Store embryos at 29 °C with a light cycle of 14 h light; 10 h dark.
5. Clean embryos/larvae every other day, replacing half of the EM and removing dead embryos/larvae and chorion debris (*see* **Note 3**).

3.2 Storage of Chicken Egg Yolk

1. Chicken egg yolk is acquired by separating yolks from 1 dozen chicken eggs.
2. Pool yolks in a glass beaker.
3. Store 1 mL aliquots in 1.5 mL Eppendorf tubes at −80 °C.

3.3 Labeling Oil with Fluorescently Tagged Fatty Acids

1. Pipet desired aliquot of fluorescently tagged fatty acids into plastic 1.5 mL Eppendorf tube.
2. Using a stream of nitrogen, gently dry down fluorescently tagged fatty acids from their storage solution.
3. Resuspend in canola oil (final fluorescently tagged fatty acid concentration: 0.5–1.5 μg/μL).

3.4 Labeling Oil with Radiolabeled Fatty Acids

1. Pipet desired aliquot of radiolabeled fatty acids into plastic 1.5 mL Eppendorf tube.

2. Using a speed vacuum, dry down Tritiated fatty acids from their storage solution.
3. Resuspend in canola oil (final radiolabeled fatty acid concentration: 16.77 uCi/uL).

3.5 Injecting Labeled Oil Droplet into Yolk of Embryos or Larvae

1. Warm EM to 28.5 °C to wash the embryos/larvae after they are injected.
2. Pull a glass injection needle from a hollow glass capillary tube using a micropipette puller.
3. Use forceps to break off the tip of the injection needle to increase the size of the bore, allowing for filling the needle with viscous oil.
4. Insert the injection needle into your microinjection apparatus.
5. Attach a 50 cc syringe to the other end of the tubing of your microinjection apparatus to allow for a source of suction that will pull oil into the injection needle (Fig. 1a).
6. Pipet labeled oil (*see* Subheading 3.2, **step 1** or **2**) onto parafilm mounted to a stereomicroscope stage.
7. This step requires two people. One person will insert the injection needle into the oil droplet on the parafilm. Throughout the loading procedure, this person will observe progress through the stereomicroscope to verify that the tip of the injection needle remains constantly immersed in the labeled oil and is not broken. The second person will pull the plunger of the 50 cc syringe to create suction that will draw up the oil (Fig. 1b).
8. The 50 cc syringe is removed and the microinjection apparatus is reconstituted to allow N_2 gas pressure injection.
9. Embryos/larvae are anesthetized with 0.03% tricaine in EM.
10. Gently hold larvae with forceps to provide resistance against injection needle.
11. Inject 3–5 nL of labeled oil into zebrafish yolk. Early-stage embryos were injected vegetally to avoid the embryo proper. Embryos/Larvae > 24 hpf were injected ventrally and posteriorly to avoid the body of the embryo/larva and vasculature overlying the yolk (Fig. 1c).
12. After injections, embryos were washed with warmed EM and incubated for 0.5–24 h in EM at 28.5 °C. During this time, labeled fatty acids diffuse out of the oil droplet and are accessed by the lipid metabolic machinery and lipid transport proteins (Fig. 1d). The majority of the canola oil remains in the confines of the oil droplet for the duration of the yolk structure (Fig. 1e).

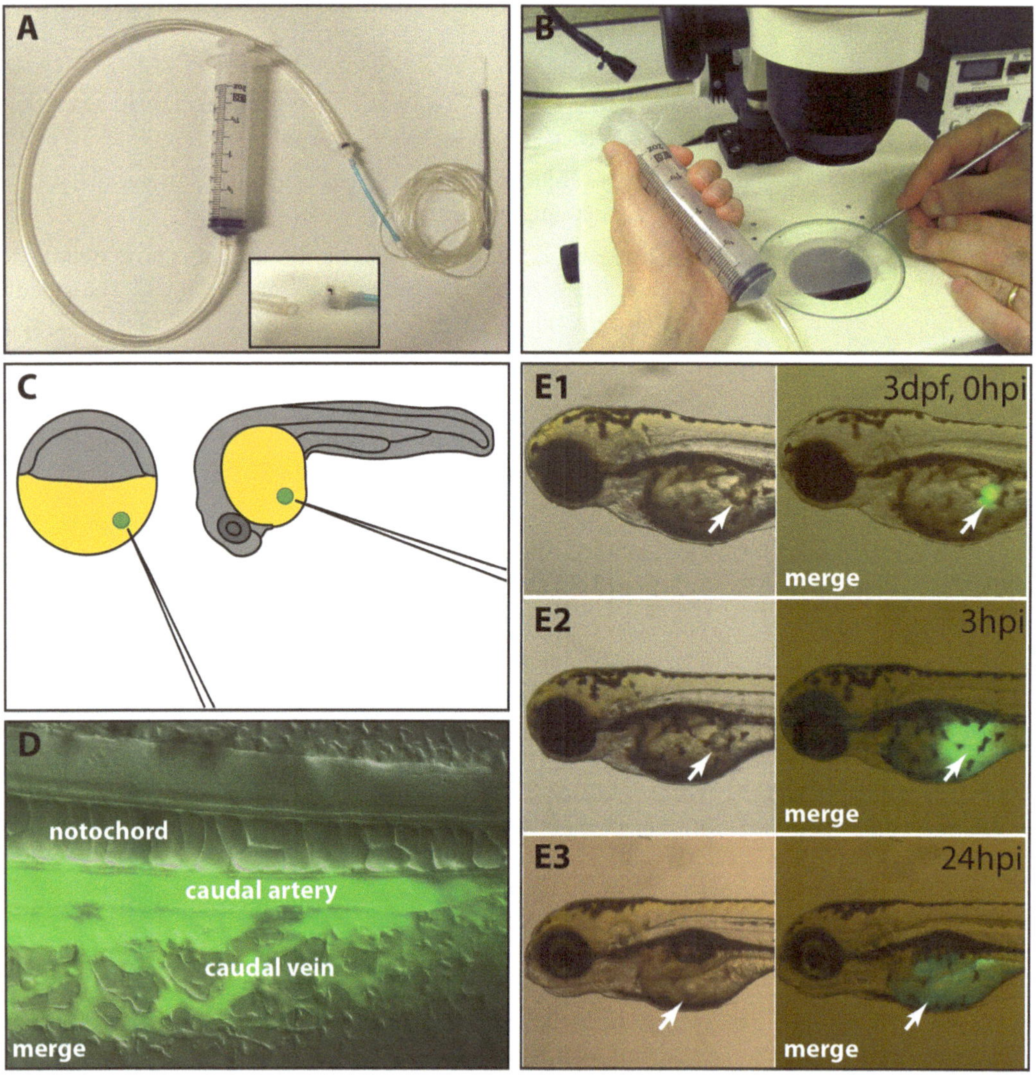

Fig. 1 Labeled Oil Droplet Injection into Yolk of Embryos or Larvae to study transport and metabolism of fatty acid analogs, as in Subheading 3.5. (**a**) Setup of apparatus to load oil into injection needle. (**b**) 2-person loading of oil into injection needle. (**c**) Location of oil injection. (**d–e**) [17] (**d**) Injected BODIPY®-c12-labeled fatty acid analogs are transported from site of the oil drop injection to distal tissues through the vasculature. Image of 1.5 dpf larvae, 2 h post injection (hpi) (**e1–e3**) Timecourse following a single fish injected with BODIPY®-c12-labeled oil at 3dpf. Different timepoints following injection are indicated in upper right corner of images. The oil droplet is indicated by the *white arrow*. Green fluorescence signal represent the BODIPY®-c12. [(**d**) and (**e**) used with permission from Miyares, R.L., V.B. de Rezende, and S.A. Farber, Zebrafish yolk lipid processing: a tractable tool for the study of vertebrate lipid transport and metabolism. Dis. Model. Mech., 2014. 7(7): p. 915–27]

3.6 Preparing BODIPY® Fatty Acid Analogs

1. Transfer desired volume of BODIPY® fatty acid analog to a 1.5 mL Eppendorf tube. The desired final concentration is 6.4 μM fatty acid for a 5 mL chicken egg yolk emulsion.
2. Remove all storage solutions by drying down BODIPY® fatty acid analog under a stream of Nitrogen gas.
3. Resuspend BODIPY® fatty acid analog in 10 μL of 200 proof ethanol. Circle the pipet tip along the wall of the Eppendorf tube as you release the ethanol to ensure that dried analog on the sides of the tube is solubilized (*see* **Note 4**).
4. Add 190 μL EM.
5. Protect BODIPY® fatty acid analog solution from light by wrapping the Eppendorf tube in aluminum foil.
6. Set aside.

3.7 Preparing Chicken Egg Yolk Emulsion and Labeling Liposomes with BODIPY® Fatty Acid Analogs

1. In a 50 mL conical tube prepare 5 % chicken egg yolk emulsion by combining 19 mL EM with 1 mL frozen chicken egg yolk aliquot that has been thawed slightly. To add the thawing egg yolk to the EM, use a metal spatula to scoop the entire semi-frozen aliquot into the EM solution.
2. Vortex until the egg yolk fully dissolves in the EM, and set on room temperature rocker until ready to sonicate.
3. Immerse 1/4th inch tapered microtip of the sonicator halfway into 5 % chicken egg yolk solution.
4. Pulse sonicate 5 % chicken egg yolk emulsion for 40 s with a programed setting (1 s on, 1 s off, output intensity: 3 W). Halfway through sonication, pour 5 % chicken egg yolk solution through a nylon mesh strainer into new 50 mL conical tube to remove any solid debris.
5. Immediately after sonication, pour 5 mL of 5 % egg yolk emulsion into a 15 mL conical tube.
6. Quickly add the prepared BODIPY® fatty acid analog to the sonicated chicken egg yolk emulsion.
7. Vortex at full speed for 30 s. This will incorporate the BODIPY® fatty acid analog into liposomes that are forming as a result of the sonication procedure.

3.8 Feeding Fluorescently Labeled Liposome Solution to Larvae 6 dpf and Older (Fig. 2a)

1. Place larvae in the fluorescent liposome solution in a 35 × 10 mm Petri dish or a 6-well culture dish. An optimal feeding density is 20–100 larvae per 5 mL of fluorescent liposome solution.
2. The feeding larvae are covered with aluminum foil to protect BODIPY® from light, and placed on an incubated orbital shaker (set to 29.5 °C and 30 RPM) for 1–8 h, depending on the design of the experiment.

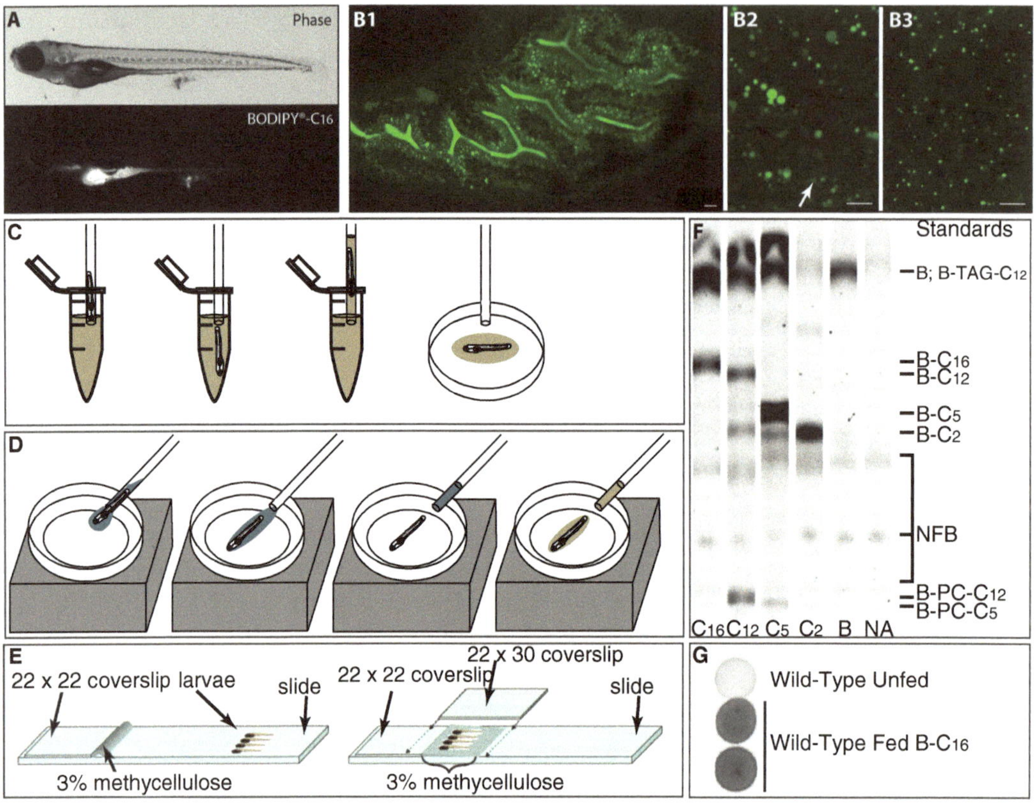

Fig. 2 Analysis of fluorescently labeled fatty acid absorption, transport, and packaging upon ingestion, as in Subheadings 3.9–3.11. (**a**) 6 dpf zebrafish larvae that has ingested BODIPY®-c16-labeled chicken egg yolk. BODIPY®-c16 fluorescence can be seen throughout the intestine. (**b1**–**b3**) Reprinted with permission from Developmental Biology, 360 (2), Juliana D. Carten, Mary K. Bradford, Steven A. Farber, Visualizing digestive organ morphology and function using differential fatty acid metabolism in live zebrafish, 276–285, 2011. Confocal images of the intestine (**b1**), liver (**b2**), and pancreas (**b3**) of 6 dpf zebrafish fed BODIPY®-c12-labeled chicken egg yolk. Scale bars = 10 μm. (**b1**) BODIPY®-c12 is taken up from the intestinal lumen by intestinal epithelial cells, and is stored in multiple lipid droplets in each cell. (**b2**) BODIPY®-c12 is transported to the liver where it accumulates in large, round lipid droplets and in canaliculi (white arrow). (**b3**) BODIPY®-c12 is transported to the pancreas and accumulates in distinct punctae. (**c**) Prepping larvae for long-term live imaging by upright microscopy using an immersion objective, as described in Subheading 3.9. (**d**) Prepping larvae for long-term live imaging by inverted microscopy using a standard objective, as in Subheading 3.10. (**e**) Image courtesy of James W. Walters. Prepping larvae for short-term live imaging by upright or inverted microscopy using a standard objective, as in Subheading 3.11. (**f**) Reprinted from Developmental Biology, 360 (2), Juliana D. Carten, Mary K. Bradford, Steven A. Farber, Visualizing digestive organ morphology and function using differential fatty acid metabolism in live zebrafish, 276–285, 2011. Different BODIPY® fatty acid carbon chain lengths are metabolized differently upon ingestion by zebrafish larvae. Thin layer chromatography analysis of total larval lipids extracted following independent chicken egg yolk feeds spiked with the BODIPY® analog indicated at the bottom of each lane. A no analog (NA) lane serves as a control and exhibits natural fluorescent lipid bands (NFB). Abbreviations for fluorescent lipid standards are as follows: BODIPY®-Cholesterol (B-CE-C12), BODIPY®-Triacylglyceride (B-TAG-C12), BODIPY® (B), BODIPY®-c16 (B-C16), and BODIPY®-Phosphatidylcholine (B-PC-C12, B-PC-C5). (**g**) Quantifying ingestion of BODIPY®-labeled liposomes using a spot assay of total larval extracts. Unfed fish exhibit a fluorescent signal, indicating natural fluorescent lipids

3. At the end of the allotted feeding time, fed larvae are washed twice in EM by moving larvae to dishes filled with fresh EM (*see* **Note 5**).
4. Larvae are anesthetized by moving to a dish with a low dose of tricaine in EM.
5. Anesthetized larvae are examined under a stereomicroscope to verify feeding. Using a poker, larvae are gently turned to allow a lateral view that clearly shows the intestine. Fed larvae will have darkened intestines, whereas unfed larvae will not (*see* **Note 6**).
6. Fed larvae will then be returned to a dish with fresh EM in the incubated shaker to recover and continue to process their meal for a designated period of time. Or larvae will be immediately processed for imaging and biochemical analysis (*see* Subheading 3.4 or 3.5, respectively).

3.9 Long-Term Live Imaging by Upright Microscopy Using an Immersion Objective (Fig. 2c)

1. 1.2% low melt agar is heated to 70 °C in a heat block. Agar should flow smoothly when tube is inverted.
2. Once the agar turns liquid it is transferred to a 42 °C heat block to maintain its fluidity during the mounting process. The agar must be reduced to 42 °C before the embryos/larvae are introduced, as higher temperatures will kill the fish.
3. Anesthetize embryos in tricaine solution.
4. Holding the tube of agar in one hand, use the other hand to pipet individual anesthetized embryos/larvae into a wide bore pipet and place them in the agar, transferring as little EM as possible.
5. Once the desired number of fish (1–10 fish) have been added to the agar, use a wide bore pipet to transfer the fish and agar to a droplet in the middle of a 35 mm Petri dish.
6. While the agar is still liquid, quickly square off the round agar droplet with a poker to equalize the height of the agar.
7. Quickly orient the fish with a poker to the desired position before the agar hardens.
8. Once the agar hardens, add a drop of EM to its surface to prevent drying. Multiple samples can be prepared consecutively in this manner and kept for 1 h. However, if fish will be in the agar drop for >1 h, cover the agar droplet with EM. When moving the dish, remove the EM so that the agar droplet does not dislodge from the plastic.
9. Move your materials to the microscope.
10. Add EM to the Petri dish with the agar-immobilized zebrafish so that it fully covers the agar droplet.
11. Place the dish on your microscope stage.
12. Position the immersion objective and focus (*see* **Note** 7).
13. Image the desired regions with appropriate channels (i.e., to image BODIPY® use a 488 laser; 498 excitation; 530 emission).

3.10 Long-Term Live Imaging by Inverted Microscopy Using a Standard Objective (Fig. 2d)

1. 1.2% low melt agar is heated to 70 °C in a heat block. Agar should flow smoothly when tube is inverted.
2. Once the agar turns liquid it is transferred to a 42 °C heat block to maintain its fluidity during the mounting process. The agar must be reduced to 42 °C before the embryos/larvae are introduced, as higher temperatures will kill the fish.
3. Anesthetize embryos/larvae in 0.03% tricaine solution.
4. Take a metal block out of the freezer and place on your bench. Cover with two folded kimwipes.
5. Place the glass-bottom dish onto the kimwipe-covered metal block.
6. Using a wide bore glass pipet, transfer a single anesthetized larvae to a glass bottom dish.
7. Slowly draw out off the EM so that the fish lays flat with its lateral side flush with the glass.
8. Use the wide bore pipet to pipet a small amount of 42 °C agar, holding it in the pipet for 5 s so that it will cool slightly.
9. Gently pipet the agar onto the fish moving from its head to its tail. The cold metal block will harden the agar from the bottom up, so that the fish will remain pressed against the glass.
10. Repeat with no more than four fish per glass bottom dish.
11. Pipet a drop of EM onto each agar droplet.
12. Move your materials to the microscope.
13. Place the dish on your microscope stage.
14. Position the objective and focus (*see* **Note** 7).
15. Image the desired regions with appropriate channels (i.e., to image BODIPY® use a 488 laser; 498 excitation; 530 emission).
16. Add drops of EM as needed to the agar to prevent drying.

3.11 Short-Term Live Imaging by Upright or Inverted Microscopy Using a Standard Objective (Fig. 2b, e)

1. Place prepared slide on a stereomicroscope.
2. Using the tip of a wide bore pipet, gently scoop a small portion of 3% methylcellulose and draw a bead across the slide immediately to the right of the glued coverslip. This will create a methylcellulose cushion.
3. Place anesthetized larvae in an EM droplet on the slide.
4. Dip a poker into the 3% methylcellulose and then use it to gently pick up an larvae from the droplet. Move the larvae into the methylcellulose. This minimizes the amount of EM carried into the methylcellulose cushion. Move all of the larvae into the methylcellulose, and then wipe off the droplet of EM using a Kimwipe.
5. Use the poker to orient the larvae with heads proximal to the glass ledge established by the glued coverslip. Use the poker to position the larvae.

6. Remove excess methylcellulose from the slide using a Kimwipe, leaving a squared off methylcellulose cushion.
7. Take a 22 × 30 mm glass coverslip and apply four beads of QuickTite® Instant Adhesive Gel, one to each corner of the 22 × 30 coverslip.
8. Attach the 22 × 30 mm glass coverslip to the prepared slide. With the 22 × 30 mm coverslip held at an angle, first adhere the leftmost edge to the top of the 22 × 22 mm coverslip, then gently press down the 22 × 30 mm coverslip to adhere it to the slide beyond methylcellulose cushion. This will cover and compress the larvae in the methylcellulose cushion (*see* **Note 8**).
9. Place the slide on your microscope stage.
10. Position the objective and focus.
11. Image the desired regions with appropriate channels (i.e., to image BODIPY® use a 488 laser; 498 excitation; 530 emission).

3.12 Isolation of Total Lipids Following the Bligh and Dyer Method [23]

1. Place ten anesthetized larvae in a 1.5 mL Eppendorf tube. Larger numbers of larvae can be used, but extraction volumes need to be adjusted accordingly. The following procedure is catered to a sample size of ten larvae.
2. All liquid was removed, leaving the larvae in the bottom of the tube.
3. Larvae were snap frozen using dry ice.
4. Larvae can be stored at −80 °C. Note that the BODIPY® signal does decrease over time, and it is recommended that replicate fluorescent samples be stored at −80 °C for identical durations of no more than 24 h.
5. Samples are removed from the −80 °C freezer and placed on ice.
6. 100 μL of H_2O is added to the frozen larvae.
7. Sample is sonicated for 4 s with a 1/4th inch tapered microtip and an output of 3 W. The tip of the sonicator should remain immersed in the solution for the duration of the pulse to prevent foam from forming.
8. Add 375 μL chloroform:methanol (1:2) to the homogenized fish.
9. Vortex 30 s.
10. Incubate at room temperature for at least 10 min. Fluorescent samples should be incubated in the dark.
11. Add 125 μL chloroform.
12. Vortex 30 s.
13. Add 125 μL 200 mM Tris-HCL, pH 7.
14. Vortex 30 s.
15. Centrifuge at 4000 rpm for 5 min.

16. Gently remove samples from the centrifuge, and carefully transfer the bottom organic phase to a clean 1.5 mL Eppendorf tube using a glass capillary pipet.
17. Samples can be stored at −80 °C, remembering that the BODIPY® signal does decrease with time. Again, we recommend that replicate fluorescent samples be stored at −80 °C for identical durations of no more than 24 h.

3.13 Running TLC Plate Using a Two-Solvent System (Fig. 2f)

1. Construct your polar and nonpolar solvent tanks, allowing the tanks 30 min to equilibrate. The polar solvent (ethanol:triethylamine:water; 27:25:6.4) will be run first. The nonpolar solvent (petroleum ether:ethyl ether: acetic acid; 64:8:0.8) will be run second.
2. Dry down lipid extracts (*see* Subheading 3.4, **step 1**) in a speed vacuum until solvent has fully evaporated.
3. Resuspend samples in 40 μL chloroform:methanol (2:1).
4. Load samples and standards onto silica gel chromatography plates. If your TLC plates do not have a loading region, be sure to load each sample at the same distance from the bottom of the plate.
5. Place the loaded TLC plate into the polar solvent tank and run the solvent halfway up the plate.
6. Remove the plate from the polar solvent tank and allow to air dry.
7. Place the loaded TLC plate into the nonpolar solvent tank and run the solvent to near the top of the plate.
8. Remove the plate from the nonpolar solvent tank and allow to air dry.
9. TLC plates loaded with fluorescent lipids should be carefully stored in the dark to protect the BODIPY® from the light.

3.14 Analyzing TLC of Fluorescent Lipids

1. To detect BODIPY®-labeled lipids, scan the dried TLC plates in a Typhoon Scanner using a blue fluorescence laser (excitation: 488 nm; emission: 520 nm Band Pass; PMT 425).
2. Fluorescence intensity of bands, as well as background fluorescence, can be quantified using ImageQuant or Image J software.
3. Background fluorescence should be subtracted before further analysis (*see* **Note 9**).

3.15 Analyzing TLC of Radiolabeled Lipids

1. To detect radiolabeled lipids, dried TLC plates were scanned for total counts across all lanes using a Bioscan radio-TLC Imaging Scanner. The scanner counts tritium's β emission across each lane of the TLC plate.
2. Export data obtained from Winscan software package of the Imaging Scanner.

3. Import data into the Peak Analyzer Pro software package to determine and subtract baselines; and find, integrate, and fit peaks.
4. Represent data as chromatograms, graphs of percent total counts, or graphs of percent total metabolites.

3.16 Quantifying Ingestion of BODIPY®-Labeled Liposomes (Fig. 2g)

1. Dry down lipid extracts (*see* Subheading 3.5, **step 1**) in a speed vacuum until solvent has fully evaporated. Lipids should be extracted from paired fed and unfed larvae (*see* **Note 10**).
2. Resuspend samples in 12 μL chloroform.
3. Pipet each sample onto a single spot on channeled TLC plate. Be careful not to touch the tip of the pipet to the surface of the silica.
4. Cover the TLC plate with aluminum foil or place in box to protect BODIPY® until scanning.
5. To detect BODIPY®-labeled lipids, scan the dried TLC plates in a Typhoon Scanner using a blue fluorescence laser (excitation: 488 nm; emission: 520 nm Band Pass; PMT 425).
6. The fluorescence intensity of each spot can be quantified using ImageQuant or Image J software.
7. Background fluorescence from unfed larvae should be subtracted before further analysis (*see* **Note 9**).

4 Notes

1. As mentioned, these methods have been successfully combined with the use of pharmacological reagents [17, 18]. Important factors to consider when using a drug for a feeding assay:
   (a) The effective dose at the stage of development you are examining. If no published dose exists in zebrafish, we recommend performing a dose response curve.
   (b) The solubility of the drug and its ability to penetrate tissue. The drug will be introduced to the zebrafish either by soaking them in EM containing the drug or by gavaging the fish. In both cases, it must be soluble in water and able to cross cell membranes into tissues.
   (c) The toxicity of the drug. Since the larvae may be soaking in the drug, it is important to consider its toxicity across multiple tissues.
   (d) Drugs can be applied once the fish are mounted in agar, as many drugs can readily diffuse through the agar.
   (e) It is possible that drugs will cause a change in the amount of food ingested. To control for this, one must perform an ingestion quantification, as described in Subheading 3.6.

2. It is important to verify that the embryos and larvae used in each experiment are developing properly. An analysis of their physical appearance should be undertaken before beginning an experiment, and all embryos and larvae exhibiting developmental abnormalities should be removed from the experiment.
3. It is important to remove the chorions before 5 dpf because the zebrafish larvae will ingest them.
4. BODIPY® fatty acid analogs are easy to solubilize in small amounts of ethanol and EM. However, larger BODIPY®-labeled lipids, such as cholesterol, can be difficult to solubilize. Contact our lab directly for helpful hints if you run into trouble.
5. We typically keep larvae in chicken egg yolk solutions no longer than 10 h, as they tend to die when kept in the solutions too long.
6. To ensure that all larvae are on the same feeding timescale during longer experiments, remove larvae from the chicken egg yolk solution at 1–2 h, examine them for feeding, and then place them in EM for the duration of the experiment.
7. Objectives with a longer working distance and a high numerical aperture are ideal for imaging tissues in a zebrafish larvae. The longer working distance helps one visualize tissues deeper within the larvae.
8. Although the larvae are somewhat protected by the methylcellulose cushion, the pressure of the coverslip does begin to kill the fish. It should be assumed that the fish would die after 30 min, which can be verified by examining the heartbeat or blood flow.
9. Zebrafish exhibit natural fluorescence that can confound analysis by fluorescence scanning. For each experiment, be sure to include an unfed control to obtain background levels of this natural fluorescence.
10. Due to variations in the extraction procedure, we recommend multiple replicates of each condition's extraction.

Acknowledgments

The authors would like to thank Drs. James Walters and Juliana Carten for the initial development of the fluorescent liposome feeding assay; Dr. Rosa Miyares and Jennifer Anderson for the initial development of the oil injection technique; Drs. Juliana Carten, Rosa Miyares, and Vitor deRezende for the initial exploration into the use of radiolabeled fatty acids and lipids in zebrafish metabolic assays; and Mahmud Siddiqi for his technical support with microscopy. This work was supported in part by the National Institute of Diabetes and Digestive and Kidney (NIDDK) [grant numbers RO1DK093399 to

S.A.F., RO1GM63904 to The Zebrafish Functional Genomics Consortium (Stephen Ekker and S.A.F.)]. The content is solely the responsibility of the authors and does not necessarily represent the official views of the National Institutes of Health (NIH). Additional support for this work was provided by the Carnegie Institution for Science endowment and the G. Harold and Leila Y. Mathers Charitable Foundation to the laboratory of S.A.F.

References

1. Bauer UE et al (2014) Prevention of chronic disease in the 21st century: elimination of the leading preventable causes of premature death and disability in the USA. Lancet 384(9937):45–52
2. Joffe BI, Panz VR, Raal FJ (2001) From lipodystrophy syndromes to diabetes mellitus. Lancet 357(9266):1379–1381
3. McNeely MJ et al (2001) Lipoprotein and apolipoprotein abnormalities in familial combined hyperlipidemia: a 20-year prospective study. Atherosclerosis 159(2):471–481
4. Watanabe S et al (2008) Liver diseases and metabolic syndrome. J Gastroenterol 43(7):509–518
5. Flegal KM et al (2010) Prevalence and trends in obesity among US adults, 1999–2008. JAMA 303(3):235–241
6. Ogden CL et al (2010) Prevalence of high body mass index in US children and adolescents, 2007–2008. JAMA 303(3):242–249
7. Anderson JL, Carten JD, Farber SA (2011) Zebrafish lipid metabolism: from mediating early patterning to the metabolism of dietary fat and cholesterol. Methods Cell Biol 101:111–141
8. Asaoka Y et al (2013) The expanding role of fish models in understanding non-alcoholic fatty liver disease. Dis Model Mech 6(4):905–914
9. Fang L, Liu C, Miller YI (2014) Zebrafish models of dyslipidemia: relevance to atherosclerosis and angiogenesis. Transl Res 163(2):99–108
10. Holtta-Vuori M et al (2010) Zebrafish: gaining popularity in lipid research. Biochem J 429(2):235–242
11. Schlegel A, Stainier DY (2007) Lessons from "lower" organisms: what worms, flies, and zebrafish can teach us about human energy metabolism. PLoS Genet 3(11):e199
12. Carten JD, Bradford MK, Farber SA (2011) Visualizing digestive organ morphology and function using differential fatty acid metabolism in live zebrafish. Dev Biol 360(2):276–285
13. Clifton JD et al (2010) Identification of novel inhibitors of dietary lipid absorption using zebrafish. PLoS One 5(8):e12386
14. Farber SA et al (2001) Genetic analysis of digestive physiology using fluorescent phospholipid reporters. Science 292(5520):1385–1388
15. Hama K et al (2009) In vivo imaging of zebrafish digestive organ function using multiple quenched fluorescent reporters. Am J Physiol Gastrointest Liver Physiol 296(2):G445–G453
16. Schlegel A, Stainier DY (2006) Microsomal triglyceride transfer protein is required for yolk lipid utilization and absorption of dietary lipids in zebrafish larvae. Biochemistry 45(51): 15179–15187
17. Miyares RL, de Rezende VB, Farber SA (2014) Zebrafish yolk lipid processing: a tractable tool for the study of vertebrate lipid transport and metabolism. Dis Model Mech 7(7):915–927
18. Babin PJ et al (1997) Both apolipoprotein E and A-I genes are present in a nonmammalian vertebrate and are highly expressed during embryonic development. Proc Natl Acad Sci U S A 94(16):8622–8627
19. Marza E et al (2005) Developmental expression and nutritional regulation of a zebrafish gene homologous to mammalian microsomal triglyceride transfer protein large subunit. Dev Dyn 232(2):506–518
20. Poupard G et al (2000) Apolipoprotein E gene expression correlates with endogenous lipid nutrition and yolk syncytial layer lipoprotein synthesis during fish development. Cell Tissue Res 300(2):251–261
21. Otis JP et al (2015) Zebrafish as a model for apolipoprotein biology: comprehensive expression analysis and a role for ApoA-IV in regulating food intake. Dis Model Mech 8(3): 295–309
22. Semova I et al (2012) Microbiota regulate intestinal absorption and metabolism of Fatty acids in the zebrafish. Cell Host Microbe 12(3):277–288
23. Bligh EG and Dyer WJ (1959) A rapid method of total lipid extraction and purification. Canadian Journal of Biochemistry and Physiology 37(8): 911–7

Part III

Neuroscience

Chapter 17

Targeted Electroporation in Embryonic, Larval, and Adult Zebrafish

Ming Zou, Rainer W. Friedrich, and Isaac H. Bianco

Abstract

This chapter describes three fast and straightforward methods to introduce nucleic acids, dyes, and other molecules into small numbers of cells of zebrafish embryos, larvae, and adults using electroporation. These reagents are delivered through a glass micropipette and electrical pulses are given through electrodes to permeabilize cell membranes and promote uptake of the reagent. This technique allows the experimenter to target cells of their choice at a particular time of development and at a particular location in the zebrafish with high precision and facilitates long-term noninvasive measurement of biological activities in vivo. Applications include cell fate mapping, neural circuit mapping, neuronal activity measurement, manipulation of activity, ectopic gene expression, and genetic knockdown experiments.

Key words Zebrafish, Gene expression, Electroporation, DNA, RNA, Optogenetics

1 Introduction

Fast and efficient methods for gene transfer in vivo provide a wealth of opportunities to study the structure, function, and development of neuronal circuits. Electroporation is an approach that is fast, cost effective, and circumvents the construction of viral vectors or other vehicles. Because electroporation relies on physical rather than molecular mechanisms it should be applicable in a wide range of cells, tissues, and organisms. Electroporation uses brief electrical pulses to transiently permeabilize the plasma membrane and transfer reagents, such as plasmid DNA, into cells [1]. Reagents can be delivered to confined brain areas by controlling the spatial extent of the electrical field. Single neurons or small clusters can be targeted by electroporation using a micropipette. Additional genetic specificity can be achieved by electroporating plasmids containing specific promoters and by the use of intersectional expression systems.

Electroporation has been used in zebrafish to introduce dyes or DNA constructs into larval neurons, adult muscle cells, adult retinal, and brain neurons [2–8]. We describe three electroporation

Koichi Kawakami et al. (eds.), *Zebrafish: Methods and Protocols*, Methods in Molecular Biology, vol. 1451,
DOI 10.1007/978-1-4939-3771-4_17, © Springer Science+Business Media New York 2016

procedures that use different approaches to apply electrical pulses: (a) through a pair of needle electrodes placed outside of the body ("external-electrode electroporation"; EEP); (b) through a pair of thin wire electrodes placed inside the tissue ("internal-electrode electroporation"; IEP); (c) through a glass micropipette ("pipette-electrode electroporation"; PEP). EEP targets relatively large brain areas such as forebrain hemispheres in adult fish; IEP confines gene delivery to smaller areas such as telencephalic nuclei in adult fish, and PEP is used to target small clusters or individual neurons. These methods have been used in different brain areas of larval and adult zebrafish to express a wide range of transgenes including fluorescent proteins, genetically encoded calcium indicators, and optogenetic probes [2, 5, 8]. Expression can last for weeks, depending on the DNA construct. In addition to nucleic acids, other reagents can be delivered to cells, including fluorescent dyes, facilitating lineage tracing or analysis of cell morphology.

2 Materials

2.1 Instrumentation

The physical arrangement of the experimental instrumentation is shown in Figs. 1 and 2.

1. A stereomicroscope (usually for EEP and IEP) or a fixed stage compound microscope (for PEP) are required to view the fish. The stereomicroscope should be tilted, usually by mounting on an arm, to provide vertical access to the preparation. The compound microscope should be equipped with long working distance objectives (air or water-immersion, 10, 20, or 40×). An appropriate stage and equipment are required to hold the fish and to mount micromanipulators (see below). Epifluorescence optics are recommended for electroporation of fluorescent dyes.
2. One or two micromanipulators with at least three movement axes are required. An axis moving in the direction of the pipette or electrode (axial direction) is recommended. Vibrations should be minimized.
3. For EEP and PEP, a stimulator capable of delivering 2–25 ms pulses of 1–70 V at a frequency of 1–200 Hz is used. This can be a stimulus isolator used for electrophysiology (e.g., Grass Technologies SD9 Square Pulse Stimulator) or a pulse generator used for standard electroporation applications such as the transformation of bacteria (e.g., Biorad Gene Pulser Xcell). For IEP, a more advanced device with flexible pulse control is used, e.g., NEPAGENE NEPA21. This electroporator allows for fine tuning of the pulse protocol based on tissue impedance (*see* **Note 1** and ref. 8).

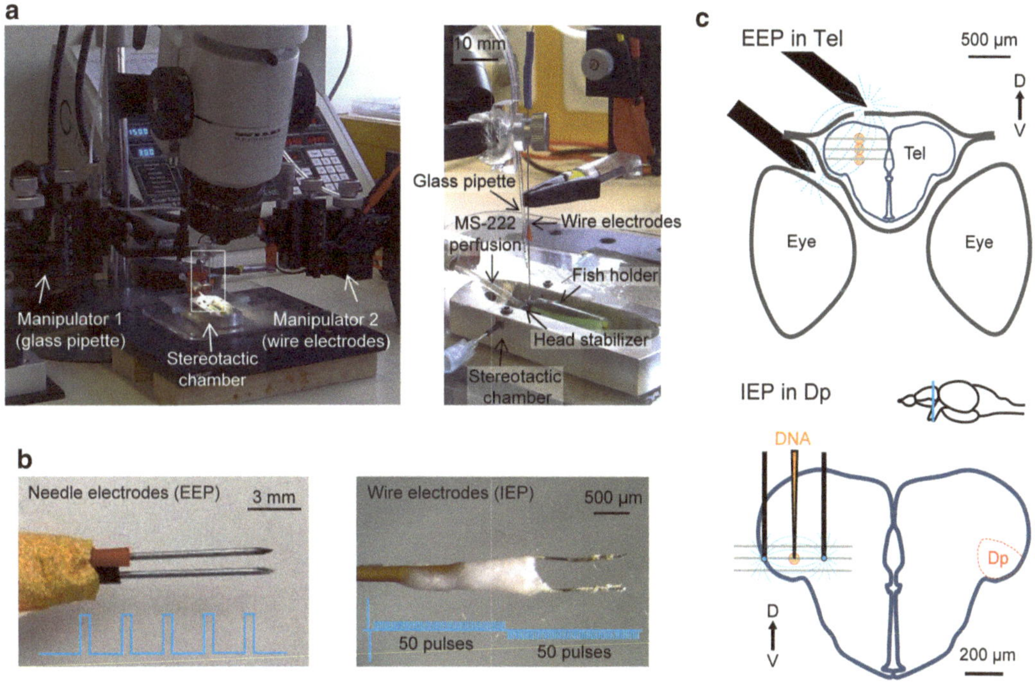

Fig. 1 External-electrode-electroporation (EEP) and internal-electrode-electroporation (IEP) in adults. (**a**) *Left*: apparatus for injection and electroporation. *Right*: Arrangement of wire electrodes and glass micropipette for targeted electroporation with internal electrodes (IEP) in the stereotactic chamber. (**b**) Needle electrodes for EEP (*left*) and wire electrodes for IEP (*right*). *Insets* show electrical pulse protocols. (**c**) Approximate positions of electrodes (*black*) and injection sites (*orange*) for EEP in the dorsal telencephalon (*top*) and targeted IEP in Dp (*bottom*). Plasmid was injected and electroporated sequentially at three different depths (*gray lines*). *D* dorsal, *V* ventral, *A* anterior, *P* posterior. Adapted from ref. 8

2.2 Small Equipment

1. *Glass micropipettes* are prepared from borosilicate capillary glass with an internal filament using a pipette puller (e.g., Sutter P-97 or P-2000). For PEP, a micropipette similar to the type used for loose-patch electrophysiology is pulled from 1.2 mm OD thin-walled capillary glass: the taper is less than 5 mm and the tip opening is 1–3 μm (*see* **Note 2**). A microelectrode holder with a silver wire connector and (optional) side port for air pressure is used for holding the micropipette (e.g., WPI MEH7W) (Fig. 2a). The side port can be connected to a 1 ml syringe by soft tubing. For EEP and IEP, the micropipette has a long taper (~1 cm). The tip is broken with a forceps to generate an opening 10–20 μm in diameter. The pipette is held almost vertically by a custom-made holder (Fig. 1a) onto one micromanipulator and the pipette is directly connected to a 60 ml syringe by soft tubing.
2. *Electrodes.* For PEP, one electrode is provided by the silver wire connected to the microelectrode holder. A second "bath electrode" consists of a short (~5 mm) piece of silver wire that can be lowered into the medium surrounding the specimen. A

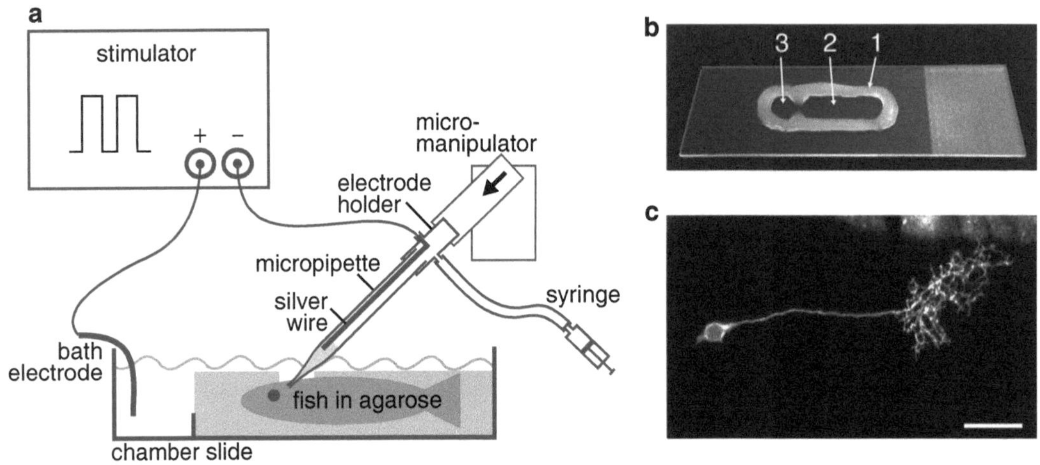

Fig. 2 Pipette electrode electroporation in embryos/larvae. (**a**) Schematic of the setup for PEP. Microscope is not shown for clarity. (**b**) Chamber slide showing (*1*) wall of epoxy resin, (*2*) region in which embryo should be mounted in agarose gel and (*3*) region which should be kept free of agarose for the bath electrode. (**c**) A single neuron in the optic tectum of a live 5 dpf larval zebrafish, which was labeled by electroporation with a plasmid encoding membrane-targeted GFP (pCS2:GAP43-eGFP). Scale bar 20 μm

simple micromanipulator can be used to position the bath electrode or it can be otherwise affixed securely to the microscope stage. EEP electrodes consist of a pair of parallel, sharp stainless steel needles with 0.5 mm diameter, separated by ~1 mm distance (Fig. 1b *left*). The needles are not insulated. Electrodes for IEP consist of a pair of parallel thin Pt electrodes with 25 μm diameter, ~400 μm distance, shank insulated, and tip exposed (modified from FHC Inc. CE2C40; Fig. 1b *right*). The IEP electrodes and the micropipette for injection of reagents are held almost vertically by two micromanipulators (Fig. 1a).

3. *Chamber slide for PEP.* For PEP, a chamber for embryos and larvae (Fig. 2b) is most easily made from a standard glass microscope slide onto which a low solid wall of epoxy resin is laid to form a rectangle of approximately 1 × 2 cm dimensions. The embryo/larva is mounted in agarose gel at one end of the chamber. The other end should be free of agarose to provide a space into which the bath electrode is placed.

4. *Holder for adult zebrafish.* A plastic tube sponge holder for adult zebrafish is used for IEP and EEP (Fig. 1a *right*). The holder can be made by gluing two small pieces of sponge on the internal walls of a ~6 mm diameter plastic tube, e.g., a disposable 2 ml plastic transfer pipette. The upper wall is cut open so that the fish can be placed in between the two pieces of sponge. This holder is placed on a small arc stage to avoid rolling. Fish water containing anesthetics is continuously supplied into the fish's mouth through a gravity-driven perfusion system (200 ml beaker, ca. 50 cm length of soft tubing, small cannula to fit into the mouth).

5. *Stereotactic chamber*. A custom-made stereotactic chamber with lateral stabilizers is used to target small areas in the brain by IEP (Fig. 1a *right*).

2.3 Reagents and Solutions

1. Standard E3 medium: 5 mM NaCl, 0.17 mM KCl, 0.33 mM $CaCl_2$, 0.33 mM $MgSO_4$; for embryos and larvae.
2. Calcium-free Ringer's solution: 119 mM NaCl, 2.9 mM KCl, 5 mM HEPES pH 7.2; or dH_2O.for dissolving DNA constructs or other molecules of interest.
3. Low melting point agarose (1–2% m/v in E3) for mounting embryos and larvae.
4. Anesthetic stock solution: 0.4% (m/v) Tricaine methanesulfonate (MS222), 20 mM Tris–HCl, pH 7. Dilute to a final concentration 0.03% in E3 for embryos and larvae or 0.03% (initial anesthesia) and 0.01% (maintenance) in fresh tank water for adults.
5. DNA constructs or other reagents of interest (*see* **Note 3**). Plasmid constructs are used at a concentration of 1 μg/μl. mRNA is used at approximately 0.4 μg/μl. Fluorescent dyes (e.g., 3 kDa tetramethylrhodamine dextran) can be used at 100 mg/ml.

3 Methods

3.1 External-Electrode Electroporation (EEP)

EEP can deliver DNA, RNA, or dyes into populations of cells. It is mainly used for simple and fast expression of transgenes without high spatial precision [4, 8]. Here we describe the procedure for gene transfer into the dorsal telencephalon of adult zebrafish. Note that electroporation using external electrodes is not equally effective throughout the brain, presumably because the electrical field inside the brain is distorted by surrounding tissues, particularly bones.

1. Anesthetize the adult fish with 0.03% MS-222 and hold the fish body dorsal side up with the sponge holder, while leaving the head free. Place the holder and its stage on a plate under the stereoscope. Put the cannula into the fish mouth (*see* **Note 4**) and perfuse the gills with tank water containing 0.01% MS-222 at a flow rate of <0.5 ml/min.
2. Make a craniotomy over the dorsal telencephalon near the midline using a dentist's drill. A small craniotomy (~250 μm) is sufficient but larger diameters are also possible. Vertically insert the micropipette containing DNA through the craniotomy into the dorsal telencephalon using a micromanipulator. Take care to avoid major blood vessels.
3. At sites 250, 350, and 450 μm below the level of the skull, slowly eject 100–300 nl of DNA solution each time by pressing the syringe for 2–5 min (*see* **Note 5**).

4. Retract the glass pipette and put the pair of parallel sharp steel electrodes at a position so that the cathode is placed on the craniotomy and the anode is located between the ipsilateral eye and the skull (Fig. 1c *top*). Electrodes can be mounted on a micromanipulator or hand-held since precise positioning of electrodes is not critical.
5. Apply a train of electrical pulses (5 × 25 ms, 70 V, 1 Hz, square) with the stimulator or electroporator. The fish should be wet but not be covered by an excess of water (*see* **Note 4**).
6. Remove the fish from the sponge holder and put it back in the home tank. The craniotomy does not need to be sealed. Inspect for fluorescence through the intact skull at successive time points, e.g., 1, 3, 5, 7 days after electroporation of DNA constructs.

3.2 Internal-Electrode Electroporation (IEP)

IEP achieves higher spatial precision than EEP because the electrical field is concentrated between small, closely spaced electrodes inside the brain. The method is therefore suitable to target gene expression to small areas of the brain (~200 μm diameter; *see* **Note 6**). Here we describe IEP in the dorsal posterior zone of the telencephalon (Dp), a higher olfactory brain area in adult zebrafish.

1. Anesthetize the adult fish with 0.03% MS-222, mount it dorsal side up in the sponge holder, insert it into the stereotactic chamber, and position the lateral stabilizers immediately above the eyes. Place the chamber on a tilted stage under a stereomicroscope so that the skull surface is horizontal. Then, put the cannula into the fish mouth (*see* **Note 4**) and perfuse the gills with tank water containing 0.01% MS-222 at a flow rate of <0.5 ml/min.
2. Make a craniotomy above the brain area of interest using a dentist's drill (*see* **Note 7**). The diameter should be sufficient to pass the IEP electrodes (~450 μm).
3. Position the micropipette above the craniotomy using a micromanipulator. Using a second manipulator, place the pair of thin wire electrodes at two sides of the micropipette, almost parallel to it (Fig. 1a *right*). Then insert the glass pipette and the pair of electrodes together into the tissue through the craniotomy, avoiding major blood vessels. Position the micropipettes and the electrodes at the desired depth (400 μm below the level of the skull to target Dp, Fig. 1c *bottom*, *see* **Note 7**).
4. Slowly eject ~70 nl of DNA solution by pressing the syringe for 2–5 min (*see* **Note 5**). Measure the tissue impedance immediately afterwards with the NEPA21 electroporator and select a set of pre-programmed two-phase square electrical pulses based on the impedance. Pulse trains consist of a brief biphasic, high-amplitude poring pulse to permeabilize the plasma membrane followed by a train of longer pulses with lower amplitude and changing polarity to transfer the DNA into the cell (Fig. 1b

right, *see* **Note 1**). Apply the train of electrical pulses 1–2 times (*see* **Note 8**).

5. Repeat **step 4** at different depths if desired (500 and 600 μm for Dp). Then, slowly retract the glass pipette and the electrodes.
6. Remove the fish from the stereotactic chamber and sponge holder, return it to the home tank. Gene expression should commence after a few days (*see* **Note 9**).

3.3 Pipette-Electrode Electroporation (PEP)

PEP has been used to electroporate neurons in various species (e.g., [9]) and the method we describe for zebrafish is similar to that developed by the Cline lab [10]. PEP enables labeling of very small groups of cells (between 1 and approximately 10) with high precision in zebrafish embryos and larvae (*see* Fig. 2c and **Note 10**) [5, 11]. It has been used to deliver a variety of reagents including nucleic acids (plasmid DNA and RNA), morpholinos, and dyes. It could also be used for single cell electroporation in adults as long as the target area is accessible with the glass micropipette.

1. Anesthetize the embryo/larva with 0.03% MS-222. Embed in low-melting temperature agarose in the chamber slide. The agarose should reach the walls of the chamber, which will help to prevent it from floating loose of the slide after gelling. Ideally, leave a region at one end of the chamber free of agarose to allow easy placement of the bath electrode. Orient the embryo/larvae such that the tissue to be electroporated is as easily visible and accessible as possible. After the agarose has gelled, add a small volume of E3 to the chamber so that it covers the agarose and fills the region where the bath electrode is to be placed.
2. Carefully remove a small amount of the agarose above the target tissue with a tungsten needle or ophthalmic scalpel to provide a route for the micropipette to approach the tissue without passing through agarose gel. Take care not to remove too much agarose as the specimen may not be held sufficiently stably.
3. Load a freshly pulled micropipette (*see* **Notes 2** and **11**) with ~2 μl of the reagent to be delivered to the cell(s) (e.g., plasmid solution), ensuring the tip is filled and there are no air bubbles. Attach the micropipette to the electrode holder on the micromanipulator, ensuring the silver wire is in contact with the solution. Optionally, apply a tiny amount of positive air pressure via a 1 ml syringe attached to the side port of the microelectrode holder (move syringe plunger just 0.1 ml) (*see* **Note 12**).
4. Place the chamber slide on the microscope stage and focus on the target region. Then insert the bath electrode into the medium surrounding the mounted embryo/larva (*see* **Note 13**).

5. Set the stimulator to deliver 2 ms square pulses at 200 Hz. For electroporation of nucleic acids, pulse amplitude is 20–50 V. The negative terminal is connected to the microelectrode and the positive terminal is connected to the bath electrode. For fluorescent dyes, pulse amplitude is considerably lower, 2–5 V. Electrode polarity should be selected depending on the net charge of the specific dye that is used. To check that polarity is set correctly, deliver a brief pulse train before inserting the micropipette into the specimen and confirm that fluorescent dye is ejected from the pipette tip.
6. Manipulate the micropipette through the skin of the embryo/larva and into the target tissue. Make sure the optics of your microscope are good enough to see precisely where the micropipette is, so you can accurately target the cells of choice (*see* **Note 14**). To penetrate the skin, first press the micropipette onto the embryo to create a depression. Then use the axis parallel to the direction of the micropipette to make a short, fast advance that breaks through the skin. Ensure that the tip of the micropipette is free of any debris before advancing it towards the target tissue. It helps to enter the embryo/larva some distance from the target so that the entry procedure causes minimal disruption to the target site. Optionally, a sharp tungsten needle can be used to make a hole in the skin before trying to insert the micropipette.
7. Once the micropipette tip is located next to the target cells, deliver two or three trains of stimulation pulses (0.25–0.5 s trains of 2 ms pulses at 200 Hz, approximately 1 s between trains) (*see* **Note 15**). Observe the cells during the stimulation: one should observe a "tip reaction" where the cells change slightly in appearance (cells may change contrast or round up) and a small bolus of reagent can sometimes be seen to leave the micropipette tip. If no tip reaction is observed it is likely the electrical circuit is incomplete and all connections should be checked. When electroporating fluorescent dyes, the success of the electroporation can be monitored directly under epifluorescence illumination (*see* **Note 16**).
8. Wait for 1 or 2 min before carefully withdrawing the micropipette from the target tissue. If the micropipette is moved too quickly after the electroporation, very often cells will adhere to the tip and be pulled out of the tissue. Carefully free the embryo/larva from the agarose using fine forceps and return to fresh E3 to recover. Following electroporation of DNA/RNA, it takes a few hours before protein expression can be observed.

4 Notes

1. Use the NEPA21 electroporator to generate pulse trains consisting of a biphasic, high-amplitude poring pulse followed by a train of lower amplitude transfer pulses. The polarity of the transfer pulses is reversed after 50% of pulses are applied (Fig. 1b *right*). The amplitude of the pulses is adjusted based on tissue impedance, which is measured using the electroporator. High cell survival and expression levels are obtained when the voltage of the poring pulse is set to yield currents of 4–6 mA and the voltage of the transfer pulses is 20% of that of the poring pulse. The pulse duration should be kept short (0.1–1 ms) to avoid accumulation of heat and formation of bubbles near the tip of the IEP electrodes. For tissue with an impedance of 16–20 kΩ, for example, the pulse train consists of one biphasic square pulse of 0.1 ms and 96 V for membrane poring followed by 50 square pulses of 1 ms, 19.2 V and 10 Hz for DNA transfer and another 50 square pulses with the same parameters but opposite polarity. In order to minimize the time delay between impedance measurements and pulse application, predefined pulse trains are stored in the memory of the electroporator. *See* ref. 8 for a table of pulse parameters depending on tissue impedance.
2. For PEP, use a new micropipette at least every three embryos/larvae. Cell debris accumulates on the micropipette tip and this inhibits movement through the tissue and delivery of reagents. The best micropipette is a new one.
3. The intensity, cell-type specificity, and time course of transgene expression may depend on the promoter in an expression construct and other factors such as the transgene, DNA concentration, electrode shapes and positions, and electrical pulse settings [8]. Plasmid preparations should be endotoxin-free. For PEP, we dissolve plasmid DNA (or fluorescent dyes) in dH_2O. We often use constructs based on the pCS2 vector (containing a strong CMV promoter). In this case we assume that focal delivery of the plasmid is the main factor restricting transgene expression to one/few cells. By mixing multiple plasmids in the micropipette, it is possible to introduce these constructs into the same cells, resulting in coexpression or intersectional expression [7, 8].
4. The cannula may be stabilized with a miniature micropositioner. The fish should be kept wet but not be submerged to minimize current flow through the surrounding water. To keep the skin wet, supply tank water manually when necessary.
5. The injection volume can be measured by monitoring the movement of the meniscus inside the capillary. For the 1 mm

diameter capillary (inside diameter, 0.58 mm), 1 μl correspond to ~3.8 mm. Small ticks can be drawn on the capillary as reference for the meniscus movement. A stopcock can be added before the syringe to hold air pressure during injection.

6. The procedure can be applied to target small body areas or tissues as long as the area is accessible. In general, it takes ~30 min to prepare the setup and 15–30 min to electroporate a fish.
7. Dp is located ~400–600 μm below the dorsal skull next to a prominent bone. The stereotactic procedure for Dp was developed based on the zebrafish brain atlas [12]. The craniotomy should be on the suture between the bones over the telencephalon and tectum. On a virtual line from the lateral edge of the telencephalon to the midline, the location should be approximately 25 % from the lateral end. *See* ref. 8 for details. The precise depths of injection points for Dp are adjusted slightly based on the size of each fish.
8. The NEPA21 electroporator is capable of reporting the actual current going through the tissue when applying the electrical pulses. If the reported current is <4 mA, the train of pulse should be applied a second time. *See* **Note 1** for the specification of pulses.
9. In most cases, the expression was completely restricted to a volume of ~200 × 200 × 200 μm^3 within Dp. Expression of transgenes lasted >1 week and depended on the promoter.
10. Electroporating plasmid DNA in embryos of 24 h or older should give a success rate (of labeling ≥ 1 cell) of about 50–75 %.
11. When pulling micropipettes, vary the size of the tip opening according to the number of cells you wish to electroporate. For single cells, use micropipettes with a small tip opening (~1 μm, high resistance); for groups of multiple cells, increase opening size (~3 μm, lower resistance).
12. This ensures the reagent is right at the tip of the microelectrode and counteracts capillary action that draws E3 into the tip. If your electrode holder does not have a side port, do not worry, as this does not appear to be a crucial part of the protocol.
13. Ideally, the bath electrode is positioned with a second micromanipulator. However, it can also be "hooked" into the chamber if it can be affixed securely to the microscope stage. It may help to arrange the bath electrode so that the target cells will lie directly between the micropipette and the bath electrode. This will align the electric fields so they pass through the target cells.
14. Although this protocol suggests using a fixed stage compound microscope with relatively high magnification objectives, if the precise location of the electroporation is not so important, then

this technique can also be used with a dissecting microscope. In this case, one is unlikely to see the exact location of the pipette tip or be able to monitor the tip reaction at the time of stimulation, but successful electroporations are none the less possible.

15. With the Grass SD9, trains can be delivered manually using the repeat mode. Alternatively, pulse trains of more precise duration can be achieved by supplying a TTL signal to the trigger input of the SD9. In either case, carefully monitor the tip reaction to assess the response of the target tissue.
16. Care should be taken when observing dye-filled cells under fluorescence illumination. We find that exposing cells (for example electroporated with tetramethlyrhodamine dextran) to intense excitation light or viewing them for an extended period often results in cell death, presumably due to phototoxicity.

References

1. De Vry J, Martínez-Martínez P, Losen M, Temel Y, Steckler T, Steinbusch HWM, De Baets MH, Prickaerts J (2010) In vivo electroporation of the central nervous system: a non-viral approach for targeted gene delivery. Prog Neurobiol 92:227–244. doi:10.1016/j.pneurobio.2010.10.001
2. Concha ML, Russell C, Regan JC, Tawk M, Sidi S, Gilmour DT, Kapsimali M, Sumoy L, Goldstone K, Amaya E et al (2003) Local tissue interactions across the dorsal midline of the forebrain establish CNS laterality. Neuron 39:423–438
3. Rambabu KM, Rao SHN, Rao NM (2005) Efficient expression of transgenes in adult zebrafish by electroporation. BMC Biotechnol 5:29. doi:10.1186/1472-6750-5-29
4. Hendricks M, Jesuthasan S (2007) Electroporation-based methods for in vivo, whole mount and primary culture analysis of zebrafish brain development. Neural Dev 2:6. doi:10.1186/1749-8104-2-6
5. Bianco IH, Carl M, Russell C, Clarke JD, Wilson SW (2008) Brain asymmetry is encoded at the level of axon terminal morphology. Neural Dev 3:9. doi:10.1186/1749-8104-3-9
6. Kustermann S, Schmid S, Biehlmaier O, Kohler K (2008) Survival, excitability, and transfection of retinal neurons in an organotypic culture of mature zebrafish retina. Cell Tissue Res 332:195–209. doi:10.1007/s00441-008-0589-5
7. Pittman AJ, Gaynes JA, Chien C-B (2010) nev (cyfip2) is required for retinal lamination and axon guidance in the zebrafish retinotectal system. Dev Biol 344:784–794. doi:10.1016/j.ydbio.2010.05.512
8. Zou M, De Koninck P, Neve RL, Friedrich RW (2014) Fast gene transfer into the adult zebrafish brain by herpes simplex virus 1 (HSV-1) and electroporation: methods and optogenetic applications. Front Neural Circuits 8:41. doi:10.3389/fncir.2014.00041
9. Kitamura K, Judkewitz B, Kano M, Denk W, Hausser M (2008) Targeted patch-clamp recordings and single-cell electroporation of unlabeled neurons in vivo. Nat Methods 5:61–67
10. Haas K, Sin WC, Javaherian A, Li Z, Cline HT (2001) Single-cell electroporation for gene transfer in vivo. Neuron 29:583–591
11. Tawk M, Bianco IH, Clarke JD (2009) Focal electroporation in zebrafish embryos and larvae. Methods Mol Biol 546:145–151
12. Wullimann MF, Reichert H (1996) Neuroanatomy of the zebrafish brain: a topological atlas […] […]. Birkhäuser, Basel [u.a.]. doi:10.1007/978-3-0348-8979-7

Chapter 18

Studying Axonal Regeneration by Laser Microsurgery and High-Resolution Videomicroscopy

Yan Xiao and Hernán López-Schier

Abstract

Heterogeneous and unpredictable environmental insult, disease, or trauma can affect the integrity and function of neuronal circuits, leading to irreversible neural dysfunction. The peripheral nervous system can robustly regenerate axons after damage to recover the capacity to transmit sensory information to the brain. The mechanisms that allow axonal repair remain incompletely understood. Here we present a preparation in zebrafish that combines laser microsurgery of sensory axons and videomicroscopy of neurons in multicolor transgenic specimens. This simple protocol allows controlled damage of axons and dynamic high-resolution visualization and quantification of repair.

Key words Zebrafish, Live imaging, Microscopy, Fluorescent protein

1 Introduction

Traumatic neuronal injury disrupts the connection between peripheral sensory organs and the brain. When these connections are not naturally repaired, they invariably result in the perpetual loss of perceptual capabilities [1–3]. Transected or crushed axons degenerate, but neuronal soma is often spared, which has encouraged the search for effectors of axonal regeneration and organ reinnervation [4–11]. Significant progress over the last few decades has advanced our understanding of intrinsic neuron-repair mechanisms and identified several of the underlying causes of failed neurorepair [12–15]. However, many fundamental questions about the capacity of regenerating neurons to reestablish neural function or the existence of critical periods for neural circuit repair remain unanswered [9, 16]. Dynamic multicellular processes that occur in the nervous system should ideally be studied in the context of the whole living animal. Evolutionary proximity makes a mammal the obvious choice as the experimental system and best proxy to human biology. Conversely, invertebrates are valuable because they enormously facilitate work for mutagenic and transgenic screens. However, they suffer from evolutionary divergence that sometimes

Koichi Kawakami et al. (eds.), *Zebrafish: Methods and Protocols*, Methods in Molecular Biology, vol. 1451,
DOI 10.1007/978-1-4939-3771-4_18, © Springer Science+Business Media New York 2016

precludes direct extrapolation of results to vertebrate-specific physiological processes. Small nonmammalian vertebrates fill this gap because they afford the experimental advantages of invertebrates and share with mammals many of the cellular and physiological mechanisms that underlie how organs form, perform, or develop disease. The zebrafish is an exceptional experimental system that compares favorably with other animals used in research by offering many technical advantages [17]. It has a functionally equivalent but anatomically simpler version of the mammalian nervous system that is amenable to manipulation of individual neurons and high-resolution structural and functional imaging in the living animal. Taking advantage of these characteristics, here we present a preparation that allows a simple implementation laser-mediated microsurgery of sensory neurons in the zebrafish lateral-line mechanosensory system [18–21]. This protocol uses 5-day postfertilization (5dpf) zebrafish larvae expressing two spectrally different fluorescent proteins: one by stable transgenesis in the entire population of lateralis neurons and a second by DNA injection in individual neurons. Axon severing is done using a nanosecond ultraviolet laser that is focused on the target neurons through a high-numerical-aperture objective lens mounted on a spinning-disk confocal microscope (Fig. 1).

In particular, the laser ablations were guided by the green fluorescence in neurons of the *Tg[HGn39D]* stable transgenic line (Fig. 2A) [22]. Using a 2-kilobase enhancer element of the zebrafish cntnap2a gene (SILL), we drove expression of a fusion transgene encoding

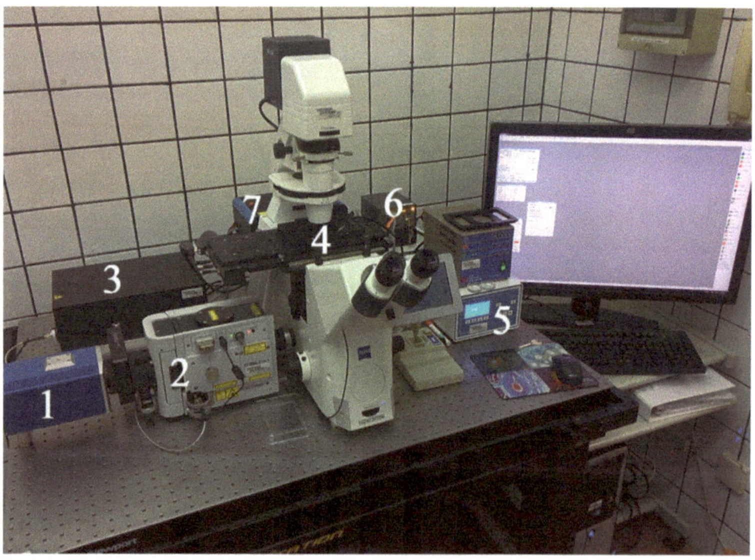

Fig. 1 The setup for laser microsurgery and live imaging. (1) Camera. (2) CSU-X1Confocal. (3) Laser power control. (4) Incubator and stage. (5) Temperature control. (6) UV laser control unit. (7) Dichroic mirror box

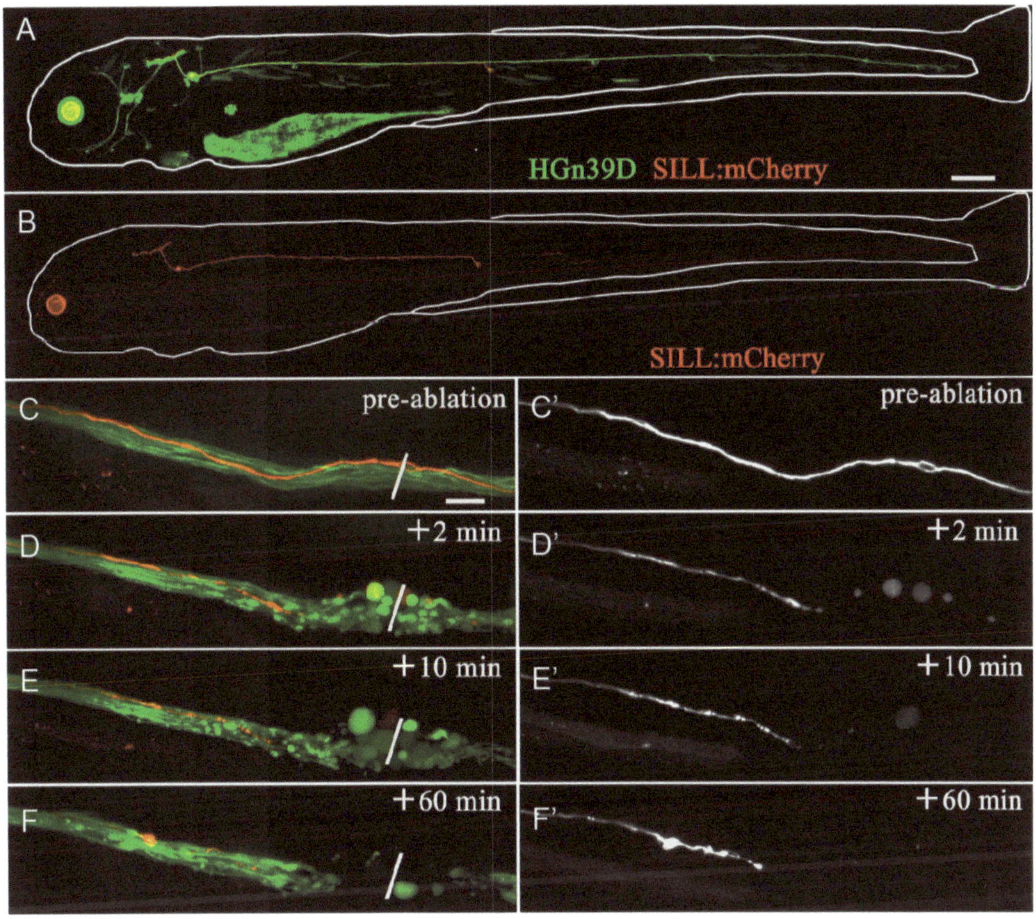

Fig. 2 Laser-mediated ablation of lateralis neurons. (**A**) Marking of individual lateralis neuron using mCherry expression in a 5 dpf by Tg[HGn39D] injected with SILL:mCherry construct. (**B**) Single neuron labeling by mCherry expression in (**A**). (**C**–**F′**) In a typical experiment, the axons are precisely and completely severed with no visible damage to the surrounding tissue. Both the proximal and distal axons experience Wallerian degeneration. White line points to the cutting site. Scale bars: 150 μm (**A**, **B**) and 10 μm (**C**–**F′**)

CAAX-mCherry in lateralis afferent neurons, which has been shown to efficiently highlight these neurons without detrimental effect (Fig. 2B) [20, 23]. Transecting axons led to acute degeneration of the proximal and distal axon ends (Fig. 2C–F′). Peripheral axons began to regenerate within 24 h-post-trauma (hpt) (Fig. 3A–C′), and their regeneration speed can be quantified (Fig. 3D). This simple and robust system has several major advantages, including its very high spatial and temporal resolution and its ability to probe regeneration in a natural context. Several functionalized proteins have been developed recently, including activity indicators and effectors [24–26]. Finally, we have employed a standard microscopy system, but implementation of more powerful microscopes, for example, lattice light-sheet imaging [27, 28], will enable the visualization of fine subcellular events, including particle transport and organellar and cytoskeletal dynamics.

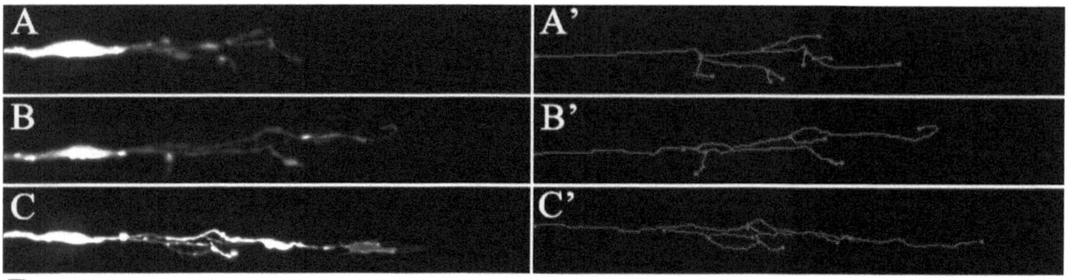

D

Time	Axon length (um)	Axon regrowth rate (um/h)
A'	107.2	/
B'	115.9	/
C'	122.5	/
B'-A'	8.7	8.7
C'-B'	6.6	6.6

Fig. 3 Regenerating neuronal-arbor tracing and calculation of regrowth rate. (**A–C**) Regenerating axon at time points 19, 20, and 21 h after axotomy. (**A′–C′**) Neurites tracing correspond to (**A–C**). (**D**) An example of the axons regrowth rate for three-dimensional quantification. To calculation of the axon regrowth, values obtained at **B**′ are subtracted from the values obtained at **A**′. The speed is the ratio of the length the axon regrows to the amount of time it takes. Here, the amount of time it regrows is 1 h between time point **A**′ and **B**′

2 Materials

2.1 Embryo Culture and Injection

1. 1× E3 embryo medium: 5 mM NaCl, 0.17 mM KCl, 0.33 mM $CaCl_2$, and 0.33 mM $MgSO_4$ in deionized water.
2. Standard equipment for raising fish and collecting eggs: fishing nylon net, breeding tank with dividers, strainer, Petri dishes, incubator.
3. Standard components for embryo injections: agarose plate, glass capillary, needle puller, microloader pipettes, forceps, micromanipulator, microinjector with air pressure, needle holder.
4. Standard components for cloning plasmid DNA and DNA extraction kit.

2.2 Embryo Selection and Mounting

1. 25× Stock solution of MS222 (also called MESAB and 3-aminobenzoic acid ethyl ester): Dissolve 400 mg MS-222 in 97.9 mL of dH_2O, and adjust the pH to 7 by adding 2.1 mL of 1 M Tris (pH 9). Store this solution at 4 °C.
2. Fluorescence stereomicroscope equipped with a filter set for RFP.
3. 1% low-melting point agarose in E3 medium to mount embryos for imaging.
4. Hair loop, plastic Pasteur pipettes, and cover-glass-bottomed culture dishes.

2.3 Laser Axotomy and Analysis of Neuron Response

1. An iLasPulse laser system (Roper Scientific SAS) mounted on a Zeiss Axio Observer spinning-disk confocal microscope that is equipped with a 63× water and oil objective lens.
2. A free cross-platform image processing software: ImageJ/Fiji.

3 Methods

3.1 Marking Single Lateralis Neurons in Zebrafish

1. Prepare plasmid DNA for injection. To obtain mosaic expression in a single afferent neuron, clone DNA expressing the red fluorescent protein mCherry under the control of hsp70 promoter and SILL enhancer to generate hsp70:mCherry-SILL (SILL:mCherry) construct [23].
2. Extract DNA using a plasmid extraction kit following the manufacturer's instructions.
3. One day prior to injections, set up the male and female Casper fish in breeding tanks with dividers in place. The following morning, pull the dividers from several tanks and allow for approximately 10 min of undisturbed mating time.
4. Circular DNA is usually diluted at a concentration of 20 ng/μL. For transient expression in single neurons, 20 pg of the SILL:mCherry construct is injected into embryos at the one- or two-cell stage (*see* Fig. 4 and **Note 1**).
5. The injected eggs are kept at 28.5 °C in an incubator to obtain a standard developmental rate until reaching the desired stage.
6. Screen and select embryos that express mCherry with a stereomicroscope under ultraviolet light for the following experiments. The onset time of expression after injection DNA depends on the promoter/enhancer used in the expression vector (*see* **Note 2**).

3.2 Laser Axotomy and Imaging

1. Add a drop of 1 % low-melting point agarose solution to a cover-glass-bottomed culture dish by a plastic Pasteur pipette. Remove the anesthetized larva directly to the cover-glass-bottomed culture dish, and carefully orientate the larva with a hair loop so that the desired side to image faces the bottom side (*see* **Note 3**).
2. Wait about 5 min to let the agarose solidify at room temperature. Carefully fill the dish with 2.5 mL of E3 medium containing MS-222 working solution, and cover the glass bottom dish with a lid (*see* **Note 4**).
3. Position the prepared larva on the microscope stage, and then use excitation light from a xenon arc lamp passed through the filter to observe red fluorescence. Further identify and sort the sample with a single mCherry-labeled afferent neuron (*see* Fig. 2A, B and **Note 5**).

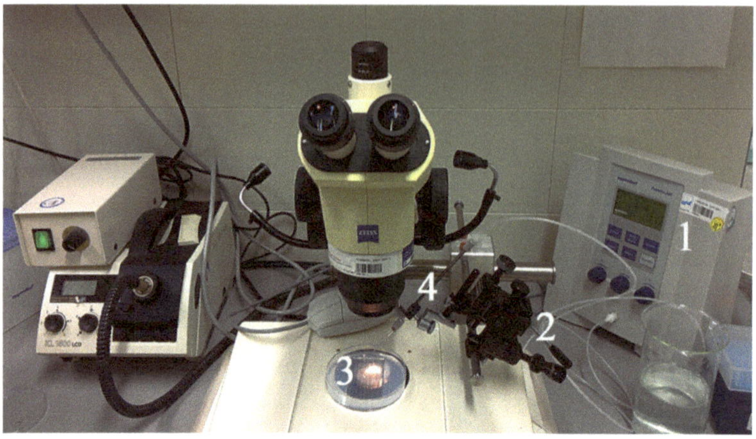

Fig. 4 The basic zebrafish microinjection system. (1) Microinjector with air pressure. (2) Micromanipulator. (3) Agarose plate with one or two-cell stage embryos. (4) Needle holder

4. Image embryos with a confocal microscope with a 20×, 40× dry objective, or a 63× water and oil immersion objective before axotomy (*see* **Note 6**). To reduce photobleaching, it is better to set the lowest laser power that allows visualizing of the single axon. Use the up and down boundaries function to set specific *Z*-stacks for each position. *Z*-stacks are acquired at 1 μm intervals (*see* Fig. 2).
5. Perform axon surgery by using a computer-controlled iLas-Pulse laser system (Roper Scientific SAS) consisting of a pulsed ultraviolet laser (355 nm, 400 ps/2.5 μJ per pulse) (*see* Fig. 1). Transect the axon with a focused laser beam coupled to a spinning-disk inverted microscope. Set the laser power to 35 mW at the sample plane (*see* **Note 7**). To sever axons, draw a region of interest (ROI) over the image of the nerve, and repeatedly apply a train of laser pulses until all fluorescence disappears within that ROI (*see* Fig. 2C–C′ and **Notes 8–10**).
6. After ablating, image the axon to evaluate the ablation result and to closely observe axon degeneration. The image settings are the same as pre-ablation imaging used (*see* Fig. 2C–F′).
7. For the following experiments, remove the larva from agarose, and leave to recover in fresh E3 embryo medium in individual Petri dishes (*see* **Note 11**).

3.3 Quantification of Axonal Regeneration

1. For regeneration experiments, larvae are remounted as the above for time-lapse imaging. A *Z* stack is made every 1 h and overnight time-lapse microscopy recordings are collected. To quantify axonal regeneration rate, we use imaging sessions at 19, 20, and 21 hpt for analysis examples (*see* Fig. 3A–C).

2. Trace the regrown axon using plug-ins in ImageJ/Fiji. Under the file menu, choose open the desired file. For quantification of axon length in three dimensions, use the stacking image instead of the *z*-projection image. Choose *Plugins*→ *Segmentation*→ *Simple Neurite Tracer* (*see* **Note 12**).
3. Select a starting point and a following point to trace the axon. For the same axon branch, trace from one point to the next point until the terminal point.
4. After completing a branch tracing, click somewhere to start a new path and repeat **step 3**.
5. Calculate the axon total length by adding the length of all the tracing paths.
6. To calculate the axon regrowth rate, it is necessary to obtain the axon regenerating length between two time points by subtracting the axon length of previous time point from the length of a specific time point. Then, the speed is the ratio of the distance the axon regenerates to the amount of time it regenerates (*see* Fig. 3A′–D).

4 Notes

1. It may be that the plasmid DNA itself determines the quantity of injected DNA. At the beginning of the experiment, it is suggested that injections with different amount of DNA should be tested to decide the proper quantity of DNA for obtaining embryos with single-labeled neurons.
2. As it is not easy to determine whether a single neuron is marked under a stereomicroscope, it is better to keep all the positive samples and screen later under a confocal microscopy and discard embryos with more than one neuron marked. Moreover, it is very important to write notes about which side of embryos is positive at this step, and it will facilitate to mount correctly the embryos in a ventral-up or dorsal-up position.
3. If larvae die during the mounting procedure, it may be caused by agarose with too high temperatures. This happens when the agarose is prepared immediately before mounting of the embryos. Therefore, wait for about 30 min for agarose to cool down.
4. The cover lid is utilized to close the embryo so it does not dry out and to act as the interface between the larva and the microscope objective. Make sure that there are no air bubbles in the imaging chamber after it is sealed; otherwise, they will interfere with imaging in the bright field.
5. If most samples are labeled by more than one afferent neuron, you try to inject DNA with lower concentrations. If none of embryos express the red marker in the afferent neurons, you

inject more DNA next time. In addition, check carefully that you are using a correct promoter/enhancer to express the fluorescent marker in the neurons of interest.

6. Before axotomy, images are taken to ensure that the axons to be cut do not have any abnormalities and as a reference to align to unperturbed control samples to compare with the images of experimental axons after severing. The decision about where along the axon axotomy is performed (near the perikaryon or near the peripheral arborization) depends on the specific aims and experiments.
7. Prior to severing, it is necessary to test different ablation laser powers and determine the optimal power to sever axon without obviously damaging surrounding tissue. If the sample is exposed to the high-power ablation beams, the damage to surrounding tissue can be observed in the bright field. In this case, the ablation requires less power. Once the optimal power has been set, it rarely needs to be adjusted.
8. When ablating the samples, simply parking the ablation beam over the target area may not always be effective. Scanning the laser beam over the ROI within a narrow range of focus might obtain better outcomes.
9. Instant loss of fluorescence does not mean that the axon has been transected, as the fluorescent marker might be only photobleached. Therefore, after performing the ablation, the axon needs to be checked whether it is completely severed in which case axonal fluorescence does not refill the ROI and axon degeneration occurs.
10. It is very important to consider eye safety and avoid directly looking at the laser beam throughout this step.
11. Be careful when you recover the larva from agarose, because the larvae are very vulnerable at this stage. Gently draw the embryo into the pipette and transfer the embryo onto a plate with fresh E3 medium. The larva should start to swim again within an hour.
12. If ImageJ does not recognize the length of your image, then it is very important to calibrate the image by setting the scale using an existing scale bar in your image.

5 Ethical Considerations

All procedures on live specimens should be performed under guidelines and approved protocols by local and general/federal agencies. The method and experiments described here were performed according to EU Directive 2010/63/EU, under which an ethics committee-approved animal protocol is required for freely feeding zebrafish larvae (after day 5.2 postfertilization). Our experiments were done using zebrafish younger than 5 days.

Acknowledgments

We thank the Central Animal Facility of the HMGU (Munich, Germany) for expert animal care. This work was supported by the Helmholtz Gemeinschaft and by a program grant of the Human Frontier Science Program (RGP0033/2014) to HL-S.

References

1. Schwab M, Strittmatter SM (2014) Nogo limits neural plasticity and recovery from injury. Curr Opin Neurobiol 27:53–60
2. Silver J, Schwab ME, Popovich PG (2014) Central nervous system regenerative failure: role of oligodendrocytes, astrocytes, and microglia. Cold Spring Harb Perspect Biol 7:a020602
3. Small SL, Buccino G, Solodkin A (2014) Brain repair after stroke—a novel neurological model. Nat Rev Neurol 9:698–707
4. Lutz BA, Barres BA (2014) Contrasting the glial response to axon injury in the central and peripheral nervous systems. Dev Cell 28:7–17
5. Ciba-Foundation (1991) Regeneration of vertebrate sensory receptor cells. Wiley, Chichester, UK
6. Cramer SC (2010) Brain repair after stroke. N Engl J Med 362:1827–1829
7. Deshmukh VA, Tardif V, Lyssiotis CA et al (2013) A regenerative approach to the treatment of multiple sclerosis. Nature 502: 327–332
8. Liu K, Tedeschi A, Park KK, He Z (2011) Neuronal intrinsic mechanisms of axon regeneration. Ann Rev Neurosci 34:131–152
9. Mar FM, Bonni A, Sousa MM (2014) Cell intrinsic control of axon regeneration. EMBO Rep 15:254–263
10. Neumann S, Bradke F, Tessier-Lavigne M, Basbaum AI (2002) Regeneration of sensory axons within the injured spinal cord induced by intraganglionic cAMP elevation. Neuron 34:885–893
11. Takeoka A, Vollenweider I, Courtine G, Arber S (2014) Muscle spindle feedback directs locomotor recovery and circuit reorganization after spinal cord injury. Cell 159:1626–1639
12. Bhatt DH, Otto SJ, Depoister B, Fetcho JR (2004) Cyclic AMP-induced repair of zebrafish spinal circuits. Science 305:254–258
13. Bradke F, Fawcett JW, Spira ME (2012) Assembly of a new growth cone after axotomy: the precursor to axon regeneration. Nat Rev Neurosci 13:183–193
14. Kilmer SL, Carlsen RC (1984) Forskolin activation of adenylate cyclase in vivo stimulates nerve regeneration. Nature 307:455–457
15. Reimer MM, Norris A, Ohnmacht J et al (2013) Dopamine from the brain promotes spinal motor neuron generation during development and adult regeneration. Dev Cell 25:478–491
16. Hubener M, Bonhoeffer T (2014) Neuronal plasticity: beyond the critical period. Cell 159:727–737
17. Santoriello C, Zon LI (2012) Hooked! Modeling human disease in zebrafish. J Clin Invest 122:2337–2343
18. Bourgeois F, Ben-Yakar A (2008) Femtosecond laser nanoaxotomy properties and their effect on axonal recovery in C. elegans. Opt Express 16:5963
19. Ghysen A, Dambly-Chaudière C (2008) The lateral line microcosmos. Genes Dev 21: 2118–2130
20. Pujol-Martí J, Faucherre A, Aziz-Bose R, Asgharsharghi A, Colombelli J, Trapani JG, López-Schier H (2014) Converging axons collectively initiate and maintain synaptic selectivity in a constantly remodeling sensory organ. Curr Biol 24:2968–2974
21. Pujol-Martí J, López-Schier H (2013) Developmental and architectural principles of the lateral-line neural map. Front Neural Circuits 7:47

22. Pujol-Martí J, Baudoin JP, Faucherre A, Kawakami K, López-Schier H (2010) Progressive neurogenesis defines lateralis somatotopy. Dev Dyn 239:1919–1930
23. Pujol-Martí J, Zecca A, Baudoin JP et al (2012) Neuronal birth order identifies a dimorphic sensorineural map. J Neurosci 32:2976–2987
24. Xiao Y, Tian W, López-Schier H (2015) Optogenetic stimulation of neuronal repair. Curr Biol 25(22):R1068–9. doi: 10.1016/j.cub.2015.09.038
25. Fenno L, Yizhar O, Deisseroth K (2011) The development and application of optogenetics. Ann Rev Neurosci 34:389–412
26. Stierl M, Stumpf P, Udwari D et al (2011) Light modulation of cellular cAMP by a small bacterial photoactivated adenylyl cyclase, bPAC, of the soil bacterium Beggiatoa. J Biol Chem 286:1181–1188
27. Chen BC, Legant WR, Wang K et al (2014) Lattice light-sheet microscopy: imaging molecules to embryos at high spatiotemporal resolution. Science 346:1257998
28. Xiao Y, Faucherre A, Pola-Morell L et al (2015) High-resolution live imaging reveals axon-glia interactions during peripheral nerve injury and repair. Dis Model Mech 8:553–564

Chapter 19

In Vivo Whole-Cell Patch-Clamp Recording in the Zebrafish Brain

Rong-wei Zhang and Jiu-lin Du

Abstract

Zebrafish (*Danio rerio*) is a newly emerged vertebrate animal model with a conserved gross architecture of the brain and a rich repertoire of behaviors. Due to the optical transparency and structural simplicity of its brain, larval zebrafish has become an ideal in vivo model for dissecting neural mechanisms of brain functions at a whole-brain scale based on a strategy that spans scales from synapses, neurons, and circuits to behaviors. Whole-cell patch-clamp recording is an indispensable approach for studying synaptic and circuit mechanisms of brain functions. Due to the small size of neurons in the zebrafish brain, it is challenging to get whole-cell recordings from these cells. Here, we describe a protocol for obtaining in vivo whole-cell patch-clamp recordings from neurons in larval zebrafish.

Key words Zebrafish larvae, Whole-cell patch-clamp recording, Perforated patch-clamp recording, Neuron

1 Introduction

Patch-clamp recording has become one of the most powerful methods in the research fields of physiology and neuroscience, since Drs. Bert Sakmann and Erwin Neher developed this technique at the end of the 1970s [1]. It is indispensable not only for studying activities of single ion channels in outside-out or inside-out mode but also for measuring macroscopic ion currents of the entire cell in whole-cell mode [2]. Whole-cell patch-clamp recording provides a high signal-to-noise measurement of synaptic activities, sensory responses, and circuit dynamics that are relevant to neural function and animal behavior [3–5].

As an elegant vertebrate animal model, zebrafish has been widely used in neuroscience research during the past decade [6, 7]. Its nervous system contains many conserved regions and structures found throughout vertebrates (Fig. 1), including the retina, olfactory bulb, habenula, optic tectum, cerebellum, hindbrain, and spinal cord [8, 9]. Both larval and adult fish exhibit a diversity of

Koichi Kawakami et al. (eds.), *Zebrafish: Methods and Protocols*, Methods in Molecular Biology, vol. 1451, DOI 10.1007/978-1-4939-3771-4_19, © Springer Science+Business Media New York 2016

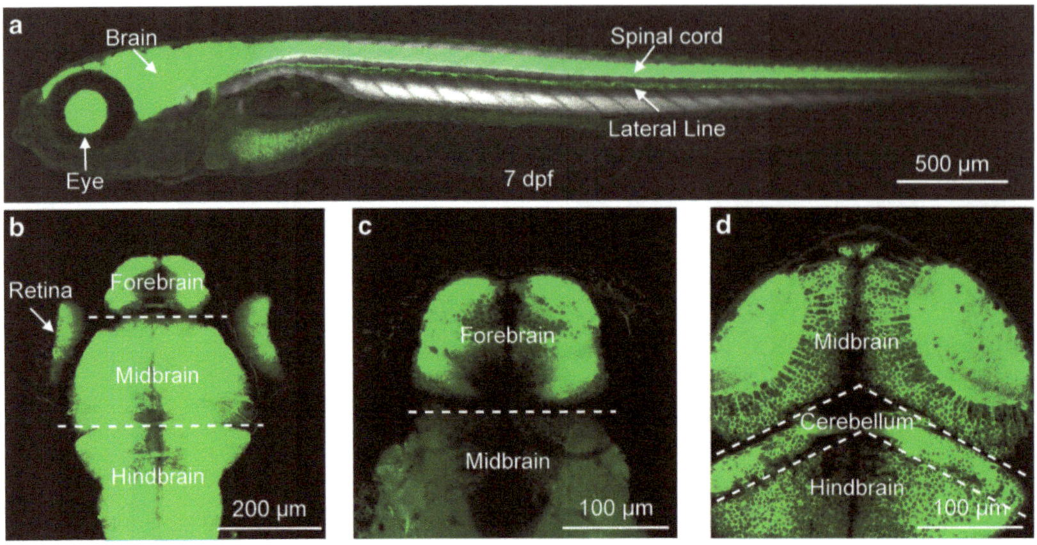

Fig. 1 The central nervous system of the larval zebrafish brain. (**a**) Merged confocal and bright-field image of an intact transgenic Tg(HuC:GFP) zebrafish larva at 7 days postfertilization (dpf), in which green fluorescent protein is specifically expressed in neurons. The fish was embedded with the left eye up. (**b**) In vivo projected confocal image of a 7-dpf Tg(HuC:GFP) zebrafish larva showing main areas of the larval brain. The fish was embedded dorsal up. (**c**, **d**) Single-plane confocal images showing individual neurons in the forebrain, midbrain, cerebellum, and hindbrain

conserved behaviors, such as the optokinetic response (OKR), prey capture, and startle/escape behaviors [10, 11]. In comparison with rodents, the brain of zebrafish is much smaller (for larva aged at 1 week, <0.4 mm in thickness, and ~0.5 mm in both length and width; for adult, <2 mm in thickness, and 2–4 mm in both length and width) and has fewer neurons (~10^{4-5} in larvae, ~10^6 in adult), which facilitates the examination of the development and function of neuronal circuits [8, 12]. Combining the advantage of a translucent brain at early larval stages and the development of advanced imaging tools, some research groups have performed calcium imaging of large-scale neuronal activities of intact larval fish to map behavior- or state-relevant brain regions or neural circuits [13–16]. Calcium imaging measurements of neuronal activity exhibit lower signal-to-noise ratio and slower kinetics than electrophysiological recordings [17]. The technique of whole-cell patch-clamp recording is indispensable to the examination of ion channel properties, subthreshold synaptic responses, or neuronal spike firing, all of which provide necessary information for dissecting synaptic mechanisms. Recently, cell-attached or whole-cell recordings were applied by several groups (including ours) in different tissues of zebrafish, in vitro or in vivo, including the retina, olfactory bulb, optic tectum, hindbrain, and spinal cord [18–23]. However, a detailed description of the technique in intact fish is still lacking.

Here, we summarize a protocol for in vivo whole-cell patch-clamp recording in the brain of intact zebrafish larvae. We first list some reagents and equipment necessary for the recording and then describe relevant dissection methods and patching techniques in detail. Finally, we also provide some critical technical suggestions and notes about common problems.

2 Materials

2.1 Reagents

1. Zebrafish: Wild-type and some transgenic lines.
2. Hank's solution: 137 mM NaCl, 5.4 mM KCl, 0.25 mM Na_2HPO_4, 0.44 mM KH_2PO_4, 1.3 mM $CaCl_2$, 1.0 mM $MgSO_4$, 4.2 mM $NaHCO_3$; adjusted to pH 7.2 by NaOH.
3. External solution: 134 mM NaCl, 2.9 mM KCl, 2.1 mM $CaCl_2$, 1.2 mM $MgCl_2$, 10 mM HEPES, and 10 mM glucose; adjusted to pH 7.8 by NaOH, 290 mOsmol/l. Store at 4 °C.
4. Low-chloride internal solution: 100 mM K-gluconate, 10 mM KCl, 2 mM $CaCl_2$, 2 mM Mg-ATP, 2 mM Na-GTP, 10 mM HEPES, and 10 mM EGTA; adjusted to pH 7.35 by KOH, 280 mOsmol/l. The internal solution should be filtered by using a 0.2-μm filter through a syringe to remove small particles and debris, then divided into 100 μl per tube, and stored at −20 °C (*see* **Note 9**).
5. High-chloride internal solution: 110 mM KCl, 6 mM NaCl, 2 mM $CaCl_2$, 2 mM $MgCl_2$, 10 mM HEPES, and 10 mM EGTA, adjusted to pH 7.4 by KOH, 270 mOsmol/l.
6. α-Bungarotoxin: Dissolve 1-mg α-bungarotoxin in 1-ml external solution to the final concentration 1 mg/ml. The solution is then divided into 20 μl per tube and stored at −20 °C.
7. Low-melting point agarose: Dissolve 0.1-g agarose in 10-ml external solution in a glass bottle to the final concentration 1 %, and then microwave until agarose is dissolved. Put the agarose bottle in water bath set at 39 °C (*see* **Note 2**).
8. Gramicidin: Dissolve 100-μg gramicidin in 100-μl DMSO to the final concentration 1 mg/ml, then aliquot the gramicidin solution into 1 μl per tube, and store at −20 °C. The stock solution should be discarded after more than 1 month because of decreased perforation efficiency. When performing gramicidin-perforated patch recording, add 100-μl high-Cl^- internal solution into one gramicidin-containing tube. The working concentration of gramicidin is 10 μg/ml (*see* **Note 17**).

2.2 Equipment for Preparation and Dissection

1. Glass-bottomed recording chamber: A hole of about 10-mm diameter is punched at the center of a 35-mm Petri dish. Then, a 20×20-mm glass slide is then glued at the back of the dish.
2. Forceps: FST, Dumont 5#.
3. Pipette holder: Narishige, H-7.
4. Dissection microelectrode: Used microelectrodes after patch-clamp recording are fine.
5. Dissection stereomicroscope: Olympus, SZX16 or SZ61.
6. Osmometer: Advanced Instruments, Model 3300.
7. pH meter: Mettler Toledo, Delta 320.
8. Electronic balance: Mettler Toledo, AL104.
9. Incubator set at 28 °C.

2.3 Equipment for Electrophysiology (Fig. 2)

1. Vibration isolator table: Meritsu, ADZ-0806.
2. Faraday cage: Custom-made.
3. X-Y stage: Custom-made.
4. Upright infrared DIC microscope: Olympus, BX51WI.
5. Objectives: Olympus Mplan 5X/0.10 and UMPlanFI/IR 60X/0.90w.
6. Amplifier: HEKA (triple patch-clamp EPC10), including electrode holders, headstages, amplifiers, and analog-to-digital converters.
7. Stimulator: A.M.P.I., Master 8.
8. X-Y translator: Siskiyou, MXMS-100.
9. Three-dimensional (3-D) micromanipulator: Burleigh, PCS-5200.
10. CCD camera: Dage-MTI, IR-1000.
11. Video monitor: SUNSPO, SP-717.
12. Electronic shutter: Uniblitz, LS6Z2.
13. Computers for electrophysiological recordings and light stimulation.
14. Perfusion system: Peristaltic pump, vacuum pump, and plastic tubing.
15. Glass microelectrode puller: Sutter Instrument, P-97.
16. Borosilicate glass capillaries: Sutter Instrument, cat. No. BF100-58-10. Recording microelectrodes have a tip opening of ~1 μm and a resistance in the range of 20–30 MΩ.
17. Visual stimulation: Portable projector (ASK, L1030) or LED, which is controlled by a computer.

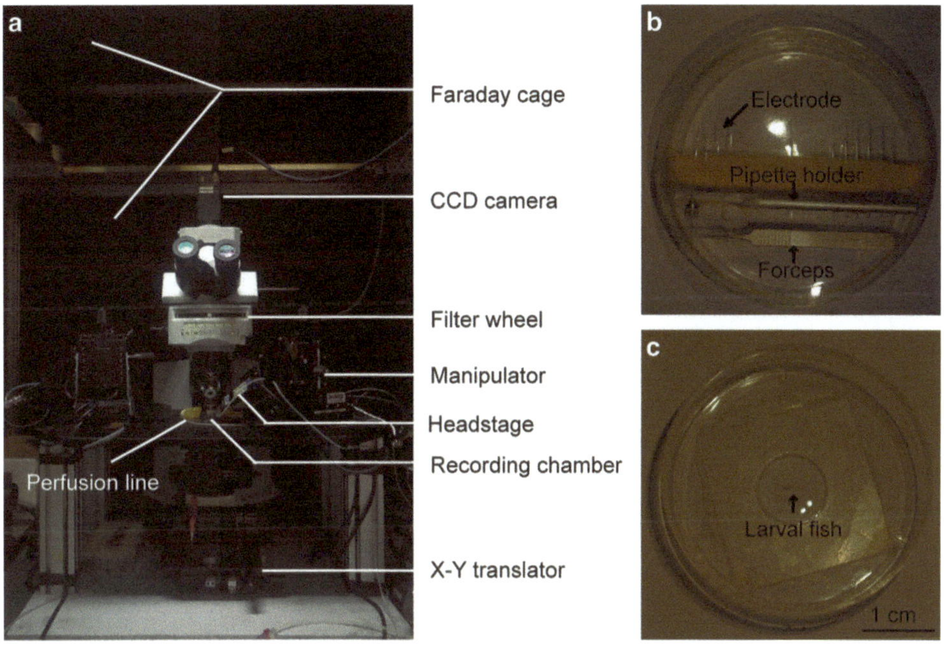

Fig. 2 Equipment for electrophysiological recording and preparation. (**a**) Electrophysiological recording setup, which is inside a Faraday cage. (**b**) Dissection tools including forceps, pipette holder, and electrode. (**c**) Custom-made recording chamber with a larval fish embedded in agarose

3 Methods

3.1 Preparation and Dissection

1. Collect several zebrafish larvae with a normal body shape and displaying active locomotion within one water drop in a 35-mm Petri dish.
2. Suck off the water using a pipette (Eppendorf) and add 20-μl α-bungarotoxin for ~15 min to block nicotinic acetylcholine receptors (nAChRs) to paralyze the larvae (*see* **Note 1**).
3. After paralysis, bathe the larvae in the external solution immediately. Select one larva and transfer it to a custom-made glass-bottomed chamber with a plastic transfer pipette.
4. Suck dry the solution around the larva, and add ~200 μl 1 % low-melting point agarose. Adjust the larva quickly to the desired body position (e.g., dorsal up when you are going to record cells in the optic tectum) with a pair of fine forceps under an upright stereomicroscope. After letting agarose set for ~5 min, immerse the larva in 3–4-ml external solution (*see* **Note 2**).
5. Mount a dissection microelectrode in the pipette holder. Cut off the agarose above the targeted brain area using the microelectrode, and then suck out the displaced agarose with a plastic pipette. Puncture a small hole using the microelectrode in the skin above the ventricle of the targeted area (e.g., optic

tectum) for the passing of recording microelectrodes. Sometimes, the skin along the midline of two tectal hemispheres may be incised to expose the brain (*see* **Notes 3** and **4**).

6. To dissect the retina, embed the larva in agarose with an eye up. Remove the cornea and lens of the upward eye by using a microelectrode after cutting off agarose. To expose retinal tissues, gently tear out the inner limiting membrane from the surface of the retina using the microelectrode. In general, recordings are easier from cells in the central retina.
7. Suck out displaced tissues or blood cells by using a microelectrode with a tip opening of ~30 μm.

3.2 Whole-Cell Patch-Clamp Recording

1. After the dissection, transfer the larva to the electrophysiological recording setup, and check whether the dissection condition is good enough for recording under a high-magnification objective. Before recording, perfuse the larva for at least 30 min with external solution at the speed of ~2 ml per min (*see* **Notes 5**–7).
2. Immerse the ground wire in the bath solution. Fill a recording microelectrode with internal solution using a 1-ml syringe and a custom-made plastic needle. For gramicidin-perforated whole-cell recording, tip-fill the microelectrode with gramicidin-free internal solution and then back-fill with high-Cl^- internal solution containing 10 μg/ml gramicidin (*see* **Notes 8** and **9**).
3. Insert the microelectrode into the electrode holder mounted on the amplifier headstage. In order to reduce the electronic noise, the tip of the chloride silver wire should be just immersed in the internal solution.
4. To avoid contamination when the electrode tip crosses the air-solution interface, apply positive pressure through a tube attached to the holder by using mouth or a 5-ml syringe. To record cells on the surface of the brain, the positive pressure should be as low as ~50 mbar. To record cells deep in the brain, the positive pressure is about 150–200 mbar (*see* **Note 10**).
5. Position the microelectrode tip above the targeted brain area under a low-magnification objective using coarse adjustment of 3-D micromanipulators. Under a high-magnification objective, adjust the position of the electrode to approach the targeted area, and then select a healthy cell to record (*see* **Notes 11–13**).
6. Null the offset potential of the electrode and apply a 5-mV, 5-ms test pulse. Approach the targeted cell slowly with the electrode using the fine adjustment until a slight dimple appears on the cell surface. Release the positive pressure immediately so that the test current will become very small due to the formation of

high-resistance seal. Sometimes a gentle but consistent negative pressure may be applied to achieve the formation of giga-ohm seal (*see* **Notes 14** and **15**).

7. Set the holding potential to –60 mV and null the fast capacitance currents. Wait for ~1 min to let the seal around the electrode tip tighter.

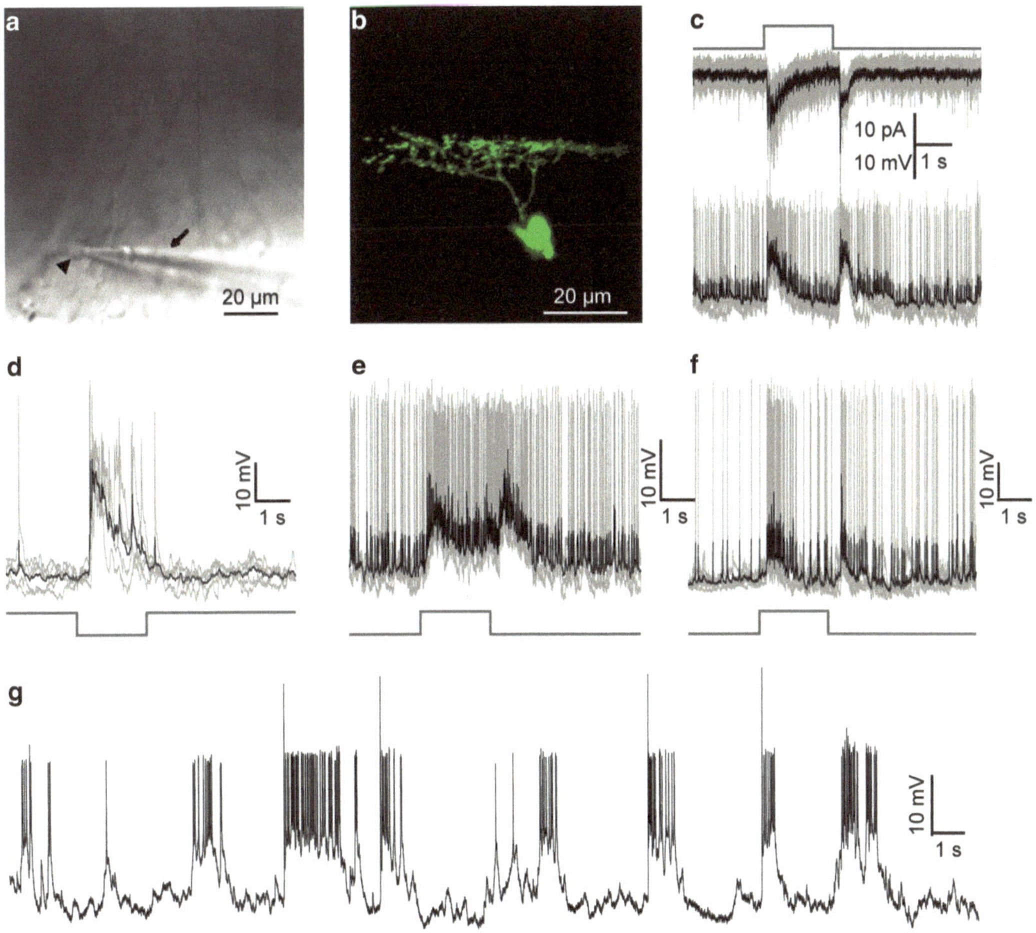

Fig. 3 Examples of in vivo whole-cell recordings of neurons in intact zebrafish larvae at 4–6 dpf. (**a**) Infrared DIC image of the ganglion cell layer surface of the retina in a 5-dpf AB wild-type larval zebrafish. Whole-cell patch-clamp recording was performed on a retinal ganglion cell (RGC). *Arrow*, recording microelectrode; *arrowhead*, recorded cell. (**b**) A typical example showing the morphology of an RGC, which was imaged after loading 1 % Lucifer yellow into the cell through the whole-cell recording microelectrode. (**c**) Representative light-evoked responses (LERs) of an ON-OFF RGC in response to a 2-s flash (*top*) when the cell was held at the equilibrium potential of Cl^- (–60 mV, middle) or recorded at current-clamp mode (*bottom*). (**d–f**) Representative examples showing OFF LERs of a forebrain neuron (**d**), ON-OFF LERs of a midbrain neuron (**e**), and ON-OFF LERs of a cerebellar neuron (**f**), respectively. The cells were recorded at current-clamp mode. The *gray* and *black* traces indicate five overlapped trials and the average, respectively. (**g**) Representative example showing spontaneous spiking and bursting activities of a cerebellar neuron

8. Apply several brief suction pulses to rupture the patch beneath the tip of the microelectrode until the capacitive transient appears. These transients arise from the membrane capacitance (C_m) of the recorded cell, which should be nulled later (*see* **Note 16**).
9. Verify the quality of whole-cell patch-clamp condition based on series resistance (R_s) and resting membrane potential (V_m). Due to the small size of zebrafish neurons, their input resistance is usually in the range of 2–10 GΩ. Recording can be accepted if the series resistance is below 100 MΩ and varied <20% during experiment. The V_m of neurons is usually more negative than –45 mV, whereas the V_m of glial cells is about –90 mV (*see* **Notes 18** and **19**).
10. Start experiment and collect data. Examples of in vivo whole-cell recordings of neurons in intact zebrafish larvae are shown in Fig. 3.

4 Notes

1. Preparation and dissection are two critical factors for a successful recording. Paralyzed zebrafish larvae should be used within 2–3 h. Strong and regular heartbeat and fast blood flow indicate that the preparation is in good condition. Discard the preparation if the larval tail is curved or the brain becomes dark.
2. Low-melting point agarose should be used within 3 days after preparation. Dirty agarose is fatal for larvae and leads to the reduction of blood flow and bad condition of brain cells.
3. Clear agarose near the targeted brain area carefully using the dissection microelectrode, because remaining agarose debris may disturb your dissection and recording.
4. Puncture the skin a little far away from your targeted region along the brain midline using the dissection microelectrode so that the targeted region will be left intact. The skin is sometimes incised a little more along the brain midline to facilitate the passing of recording electrodes. Do not insert the dissection microelectrode too vigorously since it may damage the brain.
5. Check the preparation's condition under a high-magnification objective. The brain of preparations that are in good condition should be transparent, exhibit fast blood flow, and have few damaged cells. Practice again and again until the condition of preparations is as good as desired. Sometimes vessels just beneath the skin may be punctured during dissection. Clean and suck out leaked blood cells carefully using a microelectrode before recording.

6. Experiments are performed under room temperature 22–26 °C. The condition of the preparation or whole-cell recording may become bad if the temperature is out of the optimal range.
7. Continuous perfusion of external solution is helpful for improving fish condition and prolonging recording time, because it provides a stable pH and ionic environment and prevents the accumulation of metabolites released by the preparation. Oxygenated external solution is unnecessary. Rinse the perfusion lines with 95% ethanol and deionized water in sequence after experiment.
8. Borosilicate glasses with filament are generally used to facilitate the entry of internal solution into the tip of recording microelectrodes. The microelectrode should be pulled freshly and stored in a closed dish to prevent the contamination of the tip by dust particles. Be sure there is no bubble in the electrode tip after filling of internal solution.
9. Internal solution should be filtered in advance through a 0.2-μm filter to prevent debris and particles from entering the electrode and blocking its tip. Use a fresh tube of internal solution each day. Do not reuse the internal solution and custom-made filling needle.
10. Be sure that there is no pressure leakage in the plastic tubing connected to the electrode holder. The stability of the positive pressure is very important for the formation of giga-ohm seal. Check the pressure value before advancing the electrode into the recording chamber through a barometer connected to the tubing. Clear outflow of internal solution from the microelectrode tip can be observed, when it blows away tissue debris around the tip of microelectrode.
11. Lower the microelectrode quickly and position the tip near the dissected hole. Avoid touching displaced tissues or blood cells when the tip gets into the hole. Slowly advance the tip orthogonally toward the targeted cells using the fine adjustment of micromanipulators. Avoid breaking blood vessels when getting into deep brain regions, because it will not only clog the electrode tip but also cause massive hemorrhage and bad brain condition.
12. Monitor the change of microelectrode resistance by software when advancing the microelectrode. If the positive pressure is proper, the resistance should be ~20–30 MΩ during approaching. Discard the electrode if the resistance increases to hundreds of MΩ or even several GΩ because of blockade of the tip. When the resistance is increasing, you can sometimes retreat the tip a little backward and advance it through another pathway to avoid debris.

13. Choose a healthy cell that looks clean and smooth with a clear membrane edge and oval shape. Discard dark and round cells with a rough appearance and granular particles in the cytoplasm. If the giga-seal is lost, it is most likely because the recorded cell is unhealthy or the tip of the electrode is clogged.
14. It is important to advance the electrode and obtain a giga-seal as fast as possible after the electrode tip enters the bath solution. The success rate of giga-seal formation is inversely proportional to the time the tip is immersed in the bath.
15. Be sure that the optics of the microscope is well adjusted to visualize the targeted cell and the electrode tip clearly. When the tip and cell membrane are clearly focused at the same *z*-depth, advance the tip until the cell membrane exhibits a dimple. A giga-seal usually forms when you release positive pressure, or you may sometimes apply a little negative pressure.
16. In our experience, the most effective method to rupture cell membrane is to apply a negative pressure by mouth. Short voltage pulses, or "zaps," as provided by some amplifiers may often cause cell damage and rundown and are rarely used.
17. Large insoluble particles may interfere with the formation of a giga-seal, particularly when you perform gramicidin-perforated whole-cell recording and use internal solution containing fluorescent dyes. Make sure that the drugs or dyes are vibrated, sonicated, and centrifuged before recording. Lower the positive pressure to ~50 mbar to avoid blockade of electrode tip.
18. Avoid long exposure to fluorescence light sources when recording GFP-expressing cells. Both long-pass GFP filters for emitted fluorescence and DIC light are useful for visualizing the edge and shape of GFP-expressing cells. It is very critical for successful formation of a giga-seal to make sure of the appearance of a fluorescent dimple when the tip touches on the fluorescent cell. Fluorescent dye may be sometimes added in internal solution for the confirmation of tip position.
19. The gramicidin-containing solution should be used within 2–3 h after thawing. No membrane rupture by negative pressure is needed when performing gramicidin-perforated recording. The formation of gramicidin pores on cell membrane usually takes 5–10 min. During this process, R_s will decrease gradually from several GΩ to less than 200 MΩ. The cells are discarded if the measured reversal potential for GABA-induced currents tends more and more depolarized.

Acknowledgment

This work was supported by grants of the Strategic Priority Research Program from the Chinese Academy of Sciences (XDB02040003), "973" Programs from the Ministry of Science and Technology of China (2011CBA00400, 2012CB945101), the National Outstanding Young Scientist Program from the National Natural Science Foundation of China (31325011), and the Shanghai Subject Chief Scientist Program and Shanghai Natural Science Foundation from the Science and Technology Commission of Shanghai (14XD1404100, 13ZR1446800).

References

1. Hamill OP, Marty A, Neher E et al (1981) Improved patch-clamp techniques for high-resolution current recording from cells and cell-free membrane patches. Pflugers Arch 391:85–100
2. Sakmann B, Neher E (1995) Single-channel recording, 2nd edn. Plenum Press, New York
3. Lee AK, Manns ID, Sakmann B et al (2006) Whole-cell recordings in freely moving rats. Neuron 51:399–407
4. Domnisoru C, Kinkhabwala AA, Tank DW (2013) Membrane potential dynamics of grid cells. Nature 495:199–204
5. Chorev E, Epsztein J, Houweling AR et al (2009) Electrophysiological recordings from behaving animals—going beyond spikes. Curr Opin Neurobiol 19:513–519
6. Bradbury J (2004) Small fish, big science. PLoS Biol 2:E148
7. Friedrich RW, Genoud C, Wanner AA (2013) Analyzing the structure and function of neuronal circuits in zebrafish. Front Neural Circuits 7:71
8. Rupp B, Wullimann MF, Reichert H (1996) The zebrafish brain: a neuroanatomical comparison with the goldfish. Anat Embryol 194:187–203
9. Wullimann MF, Mueller T (2004) Teleostean and mammalian forebrains contrasted: evidence from genes to behavior. J Comp Neurol 475:143–162
10. Portugues R, Engert F (2009) The neural basis of visual behaviors in the larval zebrafish. Curr Opin Neurobiol 19:644–647
11. Neuhauss SC (2003) Behavioral genetic approaches to visual system development and function in zebrafish. J Neurobiol 54: 148–160
12. Friedrich RW, Jacobson GA, Zhu P (2010) Circuit neuroscience in zebrafish. Curr Biol 20:R371–R381
13. Ahrens MB, Li JM, Orger MB et al (2012) Brain-wide neuronal dynamics during motor adaptation in zebrafish. Nature 485:471–477
14. Ahrens MB, Orger MB, Robson DN et al (2013) Whole-brain functional imaging at cellular resolution using light-sheet microscopy. Nat Methods 10:413–420
15. Kubo F, Hablitzel B, Dal Maschio M et al (2014) Functional architecture of an optic flow-responsive area that drives horizontal eye movements in zebrafish. Neuron 81:1344–1359
16. Portugues R, Feierstein CE, Engert F et al (2014) Whole-brain activity maps reveal stereotyped, distributed networks for visuomotor behavior. Neuron 81:1328–1343
17. Scanziani M, Hausser M (2009) Electrophysiology in the age of light. Nature 461:930–939
18. Zhang RW, Wei HP, Xia YM et al (2010) Development of light response and GABAergic excitation-to-inhibition switch in zebrafish retinal ganglion cells. J Physiol 588:2557–2569
19. Wei HP, Yao YY, Zhang RW et al (2012) Activity-induced long-term potentiation of excitatory synapses in developing zebrafish retina in vivo. Neuron 75:479–489
20. Mu Y, Li XQ, Zhang B et al (2012) Visual input modulates audiomotor function via hypothalamic dopaminergic neurons through a cooperative mechanism. Neuron 75:688–699
21. Gabriel JP, Trivedi CA, Maurer CM et al (2012) Layer-specific targeting of direction-selective neurons in the zebrafish optic tectum. Neuron 76:1147–1160
22. Bagnall MW, McLean DL (2014) Modular organization of axial microcircuits in zebrafish. Science 343:197–200
23. Zhu P, Frank T, Friedrich RW (2013) Equalization of odor representations by a network of electrically coupled inhibitory interneurons. Nat Neurosci 16:1678–1686

Chapter 20

Quantifying Aggressive Behavior in Zebrafish

Magda C. Teles and Rui F. Oliveira

Abstract

Aggression is a complex behavior that influences social relationships and can be seen as adaptive or maladaptive depending on the context and intensity of expression. A model organism suitable for genetic dissection of the underlying neural mechanisms of aggressive behavior is still needed. Zebrafish has already proven to be a powerful vertebrate model organism for the study of normal and pathological brain function. Despite the fact that zebrafish is a gregarious species that forms shoals, when allowed to interact in pairs, both males and females express aggressive behavior and establish dominance hierarchies. Here, we describe two protocols that can be used to quantify aggressive behavior in zebrafish, using two different paradigms: (1) staged fights between real opponents and (2) mirror-elicited fights. We also discuss the methodology for the behavior analysis, the expected results for both paradigms, and the advantages and disadvantages of each paradigm in face of the specific goals of the study.

Key words Aggression, Social dominance, Behavior, Ethogram, Event recorder, Zebrafish

1 Introduction

Aggression can be defined as any behavior directed toward another individual with the intention to cause harm [1]. It is usually seen as an adaptive behavior expressed throughout most animals' lives, which has evolved in the context of intraspecific competition for resources, such as food, shelter, mating opportunities, or social status. However, heightened aggression levels may become maladaptive, and in humans they are often associated with psychiatric disorders [2]. Therefore, the study of aggression has been prompted both by fundamental and by applied questions. Despite significant progress in the identification of the neurobiological factors associated with aggression, there is still a need to understand in more detail the neural circuits and the active molecules that control this behavior. Similar to other complex behaviors, aggression is induced by the interplay of genes, neurotransmitters, and hormones, in the building and regulation of neural circuits that appear to be conserved across vertebrate species [3, 4]. Thus, progress in this area needs a model organism with a genetic toolbox

Koichi Kawakami et al. (eds.), *Zebrafish: Methods and Protocols*, Methods in Molecular Biology, vol. 1451,
DOI 10.1007/978-1-4939-3771-4_20, © Springer Science+Business Media New York 2016

available that allows for real-time visualization of brain activity and for the precise manipulation of specific neural circuits, in order to enable the mapping of behavior into neural circuits [5].

Zebrafish have already proven to be a powerful animal model for the study of complex cognitive disorders like depression, autism spectrum disorder (ASD), drug abuse, cognitive deficits, and psychoses [6]. Several behavioral paradigms used in rodents to study these disorders have already been successfully developed in zebrafish, such as exploration (open field), anxiety-like (light-dark and alarm substance), locomotion (novel tank), and social and cognitive (shoaling, social preference, predator avoidance, and T-maze) tests [6]. The utility of this species in behavioral neuroscience has grown markedly because of its available molecular (forward and reverse genetic methods [7, 8]), electrophysiological [9], and optogenetic [10] tools, the variety of wild-type lines with distinct behavioral phenotypes [6], conditional transgenic lines [11], and the similarity its genome presents with the human genome, where approximately 70% of the genes have human orthologs [12]. All these features make zebrafish an ideal model for translational neuroscience.

Although zebrafish is a gregarious species that in nature form shoals [13], when allowed to interact in pairs, both males and females express aggressive behavior and establish dominance hierarchies [14–16]. In this species, aggression is commonly used by dominant individuals to get access to spawning sites and protect their social status from competitors [16]. Similarly to other species, the repertoire (i.e., ethogram) of zebrafish agonistic behavior consists of a series of stereotype body postures and movements that have been previously characterized (Table 1) [15]. In dyadic male fights, two distinct phases have been described: (1) A pre-resolution phase, where both fish exhibit the same repertoire of behaviors (display, circle, and bite); this phase lasts until the first chase or flee is observed, which marks the establishment of a behavioral asymmetry between the contestants (i.e., fight resolution); (2) A post-resolution phase, characterized by an asymmetry of expressed behaviors, where all agonistic behaviors are initiated by the dominant fish, whereas the subordinate only displays submissive behaviors. Therefore, the expression of the different aggressive behavior action patterns has a specific temporal structure (Fig. 1). An agonistic interaction usually starts with both opponents exhibiting lateral displays in an antiparallel position and circling each other. Then, it progresses to mutual bites, still in the pre-resolution phase. Finally, in the post-resolution phase, dominant individuals bite, chase, and strike toward subordinates, whereas the latter flee, freeze, and retreat.

Given that fish lack visual self-recognition, when exposed to a mirror, they usually display aggressive behavior toward their mirror image [17]. Therefore, aggressive behavior in fish has been quantified using either their response toward real opponents [14, 15] or toward their own mirror images [17–22]. However, recent studies have questioned

Table 1

Ethogram of zebrafish aggressive behavior (adapted from Oliveira et al. [15])

Behavioral patterns	Description
Displays	In short distance of the opponent, usually less than one body length, fish erects its dorsal and anal fins and flares its body flank toward the opponent
Circle	Two fish approach each other in antiparallel positions with their fins erected and circle one another ascending in the water column. It can last from a few seconds to minutes
Strike	The fish swims rapidly toward the opponent, but no physical contact occurs
Bite	Fish opens and closes its mouth in contact with the body surface of its opponent, usually directed toward the ventral or the posterior parts of the body of the target fish
Chase	Similar to strike but with an active pursuit by the aggressor. This behavior stops when one fish stops chasing and/or the other fish adopts a freeze behavior
Retreat	Fish swims rapidly away from the opponent in response to a strike or a bite
Flee	Continued escape reaction in response to a chase. Fish swims rapidly away from the aggressor
Freeze	Fish stays immobile with all fins retracted and the caudal region downward near the bottom or the surface of the aquaria

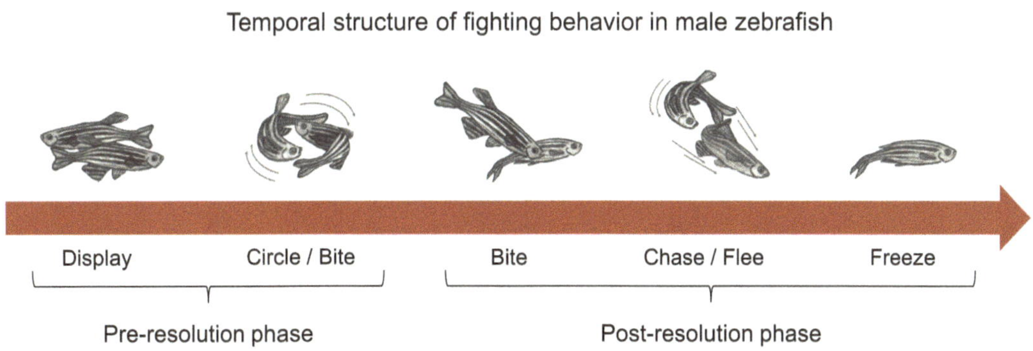

Fig. 1 Zebrafish male fights exhibit a typical temporal structure. Fights can be divided into a pre-resolution phase and a post-resolution phase. The pre-resolution phase is defined by the expression of symmetric behaviors by both contestants, and behaviors such as displays, circles, and mutual bites occur. The post-resolution phase is characterized by a transition to asymmetric expression of behaviors between the opponents, where bites, chases, and strikes are performed by the dominant individual, whereas retreat, flee, and freeze are expressed by the subordinate. The arrow represents the temporal occurrence of each type of behavior in the respective phase (adapted from Oliveira el al. [15])

whether these two tests of aggression are measuring the same aspects of behavior, since they elicit different hormonal responses in cichlid fish [17, 22]. In zebrafish, mirror-elicited fights also failed to arouse the same brain responses as real opponents in gene expression [23] and in the monoaminergic activity [19]. Despite these physiological differences elicited by the two protocols, there are no significant differences between the levels of overt aggression exhibited toward a mirror image or a real opponent [19, 20]. Thus, both protocols seem suitable for quantifying overt aggression measures, but the decision to use one or the other should take into consideration known differences between the two (Table 2), which may be advantageous or disadvantageous, depending on the specific goals of the study. Here, we describe two protocols that can be used to quantify aggressive behavior in zebrafish, using each of these two paradigms: (1) staged fight test, between real opponents, and (2) mirror-elicited aggression test.

2 Materials

1. Electronic balance.
2. Ruler/caliper.
3. Buffered tricaine methanesulfonate (MS222; *see* **Note 1**).
4. Spring scissors.
5. Forceps.

Table 2

Advantages and disadvantages between real-opponent and mirror-elicited fights as tests of aggression in zebrafish

	Real-opponent fight	Mirror-elicited fight
Advantages	– Provide the most natural social stimulus – Promote the establishment of social dominance with the emergence of dominant and subordinate phenotypes	– The opponent's behavior is standardized to that of the focal fish (i.e., it is the same) – Fighting individuals are not exposed to physical injuries, which makes it ethically more acceptable
Disadvantages	– The researcher has no or limited control of the stimulus fish, and the behavior of the focal fish depends to a great extent on the behavior of the opponent – Fighting individuals can be physically injured, and thus it is less acceptable from an ethical perspective	– The fights are unsolved and therefore the focal fish never experiences either a victory or a defeat [21] – Prevents the expression of lateral display in an antiparallel position, which is a common action pattern in real-opponent fights – The dynamics of the fight are atypical, since the opponent never initiates behavior and never displays submissive behavior

6. Fish-holding support (*see* **Note 2**).
7. 27G needle (internal diameter 0.210 mm).
8. Nylon monofilament 0.14 mm.
9. Povidone-iodine (Betadine®) or any other microbicide-like chlorhexidine to disinfect the material.
10. Nail polish.
11. Zebrafish maternity tanks (18 × 10 × 9 cm).
12. Video camera.
13. Multievent recorder software for behavior recording and analysis (Observer XT).

3 Methods

3.1 Animal Housing

The protocols described here were developed using adult wild-type zebrafish of the AB strain (*see* **Note 3**). Fish are kept in a recirculating housing system (ZebTEC Multilinking System, Tecniplast, Italy), at 28 °C with a 14 L:10D photoperiod. The water is monitored for nitrites (<0.2 ppm), nitrates (<50 ppm), and ammonia (0.01–0.1 ppm), and pH and conductivity are maintained at 7 and 700 μSm, respectively. Fish are fed twice a day, except on the day of the experiments.

3.2 Individual Tagging

In staged fights, it is important to identify each individual during the whole interaction, such that the behavior of each opponent can be quantified separately. For this purpose, individuals need to be individually tagged. There are three commonly used procedures to tag zebrafish: fin clipping [15], color tagging with nylon monofilament [24, 25], and color tagging with implanted elastomers [26] (*see* **Note 4**). Here we describe the two methods that are currently used in our lab.

3.2.1 Fin Clips

1. Anesthetize the fish by immersion in tricaine solution (160 mg/L) in a petri dish (*see* **Note 5**).
2. Use the spring scissors to clip the extremities of the caudal, dorsal, or anal fins in different combinations between pairs of opponents.

3.2.2 Color Tagging with Nylon Monofilament

1. Prepare the nylon monofilament by cutting approximately 5 cm; give three or four knots with the help of the forceps in one tip and paint the knots with nail polish (*see* **Note 6**).
2. Cut the other tip of the nylon monofilament in diagonal, in order to be pointed.
3. Place all materials, including the painted nylon monofilament previously prepared, in povidone-iodine (Betadine®) or any other microbicide solution.

4. Anesthetize the fish by immersion in tricaine solution (320 mg/L) in a petri dish.
5. Place the fish in an appropriate bedding (*see* **Note 2**).
6. Insert the hypodermic needle (27G) through the dorsal musculature immediately below the posterior insertion of the dorsal fin.
7. Insert the pointed nylon monofilament already tagged through the needle hole (Fig. 2a).
8. Remove the needle out of the fish body leaving the monofilament behind (Fig. 2b).
9. Give three or four knots, with the help of the forceps, on this tip and paint with nail polish (Fig. 2c, d) (*see* **Note 7**).

3.2.3 Recovery from Anesthesia

1. Fill a zebrafish maternity tank with water (approximately 800 mL) and place the fish to recover after any of the tagging procedures described above. Do not use more than five animals per tank to mitigate stress [27].

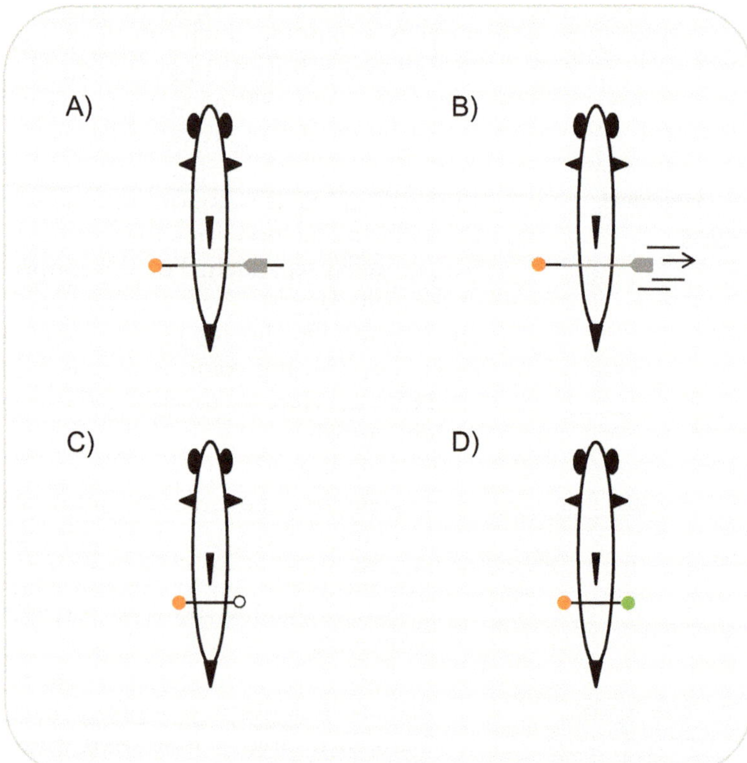

Fig. 2 Color tagging with nylon monofilament. The fish is represented in a top view: (**a**) Insertion of the hypodermic needle through the dorsal musculature of the fish and guiding the nylon monofilament already tagged through the needle hole. (**b**) Removal of the needle leaving the monofilament in place. (**c**) Giving knots on one side of the monofilament. (**d**) Painting it with nail polish (reproduced with permission from Patzner [24])

2. Animals will recover very fast from the anesthesia (in minutes); however, in order to maximize the anesthetic withdrawal, keep animals in the recovery tank for 1 h before moving them back to the home tank (*see* **Note 8**) [28].

3.3 Behavioral Recording

1. We typically use an experimental tank of 32 cm × 20 cm × 15 cm divided into two parts: (1) the posterior part (19.5 cm × 20 cm × 15 cm) containing a mechanical filter and a heater (water temperature is kept at 28 °C also during the tests) and (2) the anterior part (12.5 cm × 20 cm × 15 cm), hereafter designated as arena, where the tests take place (*see* **Note 9**).
2. Cover the back wall of the arena with white PVC, in order to improve contrast between fish and the background in video recordings.
3. Divide the arena into two parts of the same size by a removable PVC partition (Fig. 3): (a) for staged fights, the PVC partition separates the two fish in the right and left sides of the tank (Fig. 3a); (b) in mirror-elicited fights, the PVC partition contains one mirror on each side and is perforated on the sides to allow water flow between the two parts; a second removable partition should be placed in front of it to hide the mirrors from the focal fish before the start of the interaction (Fig. 3b).

3.3.1 Staged Fights

1. Pair the animals according to their weight and standard length (*see* **Note 10**).
2. Prior to the experiment, place each pair in the experimental tank, one fish on each side of the arena divided by the opaque partition, where they stay overnight in visual isolation (*see* **Note 11**, Fig. 3a, b). Before the experiment, set up a standard video camera (*see* **Note 12**) in front of the tank to record the interaction.
3. Gently remove the opaque partition and allow the two fish to interact for a period of 30 min (*see* **Note 13**, Fig. 3a′).
4. At the end of the test period (30 min), a dominant and a subordinate fish should be easily identified by the different behaviors they express (i.e., winners only express aggressive behaviors, and losers only express submissive behaviors); place the partition back into the observation tank to separate the two fish again, and note the identity of the dominant and of the subordinate fish.

3.3.2 Mirror-Elicited Fights

1. Repeat **steps 1** and **2** from the staged fights protocol in Subheading 3.3.1.
2. Gently remove the two opaque partitions that are covering the mirrors, and allow the two fish to interact with each mirror simultaneously (*see* **Note 14**, Fig. 3b′).
3. After the 30 min period, place the two opaque partitions back in place, in order to end the interaction of each fish with its own mirror image.

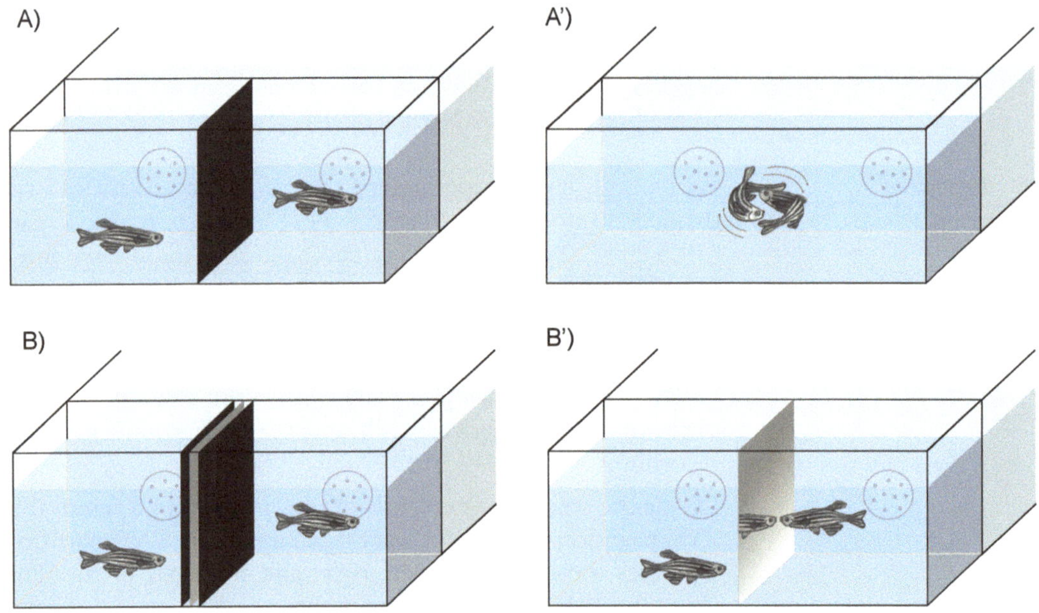

Fig. 3 Observation tanks are divided into a posterior part, which contains a mechanical filter and a heater, and an anterior part where the test takes place (the arena). Perforated plastic circles along the glass dividing the two compartments allow water exchange between the arena and the filter compartments. The arena is divided into two same-size parts by an opaque PVC partition; depending on the test (real-opponent or mirror fight), this partition can be removed or not. (**a**) For real-opponent fights, animals are separated by a removable opaque PVC partition. (**a′**) The opaque divider is removed, and the fish are allowed to interact for 30 min. (**b**) For mirror-elicited fights, the arena is divided by a PVC partition containing one mirror on each side, and a second removable partition is placed in front of each mirror to cover it. (**b′**) The two outer partitions are removed, and the fish are allowed to interact with their own mirror image throughout the test period (30 min)

3.3.3 Quantitative Behavioral Analysis

1. Analyze the video recordings using a computerized multievent recorder (Observer XT, Noldus, Wageningen, The Netherlands).
2. Use the ethogram of zebrafish agonistic behavior to identify the relevant action patterns [15], which are divided into aggressive for dominants (bite, chase, and strike) and submissive for subordinates (freeze and flee).
3. Identify the selected behaviors as states or events, and quantify the frequency or the duration of the respective behaviors (*see* **Note 15**).

3.3.4 Typical Results

For staged fights, a typical encounter starts with mutual displays (lateral displays, circling) characteristic of the pre-resolution phase. In the post-resolution phase when the dominant-subordinate status has already been established, chase and bites are the most frequent action patterns (Fig. 4a).

When comparing staged fights with mirror-elicited fights several differences can be observed (Fig. 4b–d):

1. The latency for the first attack (i.e., bite) is significantly lower in mirror fights when compared to staged fights, which may be a result of mirror "opponents" providing ambiguous information leading mirror fighters to escalate their aggressive behavior faster than individuals fighting a real opponent (*see* **Note 16**).
2. The opposite pattern is observed for the fight resolution time, with staged fights being solved more rapidly (in approximately 7 min) than mirror fights (usually still ongoing at the end of the 30 min observation period). This may result from the fact that during the pre-resolution phase, fish mutually assess their relative fighting ability and adjust their behavior accordingly.
3. Since there is no fight resolution in mirror fights, mirror fighters do not either win or lose the fight; therefore, they do not

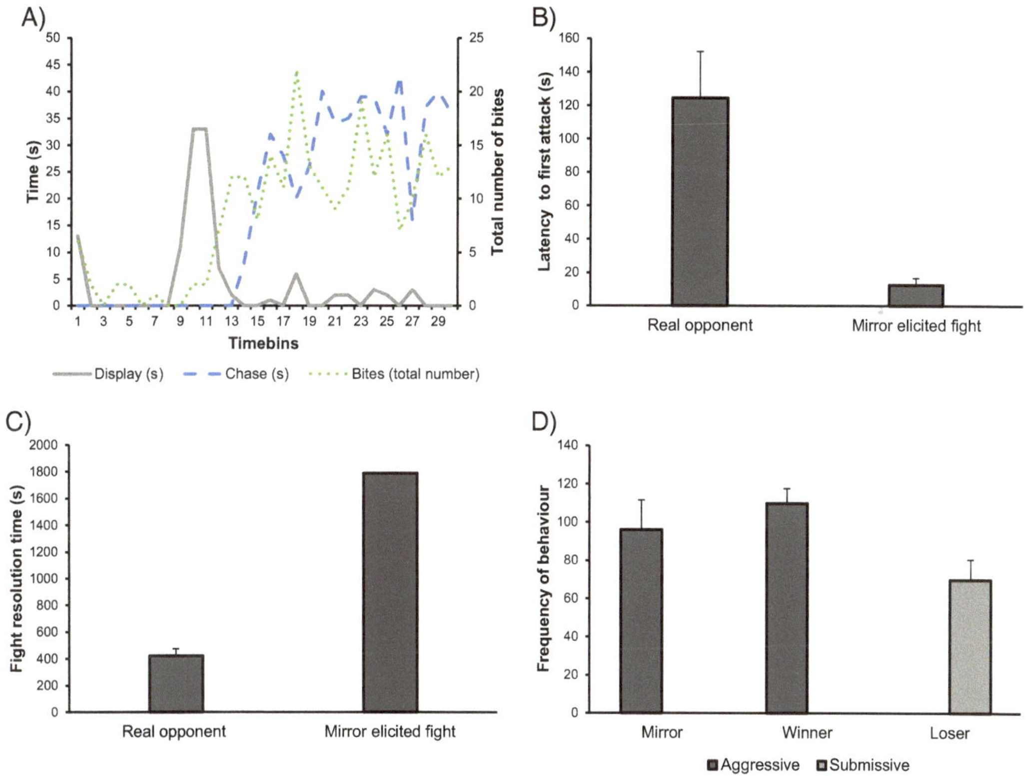

Fig. 4 Typical results for the two protocols used to quantify aggressive behavior. (**a**) Temporal dynamics of a real-opponent fight analyzed in 1 min time bins for the 30 min interaction (unpublished data). The full line represents the time in display, a typical behavior of the pre-resolution phase, and the *dashed* and *dotted lines* represent the time in chase and number of bites, respectively, behaviors typically expressed in the post-resolution phase. (**b**) Mean latencies to the first attack in real-opponent and in mirror-elicited fights (unpublished data). (**c**) Fight resolution time, measured as the time needed for a social hierarchy to be established, in real-opponent and in mirror-elicited fights (unpublished data). (**d**) Mean number of aggressive acts performed in the last 5 min of the 30 min interaction test for winners and losers of real-opponent fights and for mirror fighters; error bars represent the standard error of the mean (reproduced with permission from [19])

adopt the respective dominant or subordinate phenotype, observed in real-opponent fights, despite the expression of significant amounts of aggressive behavior.

4. Indeed, there are no significant differences in the levels of overt aggression between mirror fighters and dominants of real-opponent fights. Thus, one can conclude that a major difference between the two protocols is not so much in the behavior expressed by the focal fish, but rather in the behavior expressed by the opponent.

As a final recommendation, we suggest that researchers intending to use the mirror test to phenotype aggression should first validate it by comparing individual responses between real-opponent and mirror tests. This has been done recently for a set of different cichlid species, and the results appear to be species specific, since in some species (i.e., *Neolamprologus pulcher* and *Astatotilapia burtoni*) the results of the two tests are correlated [18, 29], whereas for other species (i.e., *Telmatochromis vittatus*, *Lepidiolamprologus elongatus*, and *Amatitlania nigrofasciata*), no relationship was found between mirror and real-opponent aggression [29].

4 Notes

1. Buffered tricaine methanesulfonate (stock solution): 4000 mg/L tricaine methanesulfonate (MS222), buffered with tris-base 1 M, pH = 9 to a final pH = 7 solution.
2. The bedding can be a small petri dish filled with aquarium-graded silicone, with a small depression in the middle to hold the fish in a dorsoventral position.
3. One should keep in mind that aggressive behavior might differ between different wild-type strains, as it has been described for other behaviors [30].
4. The choice of the tagging method depends on the experimental procedure to be used. For example, fin clips are normally used for short-term experiments since fin regeneration occurs rapidly, whereas color tagging is more appropriate for long-term experiments, despite being a more intrusive technique. Finally, visible implant elastomers are more suitable for experiments that do not require video analysis because visible implant elastomer tags may be difficult to distinguish in video images (e.g., yellow vs. orange or pink vs. red can be easily confused). Furthermore, color identification may depend on ambient light which becomes a constraint when video recordings are used [26].
5. With this dose of anesthetic a deeper anesthesia will be induced, which promotes a total loss of equilibrium and muscle tone and a very slow ventilation rate (almost absent) [31]. This will

occur very fast. As soon as these signs are present, remove the fish from the anesthetic solution.

6. Beforehand, prepare a sheet with the color combinations that you intend to use to tag the fish, to avoid repetitions of color codes.
7. Leave some clearance between the knots and the fish body to avoid skin infections and interference with body growth.
8. After tagging the animals, there must be a quarantine period before starting the behavioral tests. For fin clips, one should wait at least 24 h and, for color tagging, 10 days to guarantee wound healing. Animals should be monitored during this period for tag loss and health status.
9. The perforated plastic circles along the glass dividing these two parts of the tank allow water exchange between the two compartments (Fig. 3).
10. Since body size is highly correlated with dominance, size differences (length or weight) between opponents should not exceed 10% of total body size, in order to avoid an a priori advantage of the larger individual. Take the opportunity of having fish anesthetized for the tagging procedures to take body measurements (weight, standard length) of all individuals.
11. Previous studies had established different periods of social isolation of 5 days [14, 32] and 24 h [15] as effective to elicit aggressive behavior in zebrafish. However, overnight isolation proved sufficient to induce consistent expression of aggressive behavior for the duration of the tests (30 min) [19].
12. The camera we used had a resolution of 720×576 and frame rate of 25 frames per second; however, higher resolution cameras with higher frame rates are also appropriated.
13. In order to minimize the interaction between the observer and the focal fish, the partitions can be pulled up from a distance with the help of pulleys.
14. Subjects were also tested in pairs in the mirror-elicited test, in order to provide them with conspecific odors, which would otherwise only be present in real-opponent dyads, therefore avoiding confounding effects of putative chemical cues used in agonistic interactions.
15. For behavior quantification, it is important to distinguish between two fundamental types of action patterns, based on the time expression, because this will influence the type of measures that one should take: (1) events are action patterns that are discrete in time (i.e., have very short duration) such that it is difficult to establish their start and finish time (e.g., bites, strikes); the relevant measure of events is their frequency (number of occurrences per unit of time); (2) states are action patterns that have a significant time duration which allows to easily define their start and their

end (e.g., display, chase, freeze, and flee); states can be quantified both in terms of their frequency and their duration (e.g., percentage of time displaying). Latency, defined as the time from some specified time point (e.g., start of the test) to the first occurrence of the relevant action pattern, can also be measured, both for events and for states. Latency to initiate a fight is usually interpreted as a measure of aggressive motivation, whereas frequency and duration of events and states, respectively, reflect the engagement in the interaction. Since the engagement in the fight depends not only on the motivation of the focal fish but also on the response of the opponent, measures of latency are expected to better measure the intrinsic aggressive motivation of individuals. In our protocols, we typically analyze the latency to the first interaction and the frequency and duration of aggressive and submissive behaviors.

16. When laterally displaying to each other, as a way of assessing each other's competitive ability [33], fish can align either in a parallel (head to head) or antiparallel (head to tail) position [34]. However, since there is a left-eye bias in zebrafish for social stimuli, they prefer to display the left side of the body, making the head to tail alignment, which is not present in mirror interactions, more common during mutual displays [20]. Thus, mirror fights also change the structure of the fight making mirror fighters escalate faster than real-opponent fighters.

Acknowledgments

The authors thank Nuno Marques Pereira for helpful discussions on the tagging procedure. The writing of this book chapter was supported by a grant from Fundação para a Ciência e a Tecnologia (FCT, EXCL/BIA-ANM/0549/2012). MCT is supported by a Ph.D. fellowship from FCT (SFRH/BD/44848/2008).

References

1. Baron RA, Richardson D (1994) Human aggression. Plenum Press, New York
2. Jones LJ, Norton WHJ (2015) Using zebrafish to uncover the genetic and neural basis of aggression, a frequent comorbid symptom of psychiatric disorders. Behav Brain Res 276:171–180
3. Goodson JL (2005) The vertebrate social behavior network: evolutionary themes and variations. Horm Behav 48:11–22
4. O'Connell LA, Hofmann HA (2012) Evolution of a vertebrate social decision-making network. Science 336:1154–1157
5. Muto A, Ohkura M, Abe G et al (2013) Real-time visualization of neuronal activity during perception. Curr Biol 23:307–311
6. Kalueff AV, Stewart AM, Gerlai R (2014) Zebrafish as an emerging model for studying complex brain disorders. Trends Pharmacol Sci 35:63–75

7. Sivasubbu S, Balciunas D, Amsterdam A, Ekker SC (2007) Insertional mutagenesis strategies in zebrafish. Genome Biol 8:1–9
8. Bill BR, Petzold AM, Clark KJ et al (2009) A primer for morpholino use in zebrafish. Zebrafish 6:69–77
9. Higashijima S, Masino MA, Mandel G, Fetcho JR (2003) Imaging neuronal activity during zebrafish behavior with a genetically encoded calcium indicator. J Neurophysiol 90:3986–3997
10. Douglass AD, Kraves S, Deisseroth K et al (2008) Escape behavior elicited by single, channelrhodopsin-2-evoked spikes in zebrafish somatosensory neurons. Curr Biol 18:1133–1137
11. Kawakami K, Abe G, Asada T et al (2010) zTrap: zebrafish gene trap and enhancer trap database. BMC Dev Biol 10:105
12. Howe K, Clark MD, Torroja CF et al (2013) The zebrafish reference genome sequence and its relationship to the human genome. Nature 496:498–503
13. Spence R, Gerlach G, Lawrence C, Smith C (2008) The behaviour and ecology of the zebrafish, Danio rerio. Biol Rev Camb Philos Soc 83:13–34
14. Larson ET, O'Malley DM, Melloni RH (2006) Aggression and vasotocin are associated with dominant-subordinate relationships in zebrafish. Behav Brain Res 167:94–102
15. Oliveira RF, Silva JF, Simões JM (2011) Fighting zebrafish: characterization of aggressive behavior and winner-loser effects. Zebrafish 8:73–81
16. Paull GC, Filby AL, Giddins HG et al (2010) Dominance hierarchies in zebrafish (Danio rerio) and their relationship with reproductive success. Zebrafish 7:109–117
17. Oliveira RF, Carneiro LA, Canário AVM (2005) Behavioural endocrinology: no hormonal response in tied fights. Nature 437:207–208
18. Desjardins JK, Fernald RD (2010) What do fish make of mirror images? Biol Lett 6:744–747
19. Teles MC, Dahlbom SJ, Winberg S, Oliveira RF (2013) Social modulation of brain monoamine levels in zebrafish. Behav Brain Res 253:17–24
20. Ariyomo TO, Watt PJ (2013) Aggression and sex differences in lateralization in the zebrafish. Anim Behav 86:617–622
21. Rowland WJ (1999) Studying visual cues in fish behavior: a review of ethological techniques. Environ Biol Fish 56:285–305
22. Hirschenhauser K, Wittek M, Johnston P, Möstl E (2008) Social context rather than behavioral output or winning modulates post-conflict testosterone responses in Japanese quail (Coturnix japonica). Physiol Behav 95: 457–463
23. Oliveira RF, Simões JM, Teles MC, Oliveira CR, Becker JD, Lopes JS (2016) Assessment of fight outcome is needed to activate socially driven transcriptional changes in the zebrafish brain. Proc Natl Acad Sci 113:E654–661.
24. Patzner RA (1984) Individual tagging of small fish. Aquaculture 40:251–253
25. Dahlbom SJ, Lagman D, Lundstedt-Enkel K et al (2011) Boldness predicts social status in zebrafish (Danio rerio). PLoS One 6:e23565
26. Hohn C, Petrie-Hanson L (2013) Evaluation of visible implant elastomer tags in zebrafish (Danio rerio). Biol Open 2(12):1397–1401
27. Lawrence C (2007) The husbandry of zebrafish (Danio rerio): a review. Aquaculture 269:1–20
28. Topic Popovic N, Strunjak-Perovic I, Coz-Rakovac R et al (2012) Tricaine methane-sulfonate (MS-222) application in fish anaesthesia. J Appl Ichthyol 28:553–564
29. Balzarini V, Taborsky M, Wanner S et al (2014) Mirror, mirror on the wall: the predictive value of mirror tests for measuring aggression in fish. Behav Ecol Sociobiol 68:871–878
30. Vignet C, Bégout M-L, Péan S et al (2013) Systematic screening of behavioral responses in two zebrafish strains. Zebrafish 10:365–375
31. Ross LG, Ross B (2008) Anaesthetic and sedative techniques for aquatic animals, 3rd edn. Blackwell, Oxford
32. Dahlbom SJ, Backström T, Lundstedt-Enkel K, Winberg S (2012) Aggression and monoamines: effects of sex and social rank in zebrafish (Danio rerio). Behav Brain Res 228:333–338
33. Hurd PL (1997) Cooperative signalling between opponents in fish fights. Anim Behav 54(5):1309–1315
34. Elwood RW, Stoilova V, McDonnell A et al (2014) Do mirrors reflect reality in agonistic encounters? A test of mutual cooperation in displays. Anim Behav 97:63–67

Chapter 21

Correlating Whole Brain Neural Activity with Behavior in Head-Fixed Larval Zebrafish

Michael B. Orger and Ruben Portugues

Abstract

We present a protocol to combine behavioral recording and imaging using 2-photon laser-scanning microscopy in head-fixed larval zebrafish that express a genetically encoded calcium indicator. The steps involve restraining the larva in agarose, setting up optics that allow projection of a visual stimulus and infrared illumination to monitor behavior, and analysis of the neuronal and behavioral data.

Key words Whole-brain imaging, Behavior, 2-Photon microscopy, Zebrafish

1 Introduction

Larval zebrafish are small and transparent, and therefore, amongst vertebrate model organisms, they offer unique possibilities for studying neuronal activity using imaging techniques [1, 2]. As their brain is relatively small, roughly $800 \times 500 \times 300$ μm, it is possible to monitor, at high spatial resolution, a large fraction of their brain in a single experiment. This allows experimenters to study circuits that extend all the way from the sensory input to the motor output and to ask questions about complete neural systems. This approach has been used to study sensorimotor processing, spontaneous activity patterns, and learning [3–6]. In this chapter, we present a primer on how to perform these experiments in a restrained, but behaving, 5–7 days post fertilization (dpf) larval zebrafish using scanning 2-photon microscopy. We also describe a simple method, based on linear correlation, which allows us to understand this neuronal activity in terms of the stimulus presented and the behavior measured.

Koichi Kawakami et al. (eds.), *Zebrafish: Methods and Protocols*, Methods in Molecular Biology, vol. 1451, DOI 10.1007/978-1-4939-3771-4_21, © Springer Science+Business Media New York 2016

2 Materials

2.1 Embedding Larvae

1. 5–7 dpf. Zebrafish larvae with a fluorescent calcium indicator in some or all neurons (e.g., elavl3:GCaMP5G [4]). In this protocol, we will assume the use of a similar GFP-based indicator.
2. Fire-polished glass transfer pipette.
3. 35 mm. Petri dish.
4. Sylgard 184 silicone elastomer (Dow Corning).
5. Low melting point agarose (UltraPure LMP Agarose—Invitrogen).
6. Forceps (e.g., Fine Science Tools Dumont #5).
7. Fine scalpel (e.g., Sharpoint Stab knife, restricted depth straight; 15°, 5.0 mm blade).
8. Heating block or water bath.
9. Stereo dissecting microscope.
10. E3 embryo medium: 5 mM NaCl, 0.17 mM KCl, 0.33 mM CaCl, 0.33 mM $MgSO_4$. To make a buffered E3 solution add Tris solution, adjusted to pH 7.0 with HCl, to a final concentration of 1 mM.

2.2 Imaging Rig

1. Two-photon microscope. We use a custom-built microscope [7], but several flexible, commercially available systems would also be suitable.
2. Compact digital projector for visual stimulation.
3. Stage to hold the sample, fabricated from transparent acrylic.
4. Diffusive screen for stimulus projection (e.g., Rosco filter 3026 [8]).
5. Long pass filter for visual stimulation, e.g., Wratten #29 gel filter (Kodak).
6. 850 nm LEDs with mount and power supply.
7. High-speed infrared-sensitive camera (e.g., Pike F-032B, AVT).
8. C-mount camera objective lens that allows the appropriate field of view and working distance for your microscope (*see* **Note 1**).
9. 900 nm short-pass laser blocking filter appropriate to mount in front of the behavior camera objective.
10. 45° Hot mirror, 790 nm cut-off.
11. Silver mirror to direct the visual stimulus onto the screen.
12. Computer and software for stimulus production and camera acquisition (*see* **Note 2**).

2.3 Data Analysis

1. High performance computer.
2. Flexible software for analysis of numerical data and image processing (e.g., MATLAB from Mathworks).

3 Methods

3.1 Embedding Larvae

Zebrafish larvae in the age range of 5–7 dpf still receive nutrition from their yolk and can be embedded in low melting point agarose and survive in good health for more than a day, as long as they are immersed in well-aerated E3 embryo medium [9, 10]. With practice, it is possible to embed larvae quickly, without damage and with minimal tilt.

1. At least 2 days before starting experiments, mix the Sylgard 184 components according to the manufacturer's instructions and pour to a depth of about 3 mm into 35 mm petri dishes. Cover the dishes and allow 48 h for the Sylgard to polymerize completely.
2. Place a larva (5–7 dpf) in a drop of E3 water in the center of a Sylgard-coated 35 mm Petri dish. Carefully suck off most of the E3 with a fire-polished glass pipette (*see* **Note 3**). To allow imaging in the brain, *nacre* mutant larvae, which lack skin melanophores should be used (*see* **Note 4**) [11].
3. Cover the larva with 2 % low melting point agarose in E3 using a transfer pipette (*see* **Note 5**).
4. Melt the agarose an hour or so before, in a microwave, and then keep it in a water bath at 35° Celsius. Check that the agarose is not too hot by touching the side of the transfer pipette. Ensure that the agarose and the E3 droplet mix well. The agarose will gel in a few seconds so the next steps should be executed rapidly.
5. Use forceps to mount the larva dorsal side up (*see* **Notes 6–8**) close to the center of the dish by manipulating the agarose around the tail with the closed forcep tips: the larva will be dragged along (Fig. 1c). Pitch can be controlled by gently pressing down on the tail with the side of the forceps.
6. Let the agarose set for 10–20 min and then carefully add E3 water to completely cover the agarose (*see* **Note 9**).
7. Depending on the experiment to perform, and the behavior that will be monitored under the 2-photon microscope, it will be necessary to free the tail and/or the eyes. Do this only after adding E3 water to the dish. Using a scalpel, remove the agarose as shown in Fig. 1a. Using a scalpel, insert the blade vertically close to the fish and then cut away from the body (*see* **Note 10**). The length of the tail that is freed can be smaller for imaging versus behavioral experiments in order to minimize motion artifacts [6, 12].

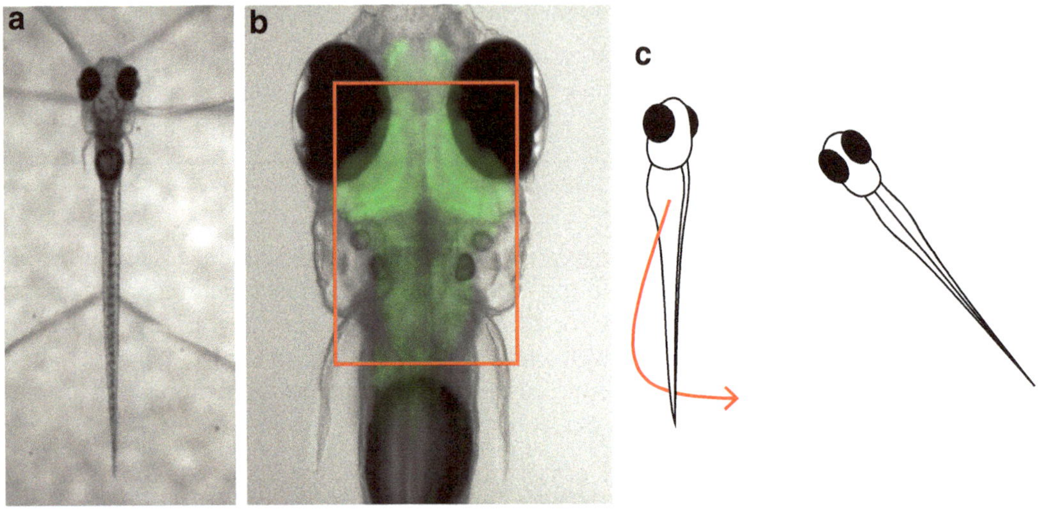

Fig. 1 Larval zebrafish embedded in agarose. (**a**) A 6 dpf larval zebrafish embedded in agarose. The agarose has been removed to free the eyes and the tail. (**b**) A *nacre* –/– fish lacking melanophores, and with pan-neuronal expression of a green-fluorescent, genetically encoded calcium indicator, Tg(elavl3:GCaMP5G). Example of the field of view that can be imaged with a 20× objective is shown in red. (**c**) Typical maneuver (*red arrow*) that is performed with the forceps without directly touching the larva to fix unwanted roll when embedding

8. Under the dissecting microscope, check that the embedded larva is healthy before transferring to the imaging rig (*see* **Note 11**).

3.2 Two-Photon Microscope

Two-photon microscopes may have widely varying capabilities and features. In this section, we describe a basic setup that may be adapted depending on the space available for the stage and the particular microscope that will be used. Such a setup is schematized in Fig. 2.

1. Present the visual stimulus on a diffusing screen below the embedded fish using a projector and mirror. The stage is cut to a custom shape from transparent acrylic.
2. In order to prevent the visual stimulus interfering with collection of green fluorescence, modify the projector so that illumination is provided only by a red LED or simply place a long-pass filter directly in front of the output beam lens of the projector [13] (*see* **Note 12**).
3. In order to track behavior, illuminate the larva using infrared LEDs (red arrows in Fig. 2), in the range of 790–850 nm. This will ensure that the illumination light is not visible to the larva but will pass through the filters used to block the laser before the high-speed camera. Place the LED as close as possible to the fish in order to use the lowest power that allows good behavioral tracking. Depending on the behavior to be tracked,

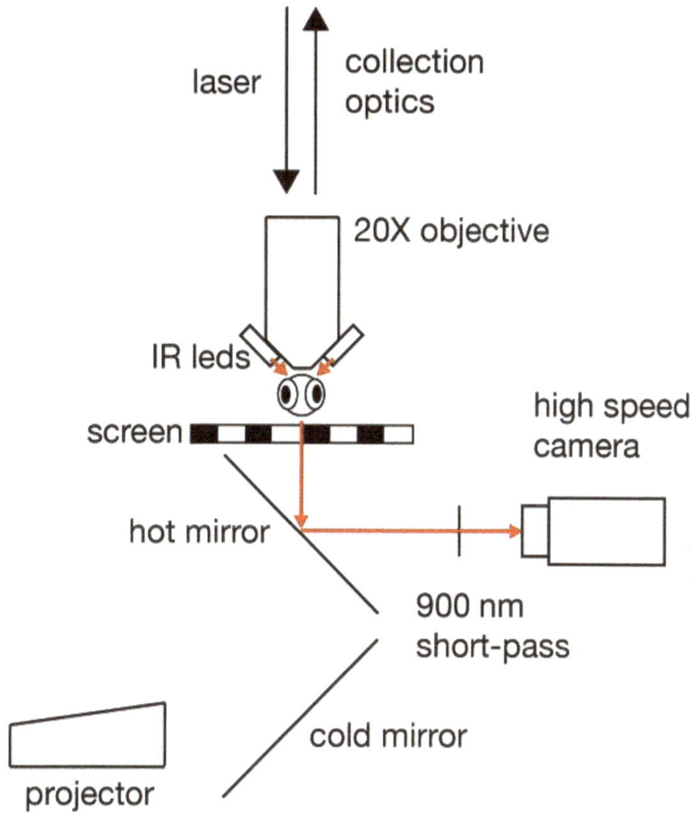

Fig. 2 Two-photon microscope setup for presenting visual stimuli and tracking behavior. The fish sits on a transparent stage in a two-photon microscope, illuminated from the side by infrared LEDs. Visual stimuli are projected on a screen below the fish. A small hole in the screen allows the tail to be imaged with an infrared-sensitive camera. A hot (infrared reflecting, visible light transmitting) mirror selectively directs infrared light to the camera

it may be convenient to use a spotlight LED placed at an angle on the side of the fish that creates a clear light/dark contrast along the length of the tail: this contrast can enable easy automatic tail tracking.

4. To combine high-speed tracking and visual stimulation from below, insert a 45° hot mirror, which transmits visible light and reflects infrared, into the optical path. Use a 900 nm short-pass filter to prevent the infrared laser light used for 2-photon excitation from reaching the camera sensor.
5. Use a laser wavelength of 950 nm for GCaMP excitation. This ensures that there is minimal bleed-through into the camera that monitors behavior, as well as reducing absorption by melanin. We recommend using a power setting that delivers around 10 mW after the objective lens. Larvae can be imaged stably under these conditions for 12 or more hours.

6. For a balanced trade-off of speed and spatial sampling, use an x–y pixel size just below 1 μm. Since a typical cell body diameter is about 5 μm, this ensures that each cell just spans sufficient pixels to allow subsequent automatic segmentation. We also suggest a similar distance between adjacent z planes. Needless to say, the higher the resolution the better, as long as the frame rate, and experimental duration remain compatible with the behavior to be studied (*see* **Note 13**).
7. For quantitative analysis, behavioral variables such as eye angle and tail curvature should be extracted from the raw images captured by the high-speed camera [6, 10, 14]. This can be done in real-time, or alternatively the raw movies can be saved for later processing. To measure eye angle, first threshold the image to identify the eye regions, and then determine their angle relative to the body axis, for example by fitting their somewhat elongated outline with an ellipse and measuring the angle of the major axis of the ellipse. The tail can be tracked by selecting a point on the center of the tail where it enters the agarose, and computing the brightness of all points in an arc at a fixed radius from this point. The peak of brightness along this arc will indicate the tail direction (we are assuming that the illumination is such that the tail appears bright on a dark background). This process can be repeated, with each new point serving as the center for the next arc, in order to measure the curvature along the whole tail.

3.3 Data Analysis

The raw imaging data recorded in the experiments consists of the fluorescence time-series for each pixel in each plane. In addition, we have time-series for the behavioral variables monitored and for the stimulus shown. Can we understand the neuronal activity that gives rise to the former in terms of the latter? Here, present a workflow, based on that in ref. [6] which allows the experimenter to gain some insight into this question. This can represent a useful starting point for further analysis tailored to the particular experiment in question.

3.3.1 Elimination of Motion Artifacts

When imaging for prolonged periods of time, motion artifacts will arise either from slow drift or from movement of the brain as the fish behaves. It is important to eliminate these as far as possible from the data in order to perform the correlation analysis presented below. There are many software packages available to do this, for example [15]. In the protocol steps below, assume that the imaging experiment has resulted in NT fluorescence frames *ftn*, where $t=1,..,T$ labels the time points of the acquisition of the T frames per plane and $z=1,\ldots,N$ labels the individual plane, out of a total of N planes.

1. For each plane z create a total fluorescence image from several consecutive images, $f_{\text{sum}}^{z} = \sum_{t=1}^{M} f_t^z$ that can serve as the initial template to align the individual fluorescence images to. Align all the individual images to this template and call the aligned images $\tilde{f}_t^z$. It is typically sufficient to perform an affine transformation, i.e., one that maintains straight lines in the image. Using a measure of how well each individual frame aligns to the average (e.g., the normalized cross-correlation), it is possible to identify and discard frames that align very poorly due to a rapid and vigorous behavioral event such as a struggle. It may be beneficial to repeat this procedure once: reaverage all the aligned frames to a new template $\tilde{f}_{\text{sum}}^{z} = \sum_{t=1}^{T} \tilde{f}_t^z$ and align the original frames to this new template to yield the final set of aligned frames $\hat{f}_t^z$. This will improve the alignment and the crispness of the anatomical image, in cases where the larva moved during the period from which the template was taken.
2. It is then important to register all planes to each other, to correct for the slow accumulation of drift across many planes. An anatomical image for each plane can be obtained by summing all the final aligned frames $f_{\text{anatomy}}^{z} = \sum_{t=1}^{T} \hat{f}_t^z$. This z-stack can then be registered just as for the time-series in the previous step. The transformations should then be applied the raw frames from each plane.

3.3.2 Unbiased Identification of Functionally Active Units

A common approach to analyzing imaging data is to group pixels in regions of interest (ROIs) and ask: What is the activity in this ROI during the experiment? ROIs can be selected manually, or chosen using automatic segmentation methods. These methods may identify neurons using morphological criteria, pixel correlations, or both together (*see* refs. [3, 16]). Uneven partitioning of the GCaMP protein between the nucleus and cytoplasm can be used to identify cell bodies (e.g., [17]), or the reverse effect can be achieved, at the expense of some temporal resolution, by specific nuclear targeting of the indicator [18]. Such approaches can identify on the order of 90,000 neurons in the brain of a 6dpf larval zebrafish. Here, we describe a method introduced in [6] that can identify, in an unbiased way, groups of voxels in the brain, corresponding either to cell bodies, or neuropil structures, that are active during an experiment, based on spatial clustering of correlated pixels.

1. Low-pass filter the fluorescence time-series of all individual pixels. The filter should take into account the dynamics of the

calcium indicator in question: for GCaMP5G, which has a half-decay time of 0.667 s [19], one can use a filter cut-off of about 2 Hz. This step makes the assumption that high frequency components in the fluorescence signal arise predominantly from photon shot noise and not from changes in indicator fluorescence.

2. Consider a specific pixel (*x*,*y*) in a given plane. If it is part of an active neuron, its fluorescence will correlate well with its neighboring pixels, some of which will also be part of this neuron. Calculate the correlation between this pixel and its neighboring pixels. Figure 3b shows the case where the neighboring pixels considered are 24 pixels that together with the pixel in question make up a 5 by 5 square. Each pixel can then be represented by its average correlation with surrounding pixels to create an anatomical correlation map, such as the one shown in Fig. 3c (*see* **Note 14**).
3. Generate a control map, to estimate the probability of observing particular correlation values by chance. Partition each pixel's time-series into a number of equally sized segments lasting ~10 s. Randomly permute these segments and concatenate them to create a shuffled time-series for the pixel. Do this for all the pixels, performing each time a new random permutation for each. Create a control anatomical correlation map by

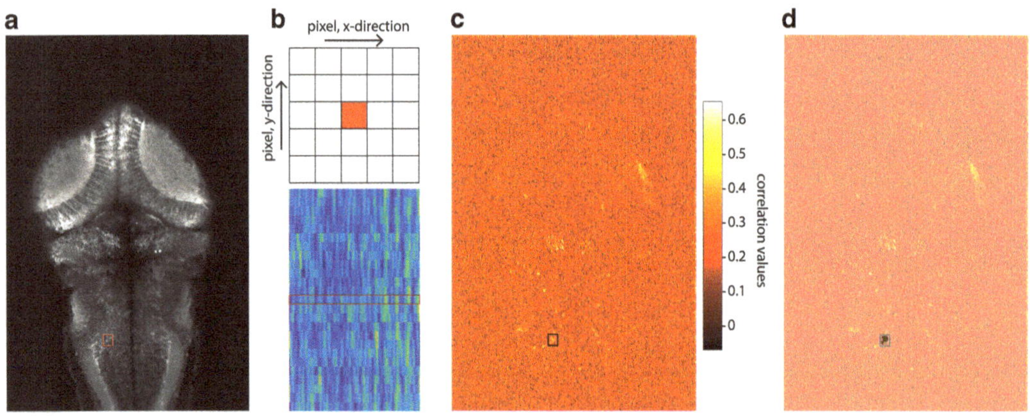

Fig. 3 Unbiased identification of functional units. (**a**) Anatomical image of a single plane created by summing all the fluorescence frames collected for this plane. The field of view corresponds to that shown in Fig. 1a. (**b**) Every pixel in the plane has an associated fluorescence time-series. For each pixel, one can ask how well it's fluorescence time-series correlates with that of its neighboring pixels. Here, we show a pixel in red and its 24 neighboring pixels in a 5 by 5 square. The time-series of all 25 pixels are shown as a color map below. (**c**) Anatomical correlation map obtained by performing the procedure depicted in (**b**) for all the pixels in plane shown in (**a**). (**d**) Starting with the pixels with highest correlation value (seeds), pixels can be aggregated to form functionally active units if their correlation with the seed exceeds a threshold value. One such functionally identified unit is overlaid on the correlation map

repeating the procedure outlined in the point above. Take care that the duration of each segment is not meaningful in the context of the task (i.e., 10 s segments should not be used if the stimulus repeats itself every 10 s).

4. Threshold the correlation map to identify putative active regions. The threshold value can be determined either by inspection, or by using the control performed above as follows: create a histogram of the distribution of correlation values for the true anatomical correlation map. Repeat this for the control map. Divide the former by the later and choose the threshold value to be that appears 20 times more often in the true versus the control correlation map.
5. Use this correlation map, together with the raw data to iteratively locate and segment ROIs. Here, we give the procedure for 3-D data, but the same approach is applicable to 2-D. Determine the location of the first ROI by choosing the voxel with the highest local average correlation and using this as a "seed." This voxel has six closest neighbors. If the correlation of the activity in any of these neighboring voxels with the activity in the "seed" voxel exceeds a predefined threshold (see below), incorporate the voxel into the ROI. Calculate the fluorescence time-series for the new ROI, and repeat the previous steps for the new set of nearest neighbors. On each iteration, perform a morphological close operation (e.g., using the Matlab *imclose* function) to remove holes within ROIs. Once no more voxels can be added, the ROI is considered segmented. Then, find the remaining voxel with the highest local correlation value and use this as the "seed" for the next ROI. The choice of threshold value determines how readily pixels are incorporated, and can be based on the distributions of true and control correlations as outlined above. This algorithm is extremely effective when the signals across planes are expected to be very stereotypical.
6. The segmentation algorithms presented above may yield ROIs that span many adjacent cells with similar functional properties or large areas of neuropil with graded responses. In order to uncover spatial variations in activity, where no clear morphological markers exist, it is useful to split large ROIs into smaller ones by repeated bisection along their longest axis until the constituent ROIs have the size of about one cell body, although, in this case, the ROI boundaries will have no particular significance.

3.3.3 Analysis of Behavior-Activity Correlations

Once we have identified ROIs that are active in our experiment we want to know what feature of the stimulus or behavior the activity reflects. A simple approach is to identify the behavioral parameters which best explain the pattern of neural activity (*see* ref. [14]). The method we present here aims to identify linear correlations between such parameters and observed activity (*see* **Note 15**). The workflow is schematized in Fig. 4.

1. First identify the set of behavioral parameters you wish to test. In this approach, neural activity is compared just with a fixed set of predictors, which reflect prior assumptions about the possible signals carried by single neurons. These could be the "raw" variables measured or controlled in the experiment, such as stimulus velocity or eye position, or "derived" variables, which involve simple combinations, or nonlinear transformations of these raw signals. For example, while we measure absolute eye position over the whole range, we can include predictors that consist only of nasal or temporal deviations from the resting position of the eye, or represent the degree of convergence between the two eyes (*see* Fig. 4b). Similarly, we can generate "derived" predictors from the stimulus velocity, such as binocular rotational or translational motion.
2. The question we ask is: is the neuronal signal we observe linearly related to one of the behavior or stimulus variables we have defined? In order to answer this question, we use correlation analysis. The first step is to convolve all the behavioral and

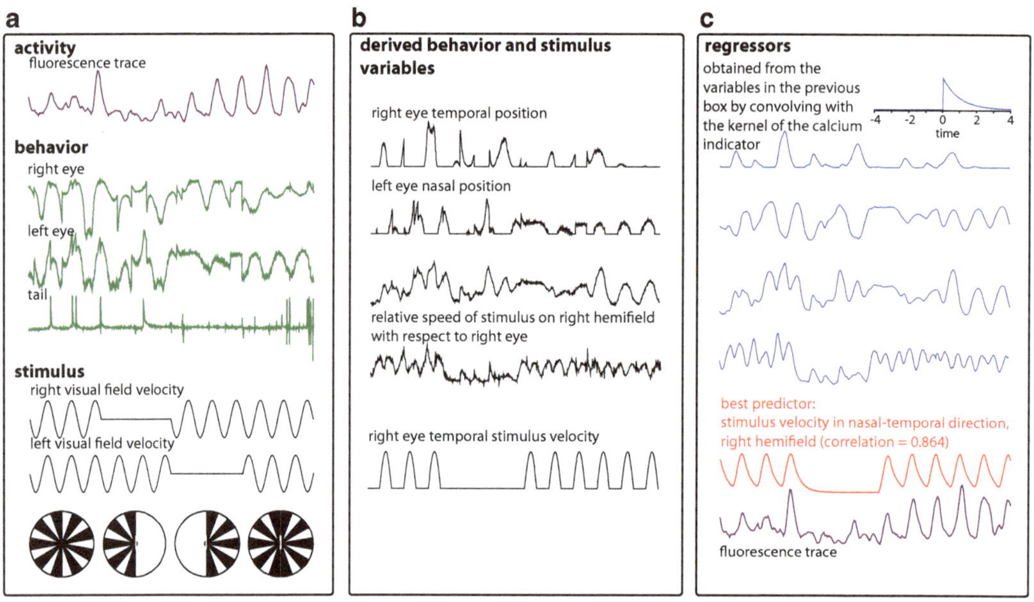

Fig. 4 Correlation analysis of neuronal activity and behavioral data. (**a**) The data collected during the experiment consists of the fluorescence traces of the units detected as described above, the monitored behavior, which in this case consists of left eye position, right eye position and tail deflection, and the stimulus presented. The trace of the unit shown in Fig. 3 is shown here in magenta as an example. (**b**) From the behavioral and stimulus variables directly measured, derived variables are defined as described in the text (*black traces*). (**c**) Each variable to be considered is convolved with the kernel of the calcium indicator (in this case GCaMP5G shown *top right*) to yield the fluorescence expected from a unit that would linearly code for this variable. Examples of these predictors from the variables in (**b**) are shown here in *blue*. For this particular unit (*magenta trace*) the predictor with the best correlation (0.864) is shown in *red*

stimulus variables with the estimated spike response function of the calcium indicator we used in our experiment. In the top of Fig. 4c we show the example of an exponential kernel with decay of 0.667 s, consistent with measured responses of GCaMP5G [19]. The resulting convolved traces are the fluorescence signals we would expect to observe if the underlying neural activity in an ROI was linearly related to the associated behavioral variable.

3. For each ROI, correlate its fluorescence trace with all the regressors. This will provide a vector of correlation values for each ROI. As a first pass classification, each ROI can be assigned to the predictor with the maximum correlation value.
4. The simple approach above may lead to some incorrect assignments, as regressors themselves can be correlated, ROIs may have similar correlation values with several different parameters or the appropriate predictors were not included in the tested set. As an alternative, the data can be clustered according to the complete pattern of behavioral correlations, and each ROI assigned to a cluster. In order to perform cluster analysis using a method such as k-means, it is helpful to reduce the dimensionality of the data. Perform principal component analysis on the correlation vectors calculated for all the ROIs. Choose the first n principal components that explain most of the variance in the data, for example 95%. To reduce the dimensionality of the data, project the correlation vector of each ROI onto this subset of principal components, which are necessarily orthogonal to each other: each ROI will now be assigned an n-dimensional vector of coefficients. Following cluster analysis, the spatial distribution of ROIs belonging to each cluster can be compared, and traces from each cluster can be averaged within or across different trials.

4 Notes

1. It is straightforward to adjust the field of view and working distance of fixed focal length camera objectives by using extension tubes to increase their distance from the camera sensor. Handy web tools for calculating camera field of view and extension tube choices can be found at http://www.1stvision.com.
2. While arbitrary stimuli can be generated directly using low-level graphics libraries such as OpenGL (http://www.opengl.org), a number of freely available packages are specifically tailored for generating visual stimuli for neuroscience, such as Psychophysics Toolbox (http://psychtoolbox.org, Matlab-based) and VisionEgg (http://visionegg.org, Python-based) [20, 21].

3. Larvae may become stuck on the plastic surface of disposable transfer pipettes, and should only be handled using a fire-polished glass pasteur pipette. Twirl a pipette in your fingers while holding the tip in a bunsen flame for a few seconds. After cooling, the pipette tip should feel smooth to the touch, but not be closed.
4. While nacre fish are particularly appropriate for visual experiments, since they have normal eye pigmentation, it is possible to suppress all pigment formation in any fish larva by treatment with 1-phenyl 2-thiourea (PTU). However it should be noted that PTU may directly affect visual processing in the retina [22].
5. Using enough agarose so that it extends to the sides of the Petri dish, can help prevent the agarose from getting unstuck from the Sylgard. However, be careful to limit the depth of agarose, since this will minimize the deviation in pitch of the larva and also ensure that the objective can get close enough to have the appropriate working distance.
6. If the larva is on its side, tap the rim of the dish with the forceps. This will typically induce it to right itself up and swim forward slightly. It is therefore useful to position the larva radially on a side of the dish facing the center and tap the edge slightly. Try to touch the larva as little as possible and avoid touching its head.
7. Make sure the larva is close to the center of the Petri dish, so that the objective does not risk touching the rim.
8. Judging that the fish is not tilted to the left or right can be tricky when using stereo microscopes that offer different views through the left and right eyepieces. To avoid this problem, turn the dish so that the long axis of the fish is oriented left to right.
9. Pass the E3 through a syringe filter, to remove any dust or other particles floating in the water that may interfere with imaging.
10. It is important to cut all the way to the Sylgard to make sure that the agarose will not flake away in pieces. Make sure a closed loop is cut around every section of agarose to be removed, and then pull the whole piece carefully away.
11. Larvae should have vigorous blood flow and a heartbeat at around 2–3 Hz. Edema around the eyes or a cloudy appearance in the brain or spinal cord are indicators of poor health.
12. Due to the very high intensity of the visual stimulus, compared with fluorescence emission, it is usually necessary to filter the projector output to remove any stray light in the emission wavelengths. It is important to consider, when choosing and

placing the filters, that the properties of interference filters are usually only specified for a small range of incidence angles, and that sometimes an absorptive filter may be a better choice.

13. It is advisable to image every plane for a period of time that ensures the occurrence of about three events of interest, to clearly distinguish relevant activity from background. These events may be stimulus presentations or particular behavioral actions. It is also important to include a sufficient "baseline" period. In order to cover most of the brain in a single experiment, we suggest using a stimulus protocol that lasts around 2 min per plane. This will ensure that 250 planes can be imaged in 500 min, just over 8 h. The frame rate should be matched to the indicator kinetics to avoid missing fast signals, at least 2 Hz for GCaMP5G.
14. There may be certain cases when the activity that we are interested in identifying is temporally reproducible from plane to plane. This is the case when the same stimulus sequence is presented in each plane and we are looking at averaged and/or highly reproducible patterns of activity. In this case, the set of neighboring pixels can be extended to include those in adjoining planes and a cube can be considered in Fig. 3b instead of a square. This is however not possible if we are seeking activity related to temporally variable features of behavior: in this case, this analysis must be performed within an individual plane.
15. In many cases, the relationship of neuronal firing to behavioral parameters will be nonlinear. For example, responses may only occur above a threshold, show saturation, or depend on conjunction of multiple variables. In cases where a particular nonlinearity is expected, e.g., threshold-linear responses to eye position, this can be explicitly included in the set of predictors [6, 23]. Alternatively, nonlinear regression methods can be used [24], including some that allow arbitrary relationships between behavioral variables and activity (e.g., *see* ref. [25]), although there will typically be a trade-off between model flexibility and interpretability.

Acknowledgments

The authors would like to acknowledge the contribution of colleagues Claudia Feierstein and Florian Engert to the development of the methods described in this protocol. MBO was supported by a Marie Curie Career Integration Grant, PCIG09-GA-2011-294049. RP was supported by the Max Planck Society.

References

1. Fetcho JR, O'Malley DM (1997) Imaging neuronal networks in behaving animals. Curr Opin Neurobiol 7:832–838
2. Kettunen P (2012) Calcium imaging in the zebrafish. Adv Exp Med Biol 740:1039–1071
3. Ahrens MB, Li JM, Orger MB et al (2012) Brain-wide neuronal dynamics during motor adaptation in zebrafish. Nature 485:471–477
4. Ahrens MB, Orger MB, Robson DN et al (2013) Whole-brain functional imaging at cellular resolution using light-sheet microscopy. Nat Methods 10:413–420
5. Wolf S, Supatto W, Debrégeas G et al (2015) Whole-brain functional imaging with two-photon light-sheet microscopy. Nat Methods 12:379–380
6. Portugues R, Feierstein CE, Engert F et al (2014) Whole-brain activity maps reveal stereotyped, distributed networks for visuomotor behavior. Neuron 81:1328–1343
7. Renninger SL, Orger MB (2013) Two-photon imaging of neural population activity in zebrafish. Methods 62:255–267
8. Niell CM, Smith SJ (2005) Functional imaging reveals rapid development of visual response properties in the zebrafish tectum. Neuron 45:941–951
9. O'Malley DM, Sankrithi NS, Borla MA et al (2004) Optical physiology and locomotor behaviors of wild-type and nacre zebrafish. Methods Cell Biol 76:261–284
10. Portugues R, Engert F (2011) Adaptive locomotor behavior in larval zebrafish. Front Syst Neurosci 5:72
11. Lister JA, Robertson CP, Lepage T et al (1999) Nacre encodes a zebrafish microphthalmia-related protein that regulates neural-crest-derived pigment cell fate. Development 126:3757–3767
12. Severi K, Portugues R, Marques J et al (2014) Neural control and modulation of swimming speed in the larval zebrafish. Neuron 83(3): 692–707
13. Orger MB, Kampff AR, Severi KE et al (2008) Control of visually guided behavior by distinct populations of spinal projection neurons. Nat Neurosci 11:327–333
14. Miri A, Daie K, Burdine RD et al (2011) Regression-based identification of behavior-encoding neurons during large-scale optical imaging of neural activity at cellular resolution. J Neurophysiol 105:964–980
15. Nestares O, Heeger DJ (2000) Robust multi-resolution alignment of MRI brain volumes. Magn Reson Med 43:705–715
16. Mukamel EA, Nimmerjahn A, Schnitzer MJ (2009) Automated analysis of cellular signals from large-scale calcium imaging data. Neuron 63:747–760
17. Romano SA, Pietri T, Pérez-Schuster V et al (2015) Spontaneous neuronal network dynamics reveal circuit's functional adaptations for behavior. Neuron 85:1070–1085
18. Kim CK, Miri A, Leung LC et al (2014) Prolonged, brain-wide expression of nuclear-localized GCaMP3 for functional circuit mapping. Front Neural Circ 8:138
19. Akerboom J, Chen TW, Wardill TJ et al (2012) Optimization of a GCaMP calcium indicator for neural activity imaging. J Neurosci 32:13819–13840
20. Straw AD (2008) Vision egg: an open-source library for realtime visual stimulus generation. Front Neuroinformatics 2
21. Brainard DH (1997) The psychophysics toolbox. Spat Vis 10:433–436
22. Page-Mccaw PS, Chung SC, Muto A et al (2004) Retinal network adaptation to bright light requires tyrosinase. Nat Neurosci 7:1329–1336
23. Kubo F, Hablitzel B, Maschio MD et al (2014) Functional architecture of an optic flow-responsive area that drives horizontal eye movements in zebrafish. Neuron 81:1344–1359
24. Bianco IH, Engert F (2015) Visuomotor transformations underlying hunting behavior in zebrafish. Curr Biol 25:831–846
25. Huber D, Gutnisky DA, Peron S et al (2012) Multiple dynamic representations in the motor cortex during sensorimotor learning. Nature 484:473–478

Chapter 22

A Practical Guide to Light Sheet Microscopy

Davis V. Bennett and Misha B. Ahrens

Abstract

Light sheet fluorescence microscopy is an efficient method for imaging large volumes of biological tissue, including brains of larval zebrafish, at high spatial and fairly high temporal resolution with minimal phototoxicity.

Here, we provide a practical guide for those who intend to build a light sheet microscope for fluorescence imaging in live larval zebrafish brains or other tissues.

Key words Light sheet, Imaging, Microscopy, Zebrafish, Calcium indicators, Systems neuroscience

1 Introduction

Fluorescence microscopy, in conjunction with fluorescent reporters of neural activity, is a useful approach for observing temporal dynamics in neural circuits in intact animals. Point-scanning approaches to imaging generate an image of a plane or volume by scanning a focused spot of light over the sample region [1]. In terms of temporal frequency, this approach scales poorly as sample volume increases because of the need to tile a three-dimensional space with a small spot. To image large volumes of tissue at higher speed requires a different imaging strategy. Light sheet imaging (Fig. 1) is one method for scaling fast imaging to large volumes. A light sheet microscope generates fluorescence by illuminating an entire focal plane simultaneously, or with a rapidly scanned line, thereby gathering fluorescence from every point in the plane at once. Volumetric imaging is thus performed by rapidly imaging adjacent planes in the sample.

The purpose of this chapter is to provide an overview of the steps required to build and operate a light sheet microscope. Although one particular microscope configuration is highlighted here (Figs. 1 and 2) [2–4], note that many different configurations exist and have been described in the literature [2–14]. Within the configuration described here, implementations can take various

Koichi Kawakami et al. (eds.), *Zebrafish: Methods and Protocols*, Methods in Molecular Biology, vol. 1451,
DOI 10.1007/978-1-4939-3771-4_22, © Springer Science+Business Media New York 2016

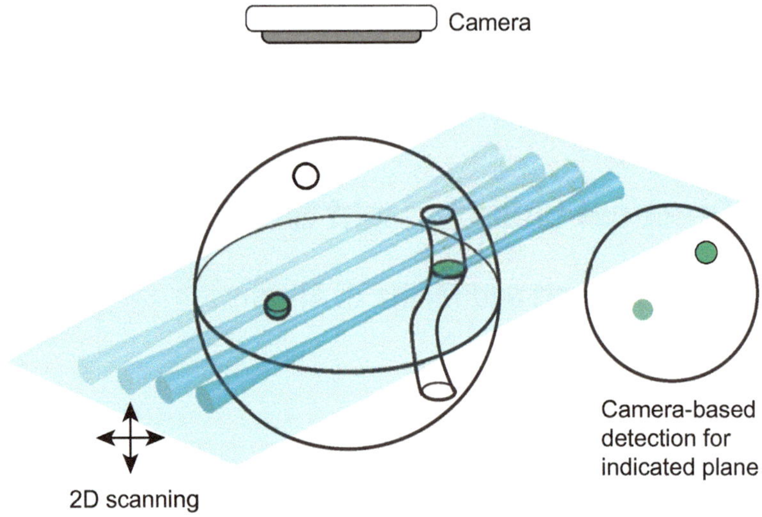

Fig. 1 Conceptual overview of described version of light sheet microscopy. In the described version of light sheet microscopy, a thin laser beam is swept through a fluorescent sample, and fluorescence is imaged using detection optics (not shown) and a camera. The laser beam is focused on the center of the sample to minimize the thickness of the light sheet in the imaged volume, and scanned in two dimensions (*x* and *z*) to cover the entire volume. At every plane, an image is collected as shown. Adapted from [14]

forms, e.g., the lenses, objectives, cameras, and mechanical hardware can be exchanged for other parts. The optimal configuration balances imaging volume, speed, cost, and other factors. The general purpose of the setup outlined here is to scan a volume of about $800 \times 500 \times 300$ μm without moving or rotating the sample using a laser beam in the visible wavelength to excite the fluorophore in one-photon excitation mode. We assume that a water-dipping objective is used for imaging, and air objectives for excitation. A light "pencil" rather than a sheet is used to scan in two directions (Fig. 1), thereby covering the entire sample about once a second (depending on the configuration); the objective is moved along with the excitation laser beam so that the excited plane is always in the focal plane of the objective. A fast camera acquires images at every plane, thus one horizontal laser sweep corresponds to one camera frame. This setup can be used for multiple purposes including imaging neural activity with, e.g., calcium indicators [5] or morphology using a variety of markers.

2 Materials

The materials listed here are those of one particular light sheet microscope design [2–4]. Note that considerable flexibility exists in the choice of optical, mechanical and electronic components, and many alternatives exist. For example, camera technology is rapidly advancing and we expect upgraded cameras to be already available around the time of publication of this chapter.

2.1 Detection Optics: Suggested Parts List

1. Fast scientific CMOS camera: Hamamatsu ORCA-Flash4.0 V2 (Hamamatsu Photonics, Shizuoka Prefecture, Japan).
2. Detection objective: CFI LWD 16XW, NA 0.8 (Nikon, Tokyo, Japan).
3. Detection tube lens: Nikon CFI second objective lens unit, 20 mm focal length (Nikon, Tokyo, Japan).
4. Fluorescence detection filters: 525/50 nm BrightLine, 32 mm diameter (Semrock, Rochester, NY, USA).
5. Neutral density filters: ND values between 1 and 3 (Melles Griot, Rochester, NY, USA).
6. Electronic filter wheel and controller (Ludl, Hawthorne, NY, USA).
7. Objective z-axis translator: 725.4 CD piezo stage (Physik Instrumente, Karlsruhe, Germany).
8. Objective z-axis translator controller: E665 piezo amplifier (Physik Instrumente, Karlsruhe, Germany).
9. Large optical rail for vertical mounting of detection optics: XT95-750 (Thorlabs, Newton, NJ, USA).

2.2 Excitation Optics: Suggested Parts List

1. Laser for fluorescence excitation: 488 nm, 80 mW peak output (Omicron, Rodgau-Dudenhofen, Germany).
2. Two broadband metallic mirrors for beam steering: PF10-03-P01 (Thorlabs, Newton, NJ, USA).
3. Two dual-axis kinematic mirror mounts: KM100 (Thorlabs, Newton, NJ, USA).
4. Electronic filter wheel and controller (Ludl, Hawthorne, NY, USA).
5. Objective translator: 725.4 CD piezo stage (Physik Instrumente, Karlsruhe, Germany) (*see* **Note 3**).
6. Objective translator controller: E665 piezo amplifier (Physik Instrumente, Karlsruhe, Germany).
7. Telecentric scan lens: 66-S80-30T-488-1100 nm, custom objective (Special Optics, Wharton, NJ, USA).
8. Excitation objective: XLFLUOR 4X, NA 0.28 (Olympus, Tokyo, Japan).
9. Tube lens: U-TLU (Olympus, Tokyo, Japan).
10. Dual-axis galvanometer scanner: Model 6215H optical scanner (Cambridge Technology, Bedford, MA, USA).
11. Galvanometer scanner driver board: MicroMax 673XX (Cambridge Technology, Bedford, MA, USA).
12. Low-profile dual-axis translation/rotation stage for galvanometer: XYR1 (Thorlabs, Newton, NJ, USA).

13. Optical rail for mounting scan lens, tube lens, and objective lens in series (Owis, Staufen, Germany).

2.3 Sample Chamber

1. Water-tight, open-top sample chamber with windows permitting lateral entry of excitation laser light and top entry of detection objective: Custom machined (*see* **Note 1**).
2. Three-axis translation stages with ~1 μm minimum step size and approximately 3 cm travel for sample alignment w.r.t. objectives (e.g., Physik Instrumente, Karlsruhe, Germany).

2.4 Control Equipment

1. Software for controlling the microscope and acquiring data (*see* Subheading 3.5 and **Note 4**).
2. Workstation for data acquisition and instrument control (e.g., Colfax International, Sunnyvale, CA, USA).
3. PC control of PXI boards: PXI-8360 MXI-Express (National Instruments, Austin, TX, USA).
4. One I/O interface board per optical axis (two total): PXI-6733 (National Instruments, Austin, TX, USA).
5. PXI chassis: NI PXI-1042Q (National Instruments, Austin, TX, USA).
6. BNC connector box: BNC-2110 (National Instruments, Austin, TX, USA).
7. Serial interface board: PXI-8432/2 (National Instruments, Austin, TX, USA).

2.5 Support Equipment

1. Large (at least 1 m × 1 m) optical table floated on air (Newport Corporation, Irvine, CA, USA) (*see* **Note 2**).
2. Right-angle prism mirror for beam alignment: Catalog no. MRA05-F01 (Thorlabs, Newton, NJ, USA).
3. Two optical iris diaphragms capable of being mounted in the excitation beam pathway for alignment: D25SZ (Thorlabs, Newton, NJ, USA).

3 Methods

The assembly of the microscope is illustrated in Fig. 2. We describe the assembly as starting from the sample (where detection and excitation converge) and then proceeding to the detection and excitation arms of the microscope. We use the following coordinate system in this guide (*see* Fig. 2):

z-axis: axis parallel to the detection objective.

y-axis: axis parallel to the excitation objective.

x-axis: axis parallel to the direction of the sweep of the light sheet.

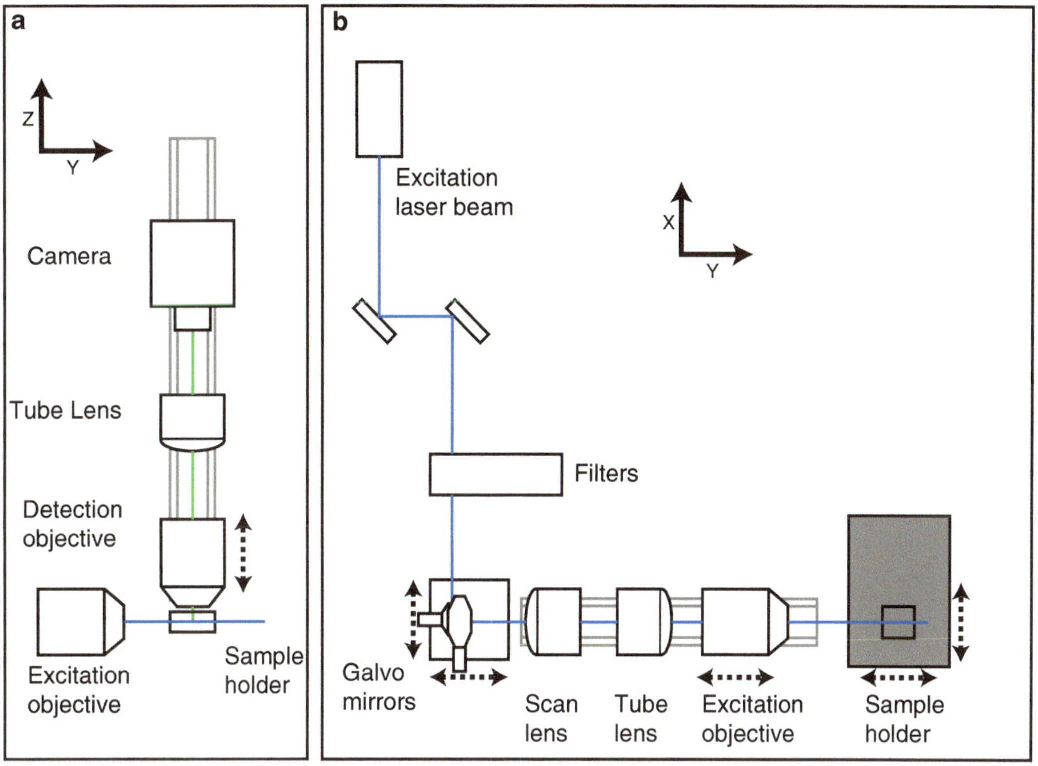

Fig. 2 Arrangement of the microscope components. Side view of the detection system (**a**) and top view of the excitation system (**b**). Drawings are not to scale

3.1 Assembly of the Sample Stage

1. Mount the *x*, *y*, and *z* stages for sample placement in a three-axis configuration (*see* **Note 1**).
2. For alignment, mount the 45° mirror (Fig. 3a′) to the floor of the sample chamber and mount the water-filled sample chamber on the top of this three-axis stage. The position of the alignment mirror in the sample chamber acts as reference for the subsequent assembly of the detection and excitation arms (*see* Subheading 3.4).

3.2 Assembly of the Detection Arm

1. *See* Fig. 2a for a schematic overview of the detection arm. Mount the 95 mm vertical rail to the optical table so that the optical devices mounted to the rail are directly above the sample area. Devices attached to this rail will form the detection arm of the microscope.
2. Mount the camera at the top of the detection rail.
3. Mount the tube lens underneath at one focal distance from the camera chip.
4. Mount the filter holder or filter wheel below tube lens.
5. Connect the detection objective to the objective *z* piezo and mount this assembly to the vertical rail. The objective should

be roughly one focal distance underneath the tube lens, but this location is flexible and the allowable range is provided by the tube lens and objective manufacturer (usually approximately 100–150 mm).

6. Position the objective so that the sample position is one focal distance away from the objective.

3.3 Assembly of the Excitation Arm

1. *See* Fig. 2b for a schematic overview of the excitation arm. Mount the optical rail on the optical table at a distance less than the focal length of the excitation objective and oriented so that optical elements placed on the rail will be aligned to the sample.
2. Connect the excitation objective on the translation stage or *y* piezo, and mount the assembly to the rail. Position the excitation objective so that the alignment mirror is roughly in focus, similar to the procedure for positioning the detection objective (*see* **Note 3**).
3. Mount the tube lens to the horizontal rail at an approximate distance from the excitation objective determined by the tube lens specifications (the setup is relatively insensitive to this; the distance is usually in the 100–150 mm range).
4. Mount the scan lens at a distance from the tube lens equal to the sum of the focal lengths of the two lenses.
5. Mount the two-axis galvo system to the *xy* translation stage, and place this so that the mirrors are one focal length of the scan lens away from the scan lens. The smaller, faster galvo mirror should be positioned so that its rotation will cause the incoming beam to be translated in *x*. This entails that the larger, slower mirror will control translation of the beam in *z*. Ensure that the height of the excitation optics (scan lens, tube lens, objective) matches the height of the second galvanometer mirror.
6. Mount any optical filters in the filter holders and mount these in the beam path leading to the first galvanometer mirror.
7. Mount the two kinematic static beam steering mirrors such that they can steer the laser beam onto the galvo mirrors. Each mirror should fold the excitation beam by 90°. The four degrees of freedom (two tilt angles per mirror) are used to ensure the laser beam is parallel to the optical elements along the excitation path (*see* Subheadings 3.3 and 3.4).

3.4 Alignment of Excitation Laser

Power on the laser at an eye-safe intensity level and use the two kinematic beam steering mirrors to guide the laser beam through the center of each optical element on the excitation pathway. The four degrees of freedom of the steering mirror should be used to position the beam at the center of the axis running from the galvo mirrors to the excitation objective. Irises mounted on the rail and centered along the axis midline can be used to ensure that the laser beam is centered.

Note that correct alignment depends on the precise default locations of the galvo mirrors — 45° relative to the horizontal (for *x*) and vertical (for *z*) axes; fine tuning is likely required. Rotating the galvos in the galvo assembly while applying the default voltage is one method for tuning the default position of the mirrors; another is using software control to identify the offset voltage that centers the laser beam on the excitation axis.

3.5 Fine Alignment and Light Sheet Generation

The excitation light should uniformly illuminate the sample at the focal plane of the detection objective. Approximating this ideal requires tuning several degrees of freedom of the excitation light path. The light sheet should strike the sample at the focal plane of the detection system, and the light sheet should pass through the sample parallel to the focal plane of the detection system (Fig. 3a). Visualizing the excitation light at the sample position is critical for properly aligning the excitation light. A 45° mirror (Thorlabs right angle prism mirror, *see* Subheading 2.4) oriented to direct excitation light into the detection objective, and thus the camera, is a simple and effective way to visualize the light sheet for tuning purposes. A schematic of this arrangement is shown in Fig. 3a′, and alignment is performed according to the following procedure.

1. Mount the 45° mirror to a sample chamber so that the center of the reflective surface is in the location where the sample would normally be (Fig. 3a′). The excitation beam should thus hit the mirror on the reflective surface. Some movement in *x*, *y*, and *z* of the translation stages may be required. Placing some scratch marks on the reflective surface of the mirror can help position the mirror into view.
2. Drive the *x*-axis galvo with a periodic waveform to generate a light sheet (see possibilities for control software, below) and image with the camera.
3. Move the detection objective until the light sheet is in focus.
4. Center the light sheet in *y* by translating the excitation objective along the *y*-axis until the light sheet is thinnest at the detection focal plane (Fig. 3b). A thin line should be visible in the center of the field of view. Some fine tuning of the galvo offsets followed by additional tuning of the excitation objective position may be required for optimal alignment. Rotate the galvo assembly until the light sheet is parallel to the *x*-axis.
5. Adjust the angle of the light sheet at the focal plane by translating the galvanometer assembly along the *y*-axis to adjust the angle at which the center of the light sheet intersects the focal plane (Fig. 3d). Ideally, the axis of propagation of the light sheet should be perfectly parallel to the focal plane of the detection

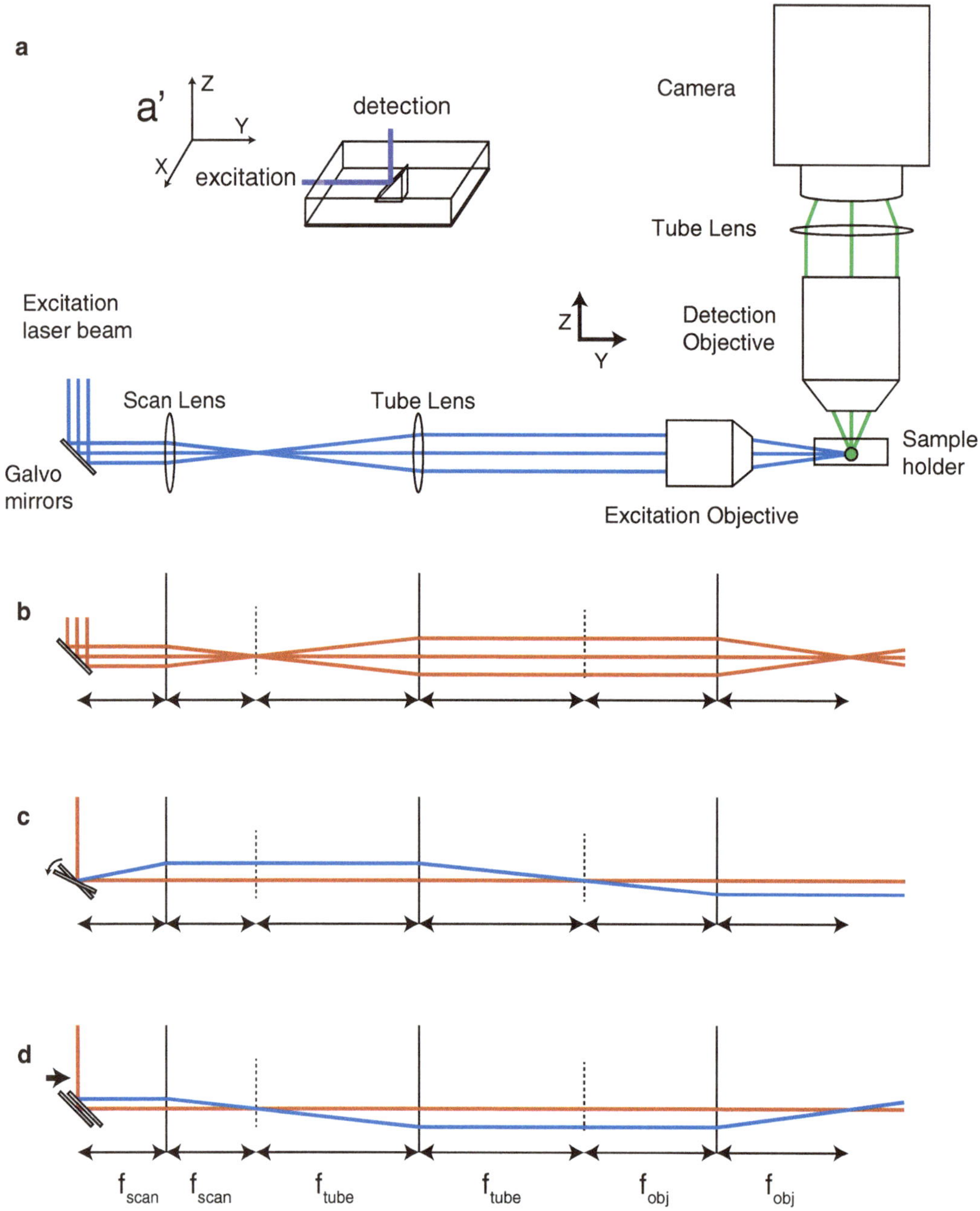

Fig. 3 Ray diagrams and alignment procedures. (**a**) Optical arrangement of the light sheet microscope, with rays shown for a static excitation laser beam (i.e., without movement of the galvo mirrors). Inset (**a**′) diagram of the sample holder with a 45° mirror used for alignment. (**b**) Ray diagram as in **a** for a static excitation beam. (**c**) Diagram of center beam position for two rotations of the galvo mirror. Rotation of the *z* galvo mirror becomes *z*-translation of the light sheet. Rotation of the *x* galvo mirror (not shown) leads to a sweep of the excitation laser beam over one plane. (**d**) Tilt alignment of the light sheet by translation of the galvo assembly. Moving the galvos closer to the excitation objective makes the beam tilt upward; moving it farther away makes it tilt downward. Other alignment procedures include moving the detection objective so that the focal plane, i.e., the point in space where the excitation beam is thinnest, is centered at the sample

objective. One way to detect tilts in the light sheet is to translate the chamber containing the 45° mirror along the *y*-axis. A horizontal light sheet will always strike the mirror in the same location, whereas a tilted light sheet will strike it at different locations when the chamber is moved in *y*. To detect such changes, scratch marks on the mirror are required as a reference, to keep track of the location of the beam on the mirror.

6. The light sheet can also be tilted in the other direction if the galvos are not mounted optimally. The best solution is to make sure the galvos are in the right position, but it is also possible to correct for such a tilt using control software, by introducing a linearly increasing *z*-offset as the laser beam sweeps in the *x* direction, with this *z*-offset resetting at the start of every plane.
7. Alternatively or additionally, one can use a solution of fluorescein to image fluorescence generated by the excitation laser beam; the laser beam profile should be brought in focus by moving the detection objective, and tilted by moving the galvanometer assembly as described above.

3.6 Operation of the Light Sheet Microscope

Light sheet imaging is based on sweeping the excitation laser beam in the *x* direction using the *x* galvo mirror while taking a camera frame, then stepping the light sheet one plane downward (or upward) using the *z* galvo and simultaneously moving the objective downward, taking the next frame, etc., so that a given number of planes covers the entire imaged volume. Thus the voltage waveform driving the *x* galvo will consist of fast oscillating signals, and the voltage waveform driving the *z* galvo of slow oscillating signals. Typically, speeds of several Hz can be achieved for volumes the size of a zebrafish brain, but much higher speeds are in principle possible. The excitation laser beam needs only a few milliseconds to scan a plane; a camera frame can be acquired in about 10 ms or less, depending on the camera. Thus when, for example, covering the brain volume with 40 optical sections, a volumetric speed of 2.5 Hz can be achieved using 100 Hz imaging.

Multiple options exist for control software, including custom-written software, published methods [2–4], open-source software such as OpenSPIM with Micro-Manager (openspim.org and micromanager.org), etc.

3.7 Alternative Approaches

Alternative implementations of light sheet microscopy include microscopes that use a cylindrical lens instead of an *x* galvo mirror [6, 7], microscopes with electrically tunable lenses before the objective that thereby do not require the detection objective to move [8], multiangle light-sheets [2, 9], objective-coupled planar illumination [7], SCAPE microscopy using a single objective for light sheet generation and imaging [10], extended depth of field light-sheet microscopy [11, 12], among others. This microscopy modality will continue to evolve and different options will best suit specific needs.

4 Notes

1. The minimal requirements for the sample chamber are that it should contain water and allow optical access to the sample from the top and side, but more elaboration on the design will likely be necessary to suit the demands of a specific experiment. Unlike some light sheet microscopes, the excitation objective in the design described here is not water dipping, which allows more flexibility in the design of the sample chamber. In order to repeatedly position the fish at the appropriate location in the sample chamber a small (~1 mm height) laser-cut acrylic pedestal mounted at the center of our sample chamber has proven very useful (3D STL design file available upon request).
2. We present a microscope that uses only a single optical breadboard, but using a large three-axis stage assembly for sample positioning or the need for an optical path below the sample may require elevating the sample to such an extent that excitation and detection optics must be mounted on a second optical breadboard, resulting in a two-tiered design (*see* ref. [4]). There is considerable flexibility in such a design as optical table manufacturers can produce custom tables with nonrectangular geometries.
3. Adjusting the *y* position of the excitation objective will translate the beam waist along the *y*-axis. For this purpose, we use the same fast piezo objective positioner as for our detection objective, but the speed and precision of this positioner are not strictly necessary unless it is desired to adjust the position of the beam waist very rapidly. Therefore it may be simpler and more economical to use a manual translation stage for the detection objective instead.
4. Light sheet microscopes generate data at high rates, and a few hours of functional data can generate terabytes of raw data. This presents challenges for acquisition, storage, and analysis. For acquisition, we recommend a powerful computer with a RAID array of hard drives for fast data writing. For storage, centralized storage including cloud storage can be used, or local file servers. Analysis presents multiple challenges, including "higher-level" problems such as model fitting, finding patterns of neuronal activation, etc., as well as "lower-level" problems such as finding regions of interests delineating neurons and regions of neuropil. Although many analyses can be run on workstations, distributed computing platforms also exist, including the Thunder library [13], available at thunder-project.org.

Acknowledgments

We thank Philipp Keller for helpful suggestions. This work was supported by the Howard Hughes Medical Institute.

References

1. Denk W, Strickler JH, Webb WW (1990) Two-photon laser scanning fluorescence microscopy. Science 248:73–76
2. Tomer R, Khairy K, Amat F et al. (2012) Quantitative high-speed imaging of entire developing embryos with simultaneous multiview light-sheet microscopy. Nat Methods 9:755–763
3. Ahrens MB, Orger MB, Robson DN et al. (2013) Whole-brain functional imaging at cellular resolution using light-sheet microscopy. Nat Methods 10:413–420
4. Vladimirov N, Mu Y, Kawashima T et al. (2014) Light-sheet functional imaging in fictively behaving zebrafish. Nat Methods 11:883–884
5. Chen T-W, Wardill TJ, Sun Y et al. (2013) Ultrasensitive fluorescent proteins for imaging neuronal activity. Nature 499:295–300
6. Voie AH, Burns DH, Spelman FA (1993) Orthogonal-plane fluorescence optical sectioning: three-dimensional imaging of macroscopic biological specimens. J Microscopy 170:229–236
7. Holekamp TF, Turaga D, Holy TE (2008) Fast three-dimensional fluorescence imaging of activity in neural populations by objective-coupled planar illumination microscopy. Neuron 57:661–672
8. Fahrbach FO, Voigt FF, Schmid B et al. (2013) Rapid 3D light-sheet microscopy with a tunable lens. Opt Express 21:21010–21026
9. Wu Y, Wawrzusin P, Senseney J et al. (2013) Spatially isotropic four-dimensional imaging with dual-view plane illumination microscopy. Nat Biotechnology 31:1032–1038
10. Bouchard MB, Voleti V, Mendes CS et al. (2015) Swept confocally-aligned planar excitation (SCAPE) microscopy for high speed volumetric imaging of behaving organisms. Nat Photonics 9:113–119
11. Tomer R, Lovett-Barron M, Kauvar I, et al. (2016) Cell 163:1796–1806
12. Quirin S, Vladimirov N, Yang CT, et al. (2016) Opt Letters 41:855–858
13. Freeman J, Vladimirov N, Kawashima T et al. (2014) Mapping brain activity at scale with cluster computing. Nat Methods 11:941–950
14. Keller PJ, Ahrens MB (2015) Visualizing whole-brain activity and development at the single-cell level using light-sheet microscopy. Neuron 85:462–483

Chapter 23

Calcium Imaging of Neuronal Activity in Free-Swimming Larval Zebrafish

Akira Muto and Koichi Kawakami

Abstract

Visualization of neuronal activity during animal behavior is a critical step in understanding how the brain generates behavior. In the model vertebrate zebrafish, imaging of the brain has been done mostly by using immobilized fish. Here, we describe a novel method to image neuronal activity of the larval zebrafish brain during prey capture behavior. We expressed a genetically encoded fluorescent calcium indicator, GCaMP, in the optic tectum of the midbrain using the Gal4-UAS system. Tectal activity was then imaged in unrestrained larvae during prey perception. Since larval zebrafish swim only intermittently, detection of the neuronal activity is possible between swimming bouts. Our method makes functional brain imaging under natural behavioral conditions feasible and will greatly benefit the study of neuronal activities that evoke animal behaviors.

Key words Calcium imaging, GCaMP, Tectum, Prey capture, Vision, Paramecium, Gal4-UAS system

1 Introduction

To understand how neuronal activity generates behavior, it is desirable to record neuronal activity from a freely behaving, unrestrained animal. Electrical activity of neurons can be indirectly measured by detecting the voltage-gated calcium influx. Development of fluorescent calcium probes and advancement in imaging technology has made it possible to simultaneously record from multiple neurons [1]. However, imaging of an animal in motion is technically challenging, for the movement of the brain results in blurred images. Therefore, most calcium imaging studies involve restraining the animal. In larval zebrafish, for example, the fish is partially restrained by embedding them in agarose [2–4]. This experimental setup raises concerns on possible stress-related activity that is irrelevant to the behavior to be studied. Another concern is that behavioral study is limited by the fact that a partially restrained larva cannot show the full spectrum of its behavioral repertoire. Thus, the use

Koichi Kawakami et al. (eds.), *Zebrafish: Methods and Protocols*, Methods in Molecular Biology, vol. 1451,
DOI 10.1007/978-1-4939-3771-4_23, © Springer Science+Business Media New York 2016

of free-swimming (i.e., unconstrained) fish in functional brain imaging is essential to study physiologically relevant neuronal signals and complement studies that used restrained larvae.

Because zebrafish larvae show only intermittent swimming activity, we can conduct calcium imaging of the brain and detect fluorescence changes during the quiescent periods. Here, we describe the detection of calcium signals in the optic tectum of the midbrain during visual perception of a paramecium (prey for the larvae). To suppress movement of the larva along the *Z*-axis (i.e., moving out of the focal plane), we used a shallow chamber (0.8 mm in depth) that we found suitable for 4- to 7-day-old larvae. The small diameter (9 or 13 mm) of the chamber provided an arena for zebrafish larvae to exhibit prey capture behavior, while reducing the need to move the *XY* stage to keep the larvae within the camera view. To maximize the size of the camera view while obtaining a fluorescent image that was bright enough, we used a 2.5×/N.A.0.12 or 5×/N.A.0.15 objective lens.

Detection of neuronal activity in a subset of cells can be observed with an epifluorescence microscope (i.e., without confocal microscopy) when these cells are specifically labeled with a functional probe. Specific labeling can be achieved by expressing genetically encoded calcium indicators such as GCaMP [5], whose expression can be driven by a specific promoter or by using the Gal4-UAS system [6–8]. A collection of Gal4FF (a variant of Gal4 [9]) driver lines with specific expression patterns has been generated, and some of them express the UAS effector in subsets of neurons in the brain [10]. Because the pigments on the body surface block both excitation and emission light, we used a mutant strain, *nacre*, which has no melanin-producing cells except for those in the retinal pigment epithelium [11]. In this chapter, we show an example of a study on vision, but the same principle may be applied to the study of other sensory modalities, such as olfaction and gustation with appropriate Gal4 lines.

2 Materials

2.1 Preparation of Zebrafish Larvae

1. UAS:GCaMP transgenic fish maintained in the *nacre* mutant background.
2. A Gal4 transgenic fish with the desired expression pattern, also maintained in the *nacre* mutant background (*see* **Note 1**).
3. E3 water (5 mM NaCl, 0.17 mM KCl, 0.33 mM $CaCl_2$, 0.33 mM $MgSO_4$, 0.00001 % Methylene Blue).
4. System water used to maintain the adult fish (Marine Biotech). (Seachem Marine Salt is automatically added to reverse osmosis water to a conductivity of 470 µs. $NaHCO_3$ is used to adjust the pH to 6.8) (*see* **Note 2**).
5. A pressed glass plate with nine concave depressions.

6. Disposable transfer pipets.
7. A fluorescence stereoscope. We use the Leica MZ16FA FLUOIII with the filter set of "GFP2" (Excitation filter: 480/40 nm (460–500 nm) and Barrier filter: 510 nm).

2.2 Preparation of Paramecia

1. Paramecia. The paramecia may be obtained from local commercial suppliers, other researchers, or the BioResource center in Japan (http://nbrpcms.nig.ac.jp/paramecium/).
2. Rice straw.
3. EBIOS (dry beer yeast, Asahi Food & Healthcare, Co. Ltd., Tokyo, Japan).
4. 1000 mL autoclavable beaker.
5. 300 mL glass beakers.
6. A squeeze bottle.
7. A funnel.
8. Stainless steel sieves with 75 μm aperture and 150 μm aperture (Tokyo Screen).
9. Nylon mesh (N-No. 508T-K (filtered object size:13 μm), and N-No. 380T (32 μm aperture), Tokyo Screen).

2.3 Recording Chamber

1. Chamber with a depth of 0.8 mm. We use Secure-Seal Hybridization Chambers (13 mm diameter × 0.8 mm depth, GRACE BIO-LABS. Item:621502), or Secure-Seal Hybridization Chamber Gasket, 8 chambers (9 mm diameter × 0.8 mm depth, Molecular Probes. S-24732).
2. A slide glass. We use Matsunami micro slide glass Superfrost (MAS-coated) or Matsunami micro slide glass (APS-coated).
3. A cover glass (e.g., Matsunami 24 × 32 mm).
4. A microblade (e.g., Feather Safety Razor Co. Ltd. Micro Feather P-715).

2.4 Calcium Imaging Equipment

1. An epifluorescent microscope and low magnification objective lenses. We use Zeiss Imager.Z1 and 2.5×/N.A.0.12 and 5×/N.A.0.15 lenses.
2. A camera suitable for fluorescence time-lapse imaging. We use a scientific C-MOS camera (ORCA-Flash 4.0V2, Model:C11440-22CU, Hamamatsu Photonics, Japan), and previously used a cooled CCD camera (ORCA-R2, Model: C10600-10B, Hamamatsu Photonics, Japan).
3. Frame grabber (Active Silicon Fire Bird, provided by the camera manufacturer as a part of the camera system).
4. PC for image acquisition. We use a Dell Precision T3610 (recommended by the camera manufacturer) with 32GB RAM and 4× 256GB solid state disks in a RAID 0 configuration.

5. Image acquisition software. We use Hamamatsu Photonics HCImage with the Hard Disc Recording module.
6. Image analysis software. We use the freeware Fiji/Image J (http://fiji.sc/Fiji).

3 Methods

3.1 Preparation of the Larvae

1. Setup crosses of the chosen Gal4 line and the UAS:GCaMP line 5 or 6 days before the imaging experiment. Collect eggs on the following morning. Raise them in E3 water at 28.5 °C (*see* **Note 3**).
2. Transfer the embryos with a disposable pipet to a pressed glass plate and view under the fluorescent stereoscope. Sort the GCaMP-expressing larvae at an early stage if it is being expressed. It is much easier to sort them before they hatch or start to swim. We maintain our Gal4FF lines with UAS:GFP as a reporter. Thus, a given clutch will contain GCaMP-positive, GFP-positive, and double-positive embryos in addition to the nonfluorescent ones. GCaMP fluorescence is weaker than GFP fluorescence, and the two could be distinguished without difficulty.
3. Raise embryos in E3 water to 4–7 days postfertilization, when they start to show prey capture behavior. Feeding of the larvae is not necessary for up to 1 week after fertilization. Remove the chorions and mold as necessary to keep the embryos clean during development.

3.2 Paramecium Culture

Note that the procedures described here are for the entire system in our lab, so the culture volume may be scaled down as necessary.

1. Autoclave 1000 mL water and 5 g of rice straw in a 1000 mL beaker covered with aluminum foil for 20 min at 121 °C. This will kill most bacteria but will allow *Bacillus subtilis*, which naturally resides on the rice straw, to survive and act as the food source for the paramecia.
2. Put 3 tablets of EBIOS into the autoclaved water with the rice straw.
3. On the following day, put 30 mL paramecium solution from the previous culture (or the original solution obtained from the supplier) into 1000 mL of the above EBIOS solution and maintain it at room temperature (approx. 25 °C) for 2 weeks. After 3 or 4 days, the solution should look cloudy and then become transparent thereafter (Fig. 1a).

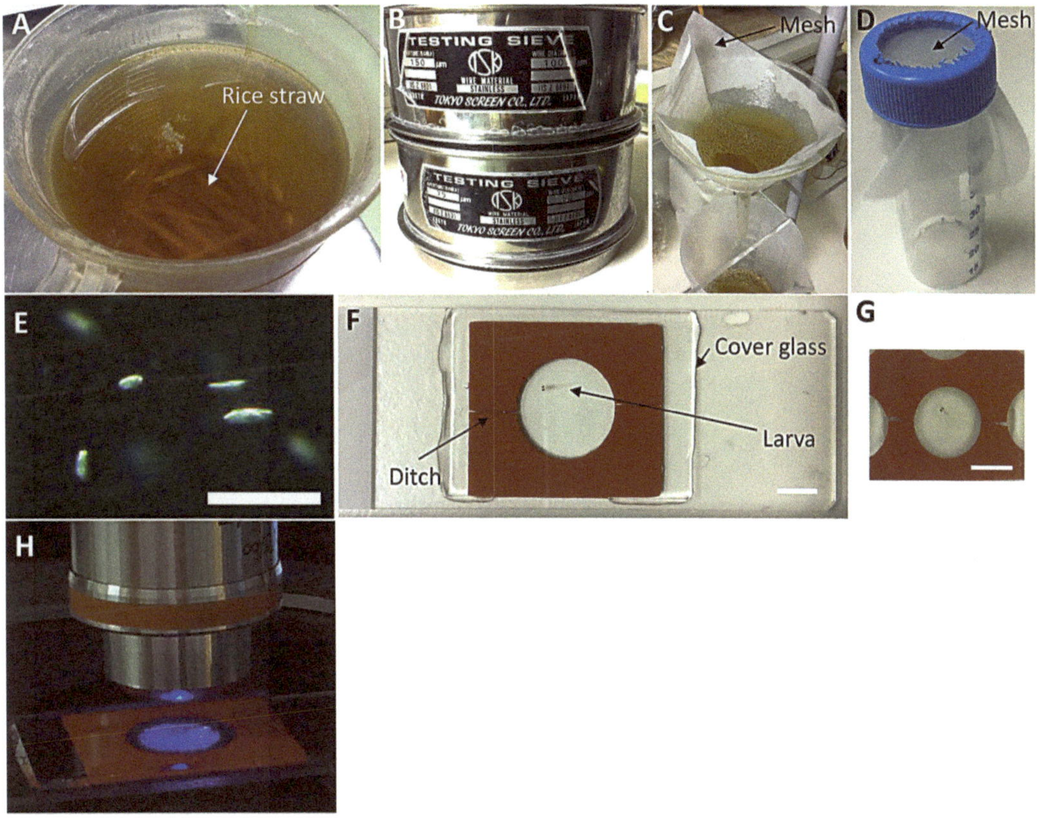

Fig. 1 Preparation of the paramecia and GCaMP imaging setup. (**a**) A 2-week culture of paramecia. The culture becomes transparent after 3 or 4 days, as shown in this picture. (**b**) Sieves to remove debris from the paramecium culture. (**c**) Collection of the paramecia in the flow-through from filtration in **b** onto a nylon mesh. (**d**) A nylon mesh to size-filter paramecia for visual stimuli. (**e**) Paramecia prepared through steps **a**–**d**, photographed in dark-field illumination. Scale bar = 0.5 mm. (**f**) A zebrafish larva in a recording chamber with a diameter of 13 mm and depth of 0.8 mm. Scale bar = 5 mm. (**g**) A zebrafish larva in a recording chamber with a diameter of 9 mm and depth of 0.8 mm. Scale bar = 5 mm. (**h**) A GCaMP-expressing zebrafish larva and paramecia in a recording chamber placed under a fluorescent microscope. Excitation light (*blue*) for GFP was used for GCaMP recording. The excitation light also serves to make visible the paramecia to both the zebrafish larva and the naked eye. Optionally, bright-field illumination can be simultaneously used to make the paramecia more easily recognizable in the camera view

3.3 Preparation of Paramecia for the Use of Prey Capture Behavioral Recording

1. Put the paramecium culture through two sieves (a 150 μm aperture, followed by 75 μm aperture) to remove debris (Fig. 1b), then collect them on a nylon mesh (N-No. 508T-K) using a funnel (Fig. 1c). Resuspended them in a 300 mL beaker using a squeeze bottle of system water. This concentrated paramecium stock solution can be kept at room temperature for 1 or 2 days.
2. Just before imaging experiments, put a few mL of paramecium stock solution into a custom-made tube with a nylon mesh (N-No. 380T; Fig. 1d) submerged in system water in a 300 mL

beaker. Rinse the paramecia in another 300 mL beaker of system water. Only the relatively large paramecia will be left on the nylon mesh. After rinsing in system water, transfer the paramecia left on the nylon mesh into a small dish using a squeeze bottle of system water. Under a stereo microscope, use a micropipette to carefully pick several paramecia that show high locomotor activity (Fig. 1e) and transfer them to the recording chamber (*see* Subheading 3.4).

3.4 Preparation of a Recording Chamber

1. Put the Secure-Seal Hybridization Chamber onto a clean slide glass, with the adhesive end attached to the slide glass (*see* **Note 4**).
2. Carefully remove the cover on the other end of the Secure-Seal Hybridization Chamber.
3. Using a microblade, cut the chamber to make a narrow ditch so that the water in the chamber is contiguous with the outer water reservoir (*see* **Note 5**).
4. Put a zebrafish larva and the paramecia into the chamber, fill the chamber with system water, and cover with a 36 × 24 mm cover glass (Fig. 1f) (*see* **Note 6**).

3.5 Calcium Imaging

1. Set the recording chamber containing a GCaMP-expressing larva and several paramecia under an epifluorescent microscope equipped with a fluorescence camera. Use an excitation/emission filter set for GFP. We use the excitation light (a mercury lamp) at maximal intensity to obtain the best possible quality of the image as long as the excitation light showed no toxic effects.
2. Let the larva adapt to this new environment. The locomotor activity may vary from individual to individual or over time. For the larva, adaptation to the new environment (i.e., the recording chamber and the excitation light illumination) may take approximately 20–30 min.
3. Focus on the larval brain and start time-lapse imaging with the selected image acquisition software (HCImage in this case) (*see* **Note 7**).
4. Place the larva at the center of the camera view by manually moving the *XY* stage of the microscope to ensure that the behavioral event (e.g., prey capture) will happen in the camera view (*see* **Note 8**).
5. After recording, open the movie file in ImageJ with the Bio-Formats plug-in (openmicroscopy.org) (*see* **Note 9**).
6. Although calcium signals may be readily recognizable in the raw fluorescence images in some cases, their visualization can be enhanced by image processing (Fig. 2b, d). To quantify the changes in fluorescence intensity, divide the indi-

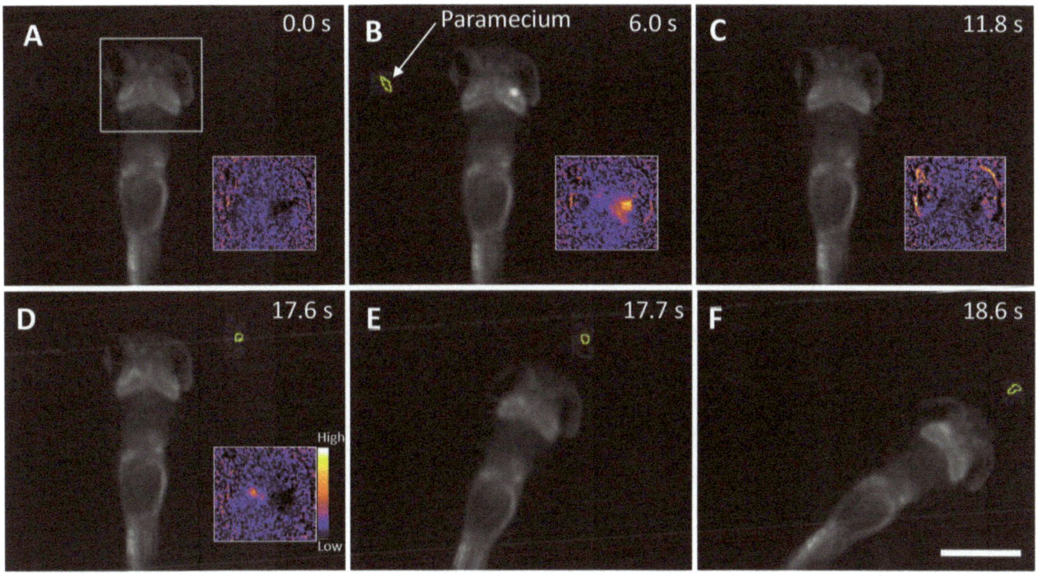

Fig. 2 Calcium imaging of free-swimming zebrafish larvae expressing GCaMP7a in the optic tectum of the midbrain. UAS:GCaMP7a transgenic fish and the gSA2AzGFF49A Gal4FF driver line were used. (**a–f**) Frames from a time-lapse recording. The paramecium is delineated by a line. Insets: Ratiometric image of the frame divided by an averaged frame. Scale bar = 0.5 mm

vidual frames by a reference image (an averaged image over all frames or an averaged image over a period with no calcium signals) (Fig. 2a–d). To do this, from the menu in ImageJ, choose Image > Stacks > Z project ... > Average Intensity, and then, Process > Image Calculator... > Divide (with the option of "32-bit (float) result"). An appropriate look-up table (LUT) can be chosen for the best presentation of the calcium signals in pseudocolor. The averaged pixel values in a region of interest (ROI) can be measured as follows: create the ROI on the image of a stack, set the measurements by choosing Analyze > Set Measurements... > Mean gray value, and select Plugins > Stacks > Measure Stack (*see* **Note 10**).

4 Notes

1. Because the pigments (melanophores) on the surface of the body block light, we use the *nacre* mutant background, which lacks melanophores. Retinal pigment epithelium is present in *nacre*, assuring normal vision. We maintain both the Gal4FF and UAS:GCaMP lines on the *nacre* background so that when mated, the clutch contains *nacre* homozygotes with Gal4FF

and UAS:GCaMP transgenes. For the study of vision-independent behavior, phenylthiourea-treatment can be used to suppress melanin formation as an alternative.

2. We use system water for behavioral recording, but E3 water (without methylene blue) also may be used instead.
3. The choice of the Gal4 line depends on the neurons from which you would like to record. A database is available at the Kawakami lab website for Gal4FF enhancer/gene trap lines that were generated in our lab [10]. The expression level of the GCaMP transgene is critical for detection of calcium signals. When the expression level of GCaMP is not high enough, one option is to make the UAS:GCaMP homozygous (i.e., express two copies of the transgene).
4. We constructed the custom-made recording chamber beforehand, and found that the depth of 0.8 mm is appropriate to minimize the movement of a 4–7 day larva along the *Z*-axis, while still allowing it to freely swim in the *XY* plane. A smaller chamber (9 mm in diameter) may also be used to increase the chance of capturing and recording behavioral events (Fig. 1g).
5. Because of the small volume of the shallow chamber, the water will rapidly evaporate, and possibly change the fluid pressure. To prevent this, a cut was made with a fine blade to make a canal to an external reservoir, so the water pressure inside the chamber will stay the same as outside.
6. Once transferred from the concentrated culture to system water, motility of paramecia decreases over time. Prepare them fresh for the imaging study.
7. Typically, the exposure time we use is 10–100 ms, with a frame acquisition rate of 10–100 fps and 2×2 binning.
8. Even with the low magnification lens (2.5× objective lens) and a wide area camera (ORCA Flash4.0), only a part of the recording chamber can be viewed and recorded. For this reason, a particular behavioral event such as prey capture occurs only occasionally in the camera view.
9. The Hamamatsu Photonics ORCA camera uses the .cxd file format to save the movie. This format can be imported into ImageJ with the Bio-Formats plugin (openmicroscopy.org).
10. With the reflected excitation light, the paramecia are visible to human eyes and probably also to the zebrafish larva. However, they are not easily recognizable in the fluorescent images, as they have no innate fluorescence nor do they have a fluorescent label. The addition of a slight amount of bright-field illumination during fluorescent imaging may help to visualize the paramecia in the image.

Acknowledgments

This work was supported by grants from the Ministry of Education, Culture, Sports, Science and Technology (MEXT) in Japan: KAKENHI Grant Number 24120521, 25290009, and 25650120.

References

1. Kettunen P (2012) Calcium imaging in the zebrafish. Adv Exp Med Biol 740:1039–1071
2. Semmelhack JL et al (2015) A dedicated visual pathway for prey detection in larval zebrafish. eLife 2014;3:e04878
3. Ritter DA, Bhatt DH, Fetcho JR (2001) In vivo imaging of zebrafish reveals differences in the spinal networks for escape and swimming movements. J Neurosci 21(22):8956–8965
4. Bianco IH, Engert F (2015) Visuomotor transformations underlying hunting behavior in zebrafish. Curr Biol 25(7):831–846
5. Nakai J, Ohkura M, Imoto K (2001) A high signal-to-noise Ca(2+) probe composed of a single green fluorescent protein. Nat Biotechnol 19(2):137–141
6. Muto A et al (2013) Real-time visualization of neuronal activity during perception. Curr Biol 23(4):307–311
7. Muto A et al (2011) Genetic visualization with an improved GCaMP calcium indicator reveals spatiotemporal activation of the spinal motor neurons in zebrafish. Proc Natl Acad Sci U S A 108(13):5425–5430
8. Asakawa K, Kawakami K (2008) Targeted gene expression by the Gal4-UAS system in zebrafish. Dev Growth Differ 50(6):391–399
9. Asakawa K et al (2008) Genetic dissection of neural circuits by Tol2 transposon-mediated Gal4 gene and enhancer trapping in zebrafish. Proc Natl Acad Sci U S A 105(4):1255–1260
10. Kawakami K et al (2010) zTrap: zebrafish gene trap and enhancer trap database. BMC Dev Biol 10:105
11. Lister JA et al (1999) Nacre encodes a zebrafish microphthalmia-related protein that regulates neural-crest-derived pigment cell fate. Development 126(17):3757–3767

Chapter 24

Fiber Optic-Based Photostimulation of Larval Zebrafish

Aristides B. Arrenberg

Abstract

The perturbation of neural activity is a powerful experimental approach for understanding brain function. Light-gated ion channels and pumps (optogenetics) can be used to control neural activity with high temporal and spatial precision in animal models. This optogenetic approach requires suitable methods for delivering light to the brain. In zebrafish, fiber optic stimulation of agarose-embedded larvae has successfully been used in several studies to control neural activity and behavior. This approach is easy to implement and cost-efficient. Here, a protocol for fiber optic-based photostimulation of larval zebrafish is provided.

Key words Optic fiber, Photostimulation, Zebrafish, Optogenetics, Laser, Agarose, Mounting

1 Introduction

The investigation of brain function in animal models depends on methods to measure and manipulate brain activity with the latter being crucial for establishing causal links. Recently developed optogenetic approaches make use of genetically encoded proteins that interact with light [1]. In particular, light-gated ion channels and pumps can be used to silence or activate expressing neurons [2, 3]. The temporal precision is high, because light sources (e.g., LEDs or lasers) can be rapidly modulated and the induced membrane potential changes are fast (<1 s). The spatial precision is dependent on both the expression pattern of the optogenetic protein and on the spread of the photostimulation light within the brain tissue. A variety of optical methods exist for shaping the stimulation light path [4–7]. This protocol describes a simple photostimulation method that utilizes blunt end optic fibers to stimulate the larval zebrafish brain [8–12]. It covers necessary equipment, the choice of the experimental light intensity, the preparation of a blunt end optic fiber, techniques to mount larval zebrafish in agarose for behavioral recordings, the positioning of the optic fiber over the skin of the animal, and the control of photostimulation. The described method is cost-efficient, easy to implement, and provides high lateral resolution of photostimulation.

Koichi Kawakami et al. (eds.), *Zebrafish: Methods and Protocols*, Methods in Molecular Biology, vol. 1451,
DOI 10.1007/978-1-4939-3771-4_24,

2 Materials

2.1 Equipment for Fiber Optic Photostimulation

This protocol utilizes low numerical aperture (NA) optic fibers that are positioned above the animal using a micromanipulator. A stereoscope is used for the positioning procedure as well as for recording animal behavior. The photostimulation is computer-controlled. Below, a parts list for the fiber optic setup is provided (Fig. 1). Manufacturers and part numbers are included here to facilitate building a setup (parts from different manufacturers can also be used).

1. The fiber-coupled light source needs to be chosen based on the required wavelength and light intensity at the desired stimulation site within the brain tissue (*see* Subheading 3.1, **Notes 1** and **2**). A pigtailed laser with analog and digital modulation and the following specifications is recommended (*see* **Note 3**): 50 mW light power, 50 μm fiber diameter, a fiber with Angled Flat (AFC) or Angled Physical Contact (APC) polishing and ferrule (FC) connector (*see* **Note 4**, e.g., a Toptica iBeam smart laser).
2. Angled Physical Contact (FC/APC) compound mating sleeve (*see* **Note 5**).
3. Custom fiber optic patch cord end A: FC/APC, end B: blunt end (*see* **Note 6**), length: 2 m. The wavelength range of the cable needs to match the wavelength of the light source. A lower NA (e.g., NA = 0.22) results in smaller lateral spread of the light in the brain tissue. The diameter of the fiber is chosen based on the experiment (i.e., 10 μm, 50 μm, or 200 μm). Custom patch cords are available at Thorlabs (e.g., part numbers HPSC10, FG050UGA, FG200UCC).
4. DAQ device (National Instruments, NI USB-6008).
5. Computer with installed drivers for the DAQ device.
6. Coarse micromanipulator (World Precision Instruments, M3301R).
7. Fine micromanipulator (Narishige, MMO-203).
8. Stereoscope (Leica MZ12) with camera (The Imaging Source, DMK 21AU04).

2.2 Reagents and Equipment for Embedding Larvae

1. E3 medium; 5 mM NaCl, 0.17 mM KCl, 0.33 mM $CaCl_2$, 0.33 mM $MgSO_4$. Adjust pH to 7.4 using sodium bicarbonate ($NaHCO_3$).
2. Low melting temperature agarose solution; 1.6% agarose in E3 medium. Heat solution in a small beaker in the microwave, stir solution and aliquot into 1.5 mL microcentrifuge tubes placed in a 39 °C block heater.
3. Platinum wire (0.127 mm diameter, e.g., article 00263 from Alfa Aesar).

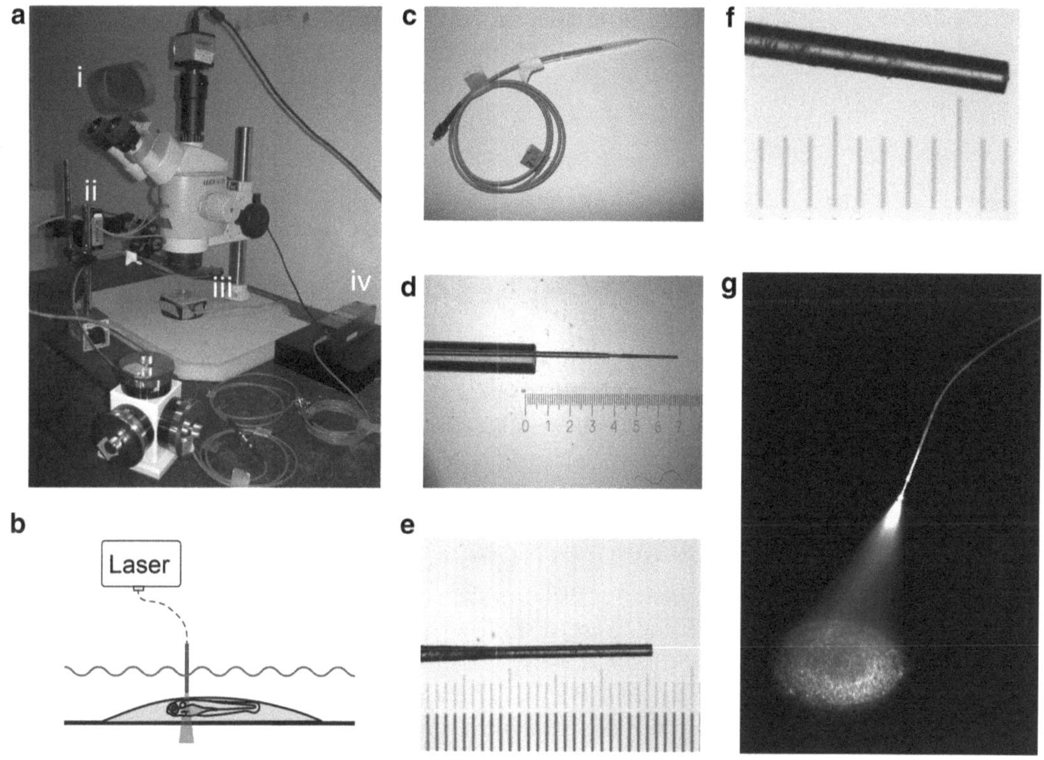

Fig. 1 Optic fiber stimulation setup. (**a**) Picture of a setup for fiber optic stimulation of zebrafish larvae. A custom-made safety shield covers the oculars (*i*) and is connected to the interlock of the laser (*iv*). The optic fiber is mounted in a glass pipette and the pipette is positioned via a fine micromanipulator mounted on a coarse micromanipulator (*ii*). In this setup, a custom backlight LED box (*iii*) is used to provide the transmitted light for the eyes (white LEDs) or the camera (850 nm LEDs, IR filter in front of the camera). In (**b**) a schematic shows a side view of a larva mounted in a drop of agarose and a fiber placed above the brain. In a flat drop of agarose the dorsal side of the animal is close to the agarose-water interface, which in turn allows the positioning of the fiber at a short distance. (**c**–**f**) A prepared 50 μm diameter optic fiber mounted in an angled glass pipette is shown at several magnifications. The numbers on the scale bar correspond to millimeters. In (**e**) the partially burnt buffer can be seen on the *left*, whereas on the *right* the fiber end is devoid of buffer. (**g**) The fiber end can be pointed at a paper screen to judge the homogeneity of the light disk. Here, fog from dry ice in water has additionally been used to visualize the light cone

3 Methods

3.1 Choice of the Illumination Intensity

1. Determine the useful illumination intensities that have been published for your optogenetic protein. For example, for halorhodopsin an intensity of 21.7 mW/mm^2 has been reported to saturate the spike inhibiting effect [2].
2. Choose an optic fiber diameter that suits your needed spatial precision (10 μm, 50 μm, or 200 μm). Large diameters can be used to stimulate complete brain regions (e.g., caudal hindbrain) and small diameters can be used to restrict activation to

a small number of expressing cells. Take into account that the light beam exiting the fiber is cone shaped, so that along the axis of stimulation, the illuminated diameter increases, and the intensity decreases (Fig. 1g).

3. Estimate the needed light power. First calculate the divergence half-angle (θ) of the emitted light, which depends on the NA (NA = 0.22) of the fiber and the refractive index of the brain tissue ($n \approx 1.36$) according to the following equation [13]: $\theta = \sin^{-1}(NA/n) \approx 9.3°$. The fractional geometric decrease along the beam path ($f(z)$) can then be calculated by dividing the fiber output area (A_0) by the illuminated area (Az) at distance z: $f(z) = A_0/Az = \pi * r^2/(\pi(r + z * \tan(\theta))^2)$. For example, a 50 μm diameter fiber ($r = 0.025$ mm) that is positioned on the brain surface, should be attenuated by a factor of 0.19 at a distance of 200 μm ($z = 0.2$ mm) into the brain: $f(0.2\ \text{mm}) = \pi * (0.025\ \text{mm})^2/(\pi(0.025\ \text{mm} + 0.2\ \text{mm} * \tan(9.3°))^2) = 0.19$. Therefore, if a light intensity of 100 mW/mm^2 is needed at 200 μm depth in this example, the light intensity at the fiber output should be 526 mW/mm^2. Given the fiber output area (A_0) of 0.00196 mm^2, this corresponds to a light power of 1.0 mW. The analog voltage of the laser can now be adjusted so that a light power of 1.0 mW exits the fiber as measured with a light power meter. Note that additional intensity loss occurs due to scattering and—to a smaller extent—absorption of light [13] (*see* **Note 7** and **step 6** in Subheading 3.4).

3.2 Preparation of the Blunt End Optic Fiber

1. Remove 10 cm of the outer PVC jacket (orange) using a wire cutter. Take care not to damage the optic fiber.
2. Remove 10 cm of the Kevlar protective threads.
3. Remove 7 cm of the polypropylene inner fiber tube using scissors. Take care not to damage the optic fiber. Now the optic fiber consisting of fiber core, cladding, and buffer is visible.
4. Hold the fiber pointing down at a 45° angle and use a lighter to ignite the tip of the fiber. The flame will burn the buffer and work its way up along the fiber. After the buffer has been burnt back 3 cm (~2 s) blow out the flame (*see* **Note 8**). This step exposes the fiber core and cladding and is needed to minimize the diameter of the fiber, which eases positioning of the fiber under the stereoscope, and also enables a clean break in the next steps [8, 10].
5. Use a diamond scribe to score the fiber at 1 cm length (*see* **Note 9**).
6. Gently bend the fiber tip until it snaps along the score.
7. Inspect the fiber tip under a stereoscope from all directions. Test the fiber light output by positioning a paper screen a few centimeters in front of the fiber tip (Fig. 1f–g, *see* **Note 10** on laser safety). The light should form a circle without inhomoge-

neities. If the cleaved surface is not completely flat or the light beam is ill-shaped, repeat **steps 5** and **6** (*see* **Note 11**).

3.3 Mounting a Larval Zebrafish in a Petri Dish

1. Keep low melting temperature agarose solution (*see* Subheading 2.2) at 39 °C in microcentrifuge tubes.
2. Select a larva (4–8 days post fertilization) with a glass Pasteur pipette and transfer it into the microcentrifuge tube in the smallest volume possible. Pipet the larva up and down once and place it (in one drop) in the lid of a petri dish under the stereoscope (*see* **Note 12**). The agarose will solidify at room temperature within approximately 20 s and practice is needed to finish the following **steps 3** and **4** in time.
3. Use a platinum wire tool (0.127 mm diameter platinum wire glued to a Pasteur pipette, Fig. 2a, b–d) to flatten the liquid agarose drop by moving the wire at the perimeter of the drop and thus increasing the agarose-covered area in the dish and reducing the height of the drop. Continue until the height of the liquid approximately matches the height of the larva, thereby reducing the amount of agarose above the head of the fish (*see* **Note 13**). This step takes about 10 s and ensures that the optic fiber can be placed close to the skin in the experiment.
4. Use the platinum wire tool to position the larva dorsal side up (*see* **Notes 14** and **15**).
5. Let the agarose solidify for 5–10 min and add E3 medium to the dish lid.
6. Remove the agarose around the tail or the eyes to enable tail or eye movements. For this step, prepare a second platinum wire tool that is flattened at the tip (*see* **Note 16**) and bent (~45° angle) about 1 mm from the tip (Fig. 2a, e–h). By moving this tool sideways, the agarose can be incised as the edge on the side of the flattened part acts like a knife. Make incisions along the borders of the agarose piece to be removed (excluding the agarose regions touching the larva). The flat side of the platinum wire can then be used to scoop out blocks of agarose after having made the incisions (Fig. 2i–j). Set the stereoscope backlight at an angle (oblique illumination) to visualize regions in which agarose has been removed. Make sure that the animal can move its eyes or tail without obstruction (*see* **Note 17**).

3.4 Positioning the Optic Fiber

1. Use a Bunsen burner to heat the tip of a Pasteur pipette and smoothly bend the last 2–5 cm of the pipette in order to achieve the desired output angle (e.g., 45°, Fig. 1c).
2. Clamp the Pasteur pipette to the micromanipulator and insert the prepared optic fiber (*see* **Note 18**). About 1.5 cm of the optic fiber should protrude from the Pasteur pipette tip. At the

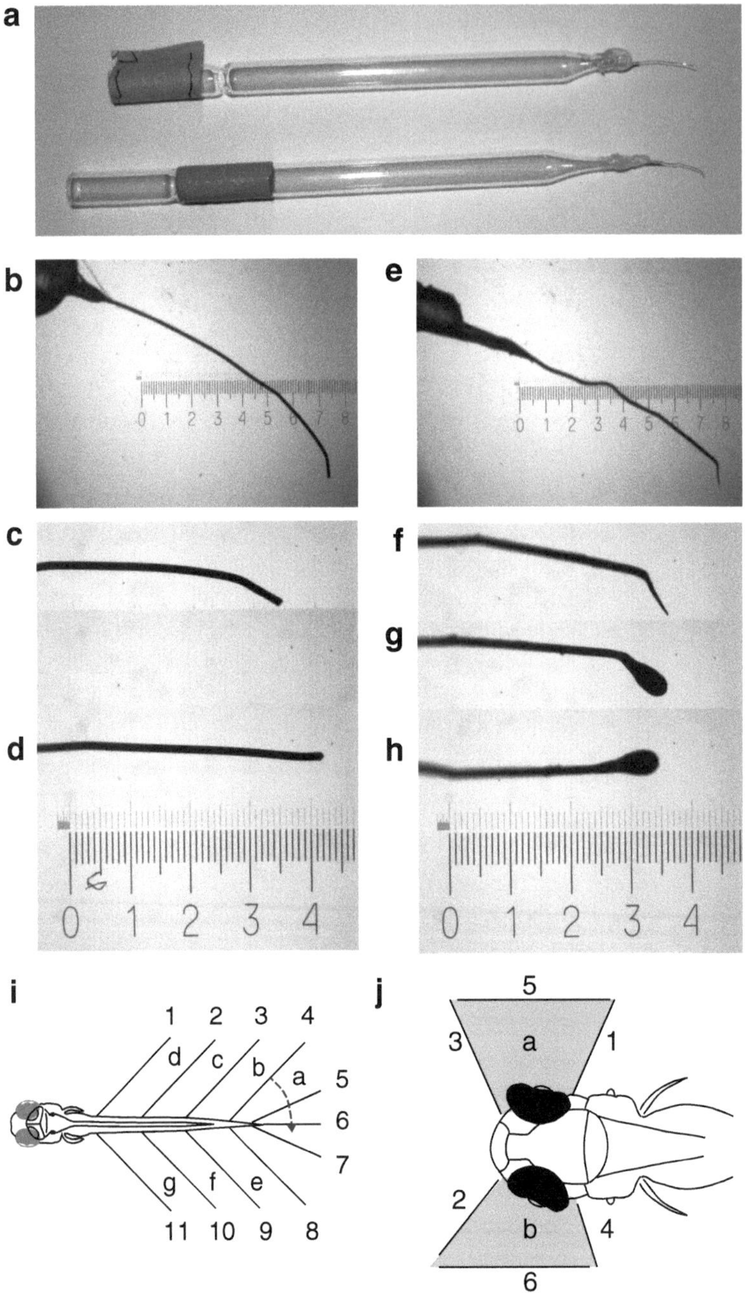

Fig. 2 Two tools for mounting zebrafish larvae in low melting temperature agarose solution. (**a**) A picture of the tools used for alignment (*top*) and cutting/scooping (*bottom*). (**b**, **e**) Side views of the platinum wire of the alignment tool (**b**) and the cutting tool (**e**). (**c**, **d**) Side view (**c**) and bottom view (**d**) of the tip of the alignment tool. (**f**–**h**) Side view (**f**), oblique view (**g**), and bottom view (**h**) of the cutting tool. The sharp side (**f**) can be used to cut blocks of agarose and the oval-shaped side (**h**) can be used to scoop these agarose blocks out of their original position. The numbers on the scale bar correspond to millimeters. (**i**, **j**) Patterns for agarose removal. (**i**) For tail movement recordings, make agarose incisions #1 through #7 using the cutting tool. Then pull the agarose stripes to the caudal end (steps a–d), starting with step a. Make incisions #8–11 and remove agarose on the *left* side (steps e–g). (**j**) For eye movement recordings, make incisions #1–6 using the cutting tool, starting directly next to the eye and moving the tool outwards. Then use the flat side of the tool to scoop out the two agarose blocks (a, b)

other end of the Pasteur pipette, fix the position of the fiber (e.g., with a piece of tape).

3. Choose the inclination angle of the fiber tip by adjusting the micromanipulator. To be able to see the tip of the fiber through the stereoscope, choose an inclination angle that is slightly smaller than 90° (*see* **Note 19**).
4. Use the coarse micromanipulator to position the fiber roughly above the fish. Use the fine micromanipulator—which is mounted on the coarse one—to position the fiber above the brain region of interest.
5. While the lateral position (x,y) can be judged easily by looking through the stereoscope, the positioning of the fiber close to the skin surface (z) is more difficult. In cases in which the position is unclear, lower the fiber tip slowly until (a) the mechanical strain on the agarose moves the skin of the fish slightly or (b) until the fish startles. Then move the fiber back up by a tiny distance (*see* **Notes 20–22**).
6. To determine the approximate spatial extent of photostimulation, animals transgenic for the photoconvertible protein Kaede (or a different photoconvertible protein) can be used. After the optogenetic experiment the light source of the optic fiber can be switched to UV light, which photoconverts Kaede (*see* **Note 23**). The animal can then be imaged using a confocal microscope to determine the stimulated volume (*see* Fig. 3). Note, however, that the optogenetic proteins and Kaede act at different wavelengths and intensities, and therefore this method can only provide an approximate estimate of the photostimulated volume.

3.5 Photostimulation Control

1. Connect the analog or digital inputs of the laser to the appropriate outputs of the DAQ box, allowing modulation of the laser intensity. The voltage can then be set using the software for the DAQ box, e.g., the NI DAQ Measurement and Automation Explorer for National Instruments devices (*see* **Note 24**). Take care to follow appropriate laser safety steps (*see* **Note 10**).
2. To control the timing of photostimulation, and synchronize with behavioral recordings, use custom software routines, written, for example, using National Instruments LabVIEW (*see* **Note 25**). Since the laser light is typically visible in the behavioral video recording, the time point of stimulation is easily determined. Interference of photostimulation and behavioral recording may be avoided by using infrared illumination (*see* **Note 26**).
3. Design the stimulation protocol. Factors to consider are the light intensity (*see* Subheading 3.1), the stimulus duration, stimulus shape, and interval. Usable stimulus durations are limited by the time it takes for the neuronal membrane poten-

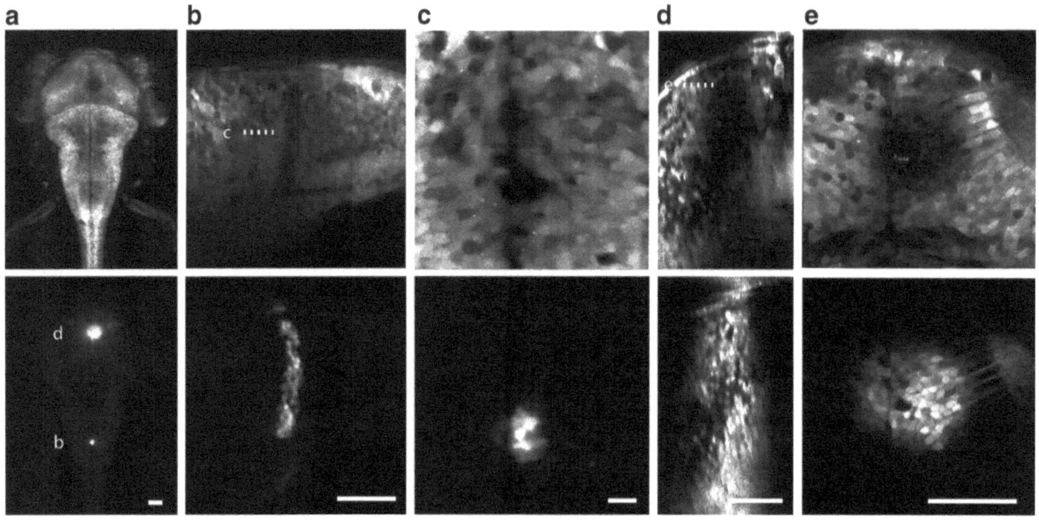

Fig. 3 Photoconversion experiment reveals fiber optic light spread [16]. (**a**) Dorsal view of a zebrafish expressing the photoconvertible fluorophore Kaede [*Et(E1b:Gal4)s1101t, Tg(UAS:Kaede)s1999t*]. In (**b**) and (**d**) 10 μm (NA = 0.1) and 50 μm (NA = 0.22) diameter fibers have been placed above the brain, respectively, and a column of cells has been photoconverted at each location using an UV laser. Side views are shown (rostral is to the *left*). The unexpected curvature of the photoconverted column in (**b**) was likely caused by cell migration after the photoconversion experiment. In (**c**) and (**e**), single confocal slices 60 μm deep (10 μm fiber diameter) and 39 μm deep (50 μm fiber diameter) are shown (dorsal views). Dashed lines in (**b**) and (**d**) indicate the level of the single confocal slices in (**c**) and (**e**). *Top*: *green* channel. *Bottom*: *red* channel. Z-projections of a selected range of confocal 3D data are shown in **a**, **b**, and **d**. Scale bars are 50 μm in (**a**, **b**, **d**, **e**) and 10 μm in **c**

tial to change after light onset (in the range of milliseconds for channelrhodopsin and halorhodopsin), by desensitization of the optogenetic protein and possibly by homeostatic effects in the stimulated neurons. Regarding the stimulus shape, constant stimulations can be used successfully, while modulated shapes (e.g., a 30 Hz rectangle profile) are preferable for controlling spike timing and frequency in channelrhodopsin experiments. The stimulation interval should be large enough to allow for resensitization of the optogenetic protein and rest of the animal.

4 Notes

1. Cost-efficient fiber-coupled high power LEDs can be used as an alternative to lasers (e.g., Thorlabs parts M470F3 and LEDD1B). However, the light intensities are typically smaller than those achievable with lasers and therefore might not be sufficient for activating some optogenetic proteins (e.g., halorhodopsin).

2. Ideally, the wavelength of the laser should match the peak of the activation spectrum of the optogenetic protein. However, costs of lasers are dependent on the wavelength; and in experiments in which more than one type of optogenetic protein is expressed, the wavelengths used should maximize the specificity of activation. For example, instead of a yellow laser, a less expensive red laser can be used to activate halorhodopsin and minimize coactivation of channelrhodopsin.
3. If no pigtailed laser is available, the laser can be coupled into a fiber using a (collimator) lens.
4. The quality of your laser should match your experimental requirements and typically a high quality laser will be needed. The laser output should be stable, the output at 0 V (analog modulation) should be minimal and there should be no delay after triggering the laser. The relation of the analog voltage and the light power needs to be measured with a power meter prior to the experiments, since it is typically not linear. In the absence of other measurement devices, the laser stability after stimulation onset can be tested with a high-speed camera.
5. Different fiber finishes are available. Both fibers to be connected via the mating sleeve should be polished at an angle to minimize back reflections and coupling loss.
6. Instead of ordering a cable with one blunt end, a double length cable with two FC/APC connectors can be ordered in order to prepare two blunt end cables after cutting it in the middle.
7. Animals used for photostimulation experiments should preferentially be devoid of pigmentation to permit the stimulation light to enter the brain. This can be achieved by working with mutants such as *nacre* [14] or by treating animals with 1-phenyl-2-thiourea (PTU). However, PTU oftentimes has adverse effects on behavior.
8. Sometimes a drop of melted buffer remains on the fiber. Such drops need to be removed, because they can block sight of the fiber tip under the stereoscope during the experiment.
9. It may be helpful to score the fiber under a stereoscope. It is important that the score runs orthogonal to the fiber axis in order to achieve an orthogonal break. If the surface of the broken fiber is not orthogonal to the fiber axis, the light beam will be angled relative to the fiber axis.
10. Lasers are dangerous for the eyes and laser safety precautions have to be arranged in accordance with a laser safety expert and existing regulations. In addition to laser goggles, we use an interlock safety system that only allows the laser to be enabled when the stereoscope's oculars are covered by a safety shield (Fig. 1a).
11. If the end of the optic fiber is chipped on one side, light will exit through the chipped area and at a high divergence angle.

12. Sometimes the agarose loses its adhesion to the petri dish lid after addition of E3 medium. Using a new petri dish lid every time ameliorates this problem.
13. Alternatively, this step (flattening the liquid agarose drop) can be omitted and the fish can be pulled upwards during the mounting. However, flattening the liquid agarose drop results in a more reliable pitch angle and elevation position of fish across experiments.
14. Take care not to poke the animal with the tip of the platinum wire. Orient the larva by (a) holding the wire shaft almost vertical and stroking the larva on its sides in rostrocaudal direction to adjust yaw and roll angles of the animal, (b) pushing the side of the wire against the nose of the animal to push the larva back into the center of the agarose drop, and (c) rotating the wire around the tail if the animal is oriented upside-down or sideways (the bent fish tail can be used as a lever).
15. If a larva has a suboptimal orientation and the agarose starts to solidify, stop manipulating the animal immediately. Otherwise you might cause harm to the animal and attempts to properly reorient the animal without contortions will likely fail. Rather take a new animal and petri dish lid and put the ill-mounted animal (with dish lid) into a beaker containing E3 medium for later euthanasia.
16. A pair of round-nosed pliers can be used to flatten the platinum wire.
17. In preparations where the agarose surrounding the eyes is removed, it is important that the agarose bridge (between cut 2 and 3 in Fig. 2j) at the nose is left intact. Otherwise, fish might start to wiggle their heads and will soon escape the agarose. After incisions around the eye have been made, use the flat side of the tool to pry the incision open and squeeze the jammed agarose block out of its original position in one piece. The agarose block should detach from the two non-incised surfaces (agarose-petri dish lid and agarose-fish eye) without leaving agarose remnants. If agarose remnants remain close to the eye, use the tool to carefully scoop the agarose out without touching or compressing the fish.
18. The more the pipette has been bent in the previous step, the more difficult it will be to insert the fiber. To avoid snapping fibers, it helps to temporarily lubricate the path by filling the pipette with water.
19. When the optic fiber is in a vertical position, it can be moved laterally to stimulate different laterally adjacent regions in the zebrafish brain expressing the optogenetic protein. If two brain regions that lie above each other need to be stimulated separately, the optic fiber can be oriented at an angle in order to target only one of the two brain regions [9, 15].

20. The depth position of the fiber end can be judged by focusing first on the fiber tip and then turning the focus wheel until the skin surface of the animal is focused. As the fiber tip gets close to the skin, it will be possible to almost get both fiber tip and skin in focus at the same time.
21. The curved surface of the water meniscus at the optic fiber shaft can obstruct the view of the optic fiber tip. Changing the water level slightly with a pipet can improve the image quality.
22. Evaporation can change the water level, resulting in loss of focus of the camera. In experiments lasting longer than 3 h, we minimize water evaporation by mounting animals in a petri dish and placing the lid (with a 1 cm diameter hole) on top. The optic fiber can then be lowered through the hole.
23. For combining lasers of different wavelengths into a single optic fiber (e.g., for optogenetic and photoconversion experiments or when using two optogenetic proteins), fiber optic couplers can be used, if no multilaser system is available.
24. In analog modulation, the analog voltage is typically not perfectly linearly related to light power and the correct setting needs to be determined empirically or measured before the experiment.
25. An alternative approach that is free to use and simple to program is the DAQtimer software available from the Burgess lab (https://science.nichd.nih.gov/confluence/display/burgess/Software). DAQtimer requires a specific DAQ device: NI PCI-6221.
26. If the laser light interferes with the recording of behavioral parameters, an IR LED panel (850 nm) can be used for illumination and an infrared transmissive filter (Schott glass RG780) can be placed in the optical path (e.g., in front of the stereoscope objective) in order to block laser light from reaching the camera.

Acknowledgments

I thank Tod R. Thiele, António M. Fernandes, and Christian Brysch for comments on the manuscript, and Tod R. Thiele for contributing his experience regarding zebrafish fiber optic experiments. I am grateful to Wolfgang Driever for support. This study was supported by the Excellence Initiative of the German Federal and State Governments DFG EXC307 (CIN - Interdisciplinary Centre for Integrative Neuroscience), DFG EXC294 (BIOSS - Centre for Biological Signalling Studies), as well as the Baden-Württemberg Stiftung (Eliteprogramme for Postdocs).

References

1. Fenno L, Yizhar O, Deisseroth K (2011) The development and application of optogenetics. Annu Rev Neurosci 34:389–412. doi:10.1146/annurev-neuro-061010-113817
2. Zhang F, Wang L, Brauner M et al (2007) Multimodal fast optical interrogation of neural circuitry. Nature 446(7136):633–639. doi:10.1038/nature05744
3. Boyden ES, Zhang F, Bamberg E et al (2005) Millisecond-timescale, genetically targeted optical control of neural activity. Nat Neurosci 8(9):1263–1268. doi:10.1038/nn1525
4. Baier H, Scott EK (2009) Genetic and optical targeting of neural circuits and behavior—zebrafish in the spotlight. Curr Opin Neurbiol 19(5):553–560. doi:10.1016/j.conb.2009.08.001
5. Vaziri A, Emiliani V (2012) Reshaping the optical dimension in optogenetics. Curr Opin Neurbiol 22(1):128–137. doi:10.1016/j.conb.2011.11.011
6. Zhu P, Fajardo O, Shum J et al (2012) High-resolution optical control of spatiotemporal neuronal activity patterns in zebrafish using a digital micromirror device. Nat Protoc 7(7):1410–1425. doi:10.1038/nprot.2012.072
7. Rickgauer JP, Tank DW (2009) Two-photon excitation of channelrhodopsin-2 at saturation. Proc Natl Acad Sci U S A 106(35):15025–15030. doi:10.1073/pnas.0907084106
8. Arrenberg AB, Del Bene F, Baier H (2009) Optical control of zebrafish behavior with halorhodopsin. Proc Natl Acad Sci U S A 106(42):17968–17973. doi:10.1073/pnas.0906252106
9. Fajardo O, Zhu P, Friedrich RW (2013) Control of a specific motor program by a small brain area in zebrafish. Front Neural Circ 7:67. doi:10.3389/fncir.2013.00067
10. Campagnola L, Wang H, Zylka MJ (2008) Fiber-coupled light-emitting diode for localized photostimulation of neurons expressing channelrhodopsin-2. J Neurosci Methods 169(1):27–33. doi:10.1016/j.jneumeth.2007.11.012
11. Schoonheim PJ, Arrenberg AB, Del Bene F et al (2010) Optogenetic localization and genetic perturbation of saccade-generating neurons in zebrafish. J Neurosci 30(20):7111–7120. doi:10.1523/JNEUROSCI.5193-09.2010
12. Thiele TR, Donovan JC, Baier H (2014) Descending control of swim posture by a midbrain nucleus in zebrafish. Neuron 83(3):679–691. doi:10.1016/j.neuron.2014.04.018
13. Aravanis AM, Wang L, Zhang F et al (2007) An optical neural interface: in vivo control of rodent motor cortex with integrated fiberoptic and optogenetic technology. J Neural Eng 4(3):S143–S156. doi:10.1088/1741-2560/4/3/S02
14. Lister JA, Robertson CP, Lepage T et al (1999) Nacre encodes a zebrafish microphthalmia-related protein that regulates neural-crest-derived pigment cell fate. Development 126(17):3757–3767
15. Kubo F, Hablitzel B, Dal Maschio M et al (2014) Functional architecture of an optic flow-responsive area that drives horizontal eye movements in zebrafish. Neuron 81(6):1344–1359. doi:10.1016/j.neuron.2014.02.043
16. Gonçalves PJ, Arrenberg AB, Hablitzel B et al (2014) Optogenetic perturbations reveal the dynamics of an oculomotor integrator. Front Neural Circ 8:10. doi:10.3389/fncir.2014.00010

Chapter 25

Genetic Ablation, Sensitization, and Isolation of Neurons Using Nitroreductase and Tetrodotoxin-Insensitive Channels

Eric J. Horstick,* Kathryn M. Tabor,* Diana C. Jordan, and Harold A. Burgess

Abstract

Advances in genetic technologies enable the highly selective expression of transgenes in targeted neuronal cell types. Transgene expression can be used to noninvasively ablate, silence or activate neurons, providing a tool to probe their contribution to the control of behavior or physiology. Here, we describe the use of the tetrodotoxin (TTX)-resistant voltage-gated sodium channel $Na_v1.5$ for either sensitizing neurons to depolarizing input, or isolating targeted neurons from surrounding neural activity, and methods for selective neuronal ablation using the bacterial nitroreductase NfsB.

Key words Ablation, Silencing, Sensitization, Isolation, Tetrodotoxin, Nitroreductase, NfsB, SCN5a, $Na_v1.5$, Zebrafish

1 Introduction

A fundamental objective in the study of neuronal circuits is to determine how activity in a given set of neurons contributes to a behavioral response. Key techniques that shed light on such questions are to either selectively inactivate or activate subsets of neurons, and determine how this influences circuit function or behavior. Here, we describe transgenic methods to manipulate targeted neurons in vivo that allow for analysis of their contribution to behavioral responses in larval zebrafish.

Several techniques have been used in zebrafish to selectively silence or ablate genetically labeled neurons. Neuronal signaling can be constitutively blocked by expression of the tetanus toxin light chain [1] or the inward-rectifier potassium ion channel KiR2.1 [2]. Light-activated membrane channels and pumps allow

*Author contributed equally with all other contributors.

Koichi Kawakami et al. (eds.), *Zebrafish: Methods and Protocols*, Methods in Molecular Biology, vol. 1451,
DOI 10.1007/978-1-4939-3771-4_25, © Springer Science+Business Media New York 2016

for reversible neuronal silencing [3–5]. These sophisticated techniques are best paired with electrical recordings to verify the extent of silencing which in many cases is partial or strongly variable between individuals [1, 2]. Ablation, which can be visually confirmed, is therefore an experimentally more tractable approach in many cases. Cell-specific ablation in zebrafish has been achieved using diphtheria toxin-A [6], the bacterial Kis/Kid system [7] and a modified caspase gene [8]. However, the best-characterized system relies on expression of the bacterial nitroreductase, *NfsB*, which metabolizes bath-applied metronidazole into a cell-impermeant cytotoxin [9–11]. This system allows temporal control of ablation through metronidazole exposure and has recently been optimized to yield robust neuronal ablation within 24 h [12, 13]. A simple way to verify that metronidazole treatment induces apoptosis is to use PhiPhiLux G1D2, a live fluorescent reporter of caspase 3-like activity (Fig. 1) [14].

Optogenetic techniques have been widely used in zebrafish to acutely depolarize and thereby activate targeted neurons upon light exposure (for example, [3, 15–17]). These are powerful methods for decoding neuronal circuits, but require illumination with intense light that may interfere with behavioral assays. Moreover, induced firing patterns may not reproduce normal patterns of neuronal activation. An alternate method is to overexpress a voltage-gated

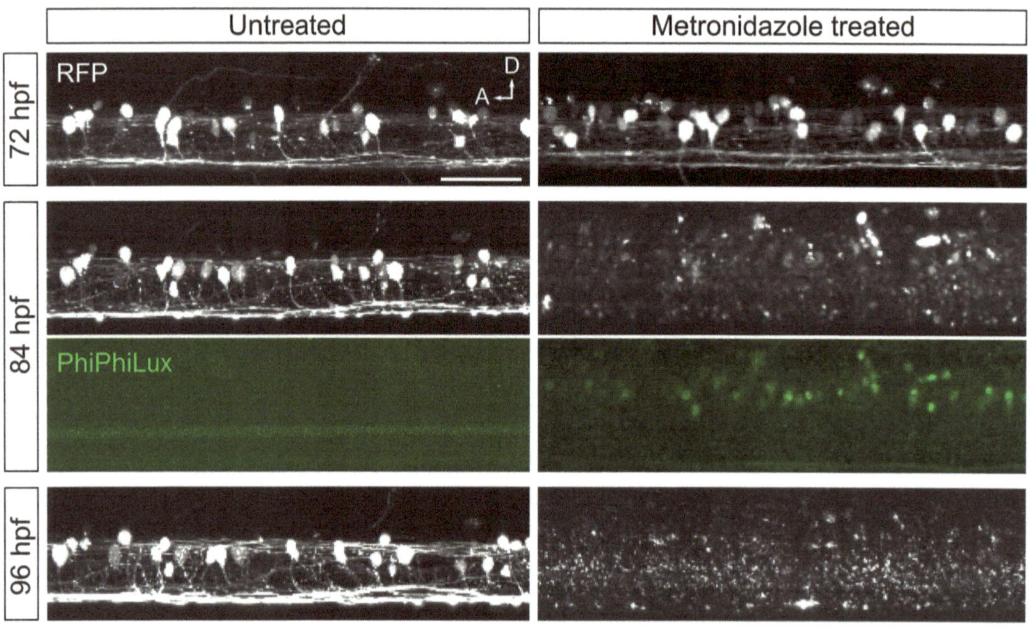

Fig. 1 Detection of apoptosis using PhiPhiLux. RFP expression in spinal cord neurons in *Et(REx2-cfos:Gal4ff) y270, UAS:epNTR-TagRFPT* control larva (*left*), and larva treated with metronidazole (*right*). Gray panels show RFP expression every 12 h from the start of metronidazole treatment at 72 hpf. At 84 hpf, larvae were treated with PhiPhiLux to label apoptotic cells (*green, middle panels*). Note fluorescent debris from ablated neurons. Scale bar 50 μm

sodium channel in neurons of interest, thereby increasing their sensitivity to endogenous patterns of depolarizing input [13]. Specific changes in behavioral measures in transgenic fish can thus reveal the contribution of sensitized neurons. The cardiac voltage-gated sodium channel α-subunit ($Na_v1.5$, encoded by the SCN5a gene) is ideally suited for this purpose, as it opens at relatively low membrane voltage and inactivates slowly [18, 19]. Finally, $Na_v1.5$ is resistant to tetrodotoxin, which blocks most central nervous system voltage-gated sodium channels. Injection of tetrodotoxin into the brain to globally suppress neuronal activity, therefore, spares neurons that express $Na_v1.5$. This feature of $Na_v1.5$ expression enables measurement of the spontaneous or stimulus evoked activity in neurons that have been isolated from synaptic input [13].

2 Materials

2.1 Embryo Rearing

1. E3 stock (60×, 5 L): 0.3 M NaCl, 10.2 mM KCl, 19.8 mM $CaCl_2$ and 19.8 mM $MgSO_4$ to 4 L water. Mix, make up to 5 L with water and store at room temperature.
2. E3h medium (1×, 10 L): Dilute E3 stock 1:60 by adding 167 mL E3 stock to 9 L water, add 15 mL 1 M HEPES pH 7.3, mix thoroughly. Add water to 10 L and store at room temperature in a carboy with a bubbling stone for aeration. We do not add methylene blue for larvae that will be tested for behavior (*see* **Note 1**).
3. Pronase: Dissolve 300 mg in 10 mL water. Make 500 μL aliquots and store at −20 °C.
4. Milked plates: Prepare a 2% solution of nonfat dry milk powder in water. Pour into clean plastic Petri dishes, leave for 10 s, then discard milk, rinse twice with purified water then once with E3h.

2.2 Ablation and Isolation Reagents

1. Metronidazole working solution (10 mM): weigh 85.6 mg of metronidazole, add to 50 mL E3h, protect from light by wrapping tube in aluminum foil and mix thoroughly at room temperature until fully dissolved (*see* **Note 2**).
2. PhiPhiLux G1D2 (8 μM): 8 μM solution is supplied by the distributor (OncoImmunin, Gaithersburg, MD). Store at 4 °C, protected from light.
3. Tetrodotoxin stock solution (500 μM): Wear personal protective equipment including gloves, mask, and eye protection when handling tetrodotoxin. Weigh 0.16 mg tetrodotoxin citrate (MW 319.27) in a chemical hood, add to 1 mL H_2O and mix thoroughly.
4. Tetrodotoxin working solution (1 μM): dilute TTX stock solution 1:500 with water and mix thoroughly. Use within 12 h.

Dispose of all solutions containing TTX in hazardous materials waste following Institute guidelines (*see* **Note 3**).

5. Lysis buffer: 10 mM Tris pH 7.5, 50 mM KCl, 0.3% Tween20, 0.3% TritonX, 1 mM EDTA. Store at 4 °C.
6. Proteinase K: Dissolve 10 mg powder in 1 mL water immediately before use. Store powder at 4 °C.

2.3 Fish Stocks

1. For ablation studies use *Tg(UAS-E1b:BGi-epNTR-TagRFPT-oPre)y268* (UAS:epNTR) which contains the highly active variant of nitroreductase epNTR fused to TagRFPT (*see* **Note 4**) [13].
2. For activation and isolation experiments use *Tg(UAS-E1b:BGi-SCN5a-v2a-TagRFPT)y266* (UAS:SCN5) which coexpresses the human voltage-gated sodium channel $Na_v1.5$ and TagRFPT [13].
3. Maintain UAS:epNTR and UAS:SCN5 together with a Gal4 line that has an easily recognized expression pattern (*see* **Note 5**).

3 Methods

3.1 Raising Larvae for Behavior Testing

1. Cross Gal4 transgenic fish to UAS:epNTR fish for ablation studies, or UAS:SCN5 for sensitization and isolation studies (*see* **Note 6**).
2. Collect embryos from crosses within 1–2 h post fertilization (hpf) and combine all individual clutches. Remove abnormal or unfertilized embryos.
3. At 6–12 hpf, sort fertilized embryos into dishes at a density of 15–25 per 6 cm Petri dish in a volume of 10 mL of E3h media. Raise and maintain larvae in a 28.5 °C light-cycle incubator (*see* **Note 7**).
4. At 1 dpf remove all dead or deformed embryos and debris.
5. At 2 dpf hatch embryos by adding 10 μL of pronase to each dish. Incubate embryos for 3 h, then pipette embryos lightly up and down to ensure all embryos are removed from their chorions. Change E3h media completely after dechorionation.
6. Sort larvae for transgene expression, keeping both transgene positive and negative groups (*see* **Notes 8** and **9**). If the UAS stocks are maintained with a different Gal4 line in the background, this pattern will need to be separated. After dechorionation, larvae tend to stick to clean plastic dishes, sometimes resulting in damage and should be sorted into milked plates (*see* **Note 10**).
7. Maintain transgenic and control nontransgenic larvae at the same density, changing medium every 2 days (*see* **Note 11**).

3.2 Ablation Using Nitroreductase and Metronidazole

1. Raise larvae for metronidazole treatment to 3 dpf (*see* **Note 12**).
2. Remove E3h media from dishes and rinse once with 5 mL metronidazole solution. Completely remove the rinse solution and add 10 mL metronidazole solution (*see* **Note 13**).
3. Place dishes with metronidazole treated and untreated larvae in a light-cycle incubator with dim light intensity (0.5 μW/cm^2) (*see* **Note 14**).
4. Optional (*see* **Note 15**): To label apoptotic cells, 12 h after starting metronidazole treatment, remove several larvae and immerse for 1 h in 8 μM PhiPhiLux G1D2 (*see* **Note 16**). Protect embryos in PhiPhiLux G1D2 solution from light. Perform three washes in E3h (20 min each) before live imaging using a 488 nm confocal laser and 500–550 nm bandpass emission filter. The fluorescence signal of PhiPhiLux G1D2 in apoptotic cells is relatively weak, but easily distinguished from the very low background fluorescence in surrounding tissue, so a relatively high laser power should be used to detect the PhiPhiLux G1D2 label.
5. After 24 h, replace the medium with freshly prepared metronidazole and return dishes to the dimly lit incubator.
6. After 48 h, remove dead or deformed individuals (*see* **Note 17**), then discard the metronidazole solution and perform three washes with fresh E3h media. Even after ablation conditions have been validated, a few larvae should be set aside and checked for loss of the fluorescent transgene-expressing cells (*see* **Note 18**).
7. Allow larvae to recover for 24 h in E3h under normal intensity light-cycle conditions before starting behavioral tests.

3.3 Sensitization of Neurons by Targeted Expression of $Na_v1.5$

1. Raise groups of larvae, changing media at every 2 days as described in Subheading 3.1.
2. Sort for TagRFPT expression, keeping nontransgenic fish as controls. Due to the large size of the $Na_v1.5$ mRNA, fluorescence is relatively dim in these fish and post-hoc genotyping may be necessary to identify transgenic larvae.
3. Establish stimulus conditions that avoid floor or ceiling effects for the behavior to be tested.
4. Measure behavior in $Na_v1.5$ expressing fish compared to transgene negative fish under normal assay conditions. Use a range of stimulus intensities that affect the aspect of behavior measured (*see* **Note 19**).
5. For post-hoc genotyping to identify $Na_v1.5$ expressing larvae, first make DNA by placing larva in 30 μL lysis buffer, incubating at 98 °C for 10 min then placing on ice. Add 5 μL proteinase K, then incubate at 55 °C for 2 h. Incubate at 98 °C for 10 min to inactivate proteinase K, then dilute 1:20 with water.

6. Setup genotyping PCR using 5 μL of diluted DNA prepared in the previous step for each of two PCRs. For the Gal4 transgene conditions will vary according to the line. For the UAS:SCN5-2a-TagRFPT transgene the primers are 5′-TCTGTGCATTGACTTGGTGAG and 5′-GGCGGTTCTACCCTGAATTA. Use standard PCR conditions with 0.2 μM each primer, 0.2 mM dNTPs and 35 cycles of 94 °C for 30 s, 53 °C for 30 s, 72 °C for 30 s. Run product on 2% agarose gel. Larvae with the UAS:SCN5 transgene will produce a 279 bp band (*see* **Note 20**).

3.4 Isolation of Neurons from Circuit Activity with $Na_v1.5$ and Tetrodotoxin

1. Raise groups of larvae with and without $Na_v1.5$ transgene expression.
2. If using a transgenic method to monitor neuronal activity, such as the GCaMP calcium sensors or arch voltage sensors, also sort for this transgene. If using a synthetic indicator to monitor neuronal activity, such as Calcium Green-1 dextran, label neurons in both $Na_v1.5$ positive and negative groups of fish (*see* **Note 21**).
3. At 5–6 dpf embed larva in 2% low melting point agarose in E3h in a chamber suitable for microscopy (*see* **Note 22**).
4. Record stimulus evoked changes in neuronal activity in neurons that express $Na_v1.5$ and in the same neurons in nontransgenic fish.
5. While still embedded in agarose, remove larva from the microscope and move to the injection stage. Backload a pulled glass

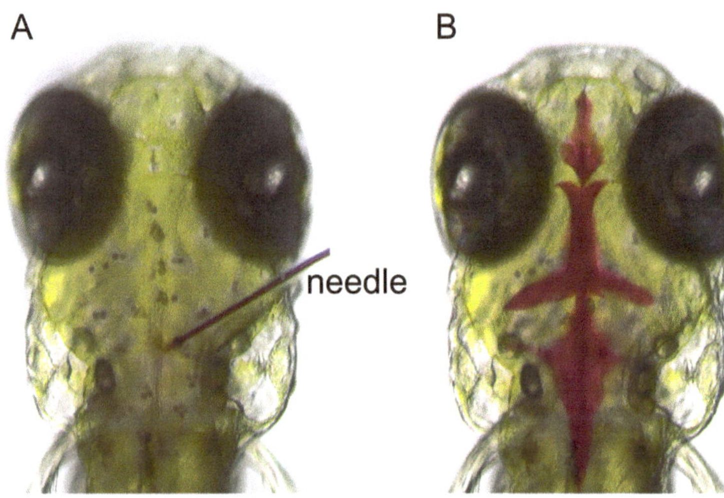

Fig. 2 Injection into the brain ventricle. (**a**) Glass injection needle (in this case filled with rhodamine-dextran for demonstration purposes) positioned to inject into the ventricle of 6 dpf larva. Needle is inserted at the midline and at the same anterior position as the otic vesicles. (**b**) Larva immediately after dye injection into the ventricle

pipette with 1 μM tetrodotoxin working solution. Attach pipette to a picospritzer to pressure inject 2–3 nL tetrodotoxin solution into the hindbrain ventricle (Fig. 2 shows how to position the needle for injection).

6. Realign embedded larvae in imaging setup and record neuronal activity-dependent fluorescence changes evoked in experimental paradigm (*see* **Note 23**).
7. To verify that TTX was active and not lethal, 10 min after TTX injection, larvae should be immobilized (completely unresponsive to a body touch); however, the heart should remaining beating. Larvae should recover sufficiently from TTX injection to resume free-swimming behavior in 3–5 h.

4 Notes

1. Although methylene blue is widely used in embryo medium to reduce fungal growth, it has potent biochemical effects, producing behavioral changes in mice [20] and molecular changes in zebrafish [21, 22].
2. 10 mM is the maximal solubility for metronidazole in water. We attach the solution to a benchtop rotator and vigorously mix for about 1 h at room temperature to fully dissolve. We make metronidazole solution fresh for each application and use within 2 h of dissolving. The powder can be stored at 4 °C, but in our laboratory we replace the stock every 12 months.
3. Tetrodotoxin is extremely hazardous. Stock powder should be stored at −20 °C in a locked container. In our laboratory we prepare stock and working solutions at BSL-2. We handle solutions and dishes that contact TTX with gloves and handle materials as medical pathological waste. Leftover TTX solution is inactivated with a 30 min exposure to 1% bleach and disposed of as chemical waste. In the United States, TTX is classified as a Select Agent toxin. Institute guidelines for using TTX will vary so we suggest you seek guidance before use.
4. Several variants of nitroreductase are available: (a) the original bacterial NfsB fused to mCherry or CFP [9, 10], (b) wild-type NfsB with amino acid mutations T41Q, N71S, F124T (mutNTR) [12], and (c) zebrafish codon optimized NfsB with amino acid mutations T41L, N71S, F124W fused to TagRFP (epNTR-TagRFP) [13]. Both mutNTR and epNTR produce more rapid and complete ablation than the original NfsB, are readily available as UAS lines and are therefore currently the best options.
5. UAS reporter lines are sensitive to transgene silencing, leading to variegated reporter expression that worsens across generations [23]. To mitigate this, UAS lines should be propagated

together with a Gal4 driver, raising only individuals with strong, complete UAS reporter expression (generally on the order of 10–20 % of larvae). Stock maintenance is facilitated by keeping the UAS and Gal4 transgenes as single-copy heterozygotes in each generation, so that strong expression is due to nonsilenced transgenes, rather than the presence of multiple copies of UAS and/or Gal4.

6. Alternatively epNTR or $Na_v1.5$ can be expressed in target neuron populations by direct expression using a characterized promoter. However because Gal4 amplifies expression, this is the recommended strategy.
7. For the light-cycle incubator, we use a 14/10 light/dark period. During the light cycle we have used light intensities between 10 and 500 $\mu W/cm^2$.
8. The UAS:epNTR and UAS:SCN5a lines coexpress TagRFPT so that transgenic larvae can be readily recognized for sorting. As silencing of the UAS reporter leads to variegated expression [23], during sorting select only larvae with robust transgenic expression in the complete expected pattern and be careful to avoid sorting larvae with weak expression into the negative control group. When sorting lines with dim fluorescence it can be helpful to screen larvae using a camera with long exposures.
9. When sorting embryos or larvae for neuronal ablation and subsequent behavioral analysis, avoid using tricaine. Exposure to tricaine, especially prolonged exposure, can alter behavior and influence subsequent analysis. When possible, sorting prior to dechorionation may be helpful, avoiding the difficulty in detecting fluorescence in moving larvae. Motile larvae can be placed in small individual drops to facilitate screening. Sorting before inflation of swim bladders is easier, as larvae do not float to the surface.
10. If embryos are resorted into the dishes they were raised in, the debris from the hatching process is sufficient to coat the bottom of the plate and prevent sticking. Otherwise, sort into milked plates.
11. Ideally maintain 15–25 larvae per 6 cm Petri dish, but in any case be sure to raise controls and experimental larvae at the same density because this can significantly alter larval behavior [24].
12. Metronidazole treatment can be applied at any time and the optimal time point to initiate ablation will need to be empirically determined for each new line. We find that starting the metronidazole treatment just before the onset of transgene expression often gives the most complete ablation. The maximal dose and treatment duration with metronidazole is 10 mM for 48 h, however if nitroreductase is highly expressed, robust ablation may be achieved with lower doses and shorter exposure times. Treatment with 10 mM for 48 h is a good starting point, because if ablation is not effective under these conditions, stronger nitroreductase expression will be needed.

13. Metronidazole treatment alone may affect behavior, for example we find that treatment of wild-type larvae sensitizes the acoustic startle response. Nontransgenic siblings, treated with metronidazole, and handled in parallel to the experimental group are an essential control for ablation experiments. We also suggest keeping a group of metronidazole untreated larvae as treatment and quality controls.
14. This can be achieved by placing the treated embryos in a container with a semitransparent lid in a regular light-cycle incubator. Inexpensive sheets of neutral density plastic can be obtained from GAM Products. Metronidazole is light sensitive and prolonged exposure will cause degradation.
15. Verification that metronidazole treatment results in complete ablation of nitroreductase-expressing cells is essential, because the effectiveness varies with cell type and levels of nitroreductase expression. Loss of neurons with the coexpressed fluorescent protein should be confirmed by an independent measure such as immunostaining for a cellular marker of the targeted neurons or detection of apoptosis in the targeted cells. For detecting apoptosis, we recommend PhiPhiLux G1D2 because of the low background in brain. However, alternate methods for detecting apoptosis include TUNEL staining [9], hoechst staining for nuclear fragmentation [9], immunolabeling of activated caspase-3 [11], and acridine orange staining [12].
16. Since PhiPhiLux treatment labels cells undergoing active apoptosis, the ideal time for labeling will need to be empirically tested for each line and cell type, however, approximately 12 h after starting metronidazole treatment is a general starting time point. We have used PhiPhiLux to detect apoptotic cells in 3 and 4 dpf larvae.
17. A common effect of metronidazole treatment is overinflation of the swim bladder which leads to abnormal balance and may preclude behavioral characterization. Larvae with overinflated swim bladders float to the surface and congregate in the meniscus. This can be minimized by using lower doses of metronidazole, or treating larvae after swim bladder inflation is complete (4 dpf). Viable lines for nitroreductase mediated ablation and subsequent behavioral testing cannot have expression of nitroreductase in tissues including skeletal muscle, heart, skin, or notochord, as the ablation will lead to severe morphological abnormalities. Expression can be suppressed outside the brain by incorporating a Neuronal Restrictive Silencing Element (NRSE) into the transgene [25, 26].
18. During postablation screening, it is common to observe fluorescent aggregates appearing as puncta that are smaller than cell bodies. Such debris does not indicate unsuccessful ablation.

19. The degree of sensitization will vary according to cell type and $Na_v1.5$ expression level. As a rough guide, larvae with Mauthner cells expressing sufficient levels of transgene that RFP was visible under epifluorescence, showed a roughly twofold increase in short latency escape responses. Larvae with motor neurons expressing the transgene at visible levels showed a 30 % increase in the bend angle during long latency escape responses [13].
20. The genotyping protocol for the UAS:SCN5 (y266) line can also be used to maintain adult transgenic fish. The primers bind in the 3'UTR and transposon arm of the transgene and do not distinguish fish with one or two copies of the transgene.
21. Several options are available to monitor neuronal activity using fluorescent indicators of calcium activity. Reticulospinal neurons can be backfilled by injection of dextran-coupled Calcium Green or Oregon Green 488 BAPTA-1 into the spinal cord [27, 28]. Injection of acetoxymethyl-esters of calcium indicators into target brain regions and pan-neuronal expression of GCaMP can also be used to monitor activity in any group of neurons (reviewed in [29]).
22. During embedding, agarose should be only slightly warm to touch to prevent damage to the larva. Submerge larva in agarose and position using a plastic pipette tip. Add E3h above the solid agarose in the chamber to prevent the larva from dehydrating.
23. If $Na_v1.5$ expressing neurons retain activity after TTX injection, this indicates that spontaneous activity or activation of these neurons by a stimulus does not require action potential dependent input from other neurons.

Acknowledgment

This work was supported by the Intramural Research Program of the *Eunice Kennedy Shriver* National Institute for Child Health and Human Development.

References

1. Asakawa K, Suster ML, Mizusawa K, Nagayoshi S, Kotani T, Urasaki A, Kishimoto Y, Hibi M, Kawakami K (2008) Genetic dissection of neural circuits by Tol2 transposon-mediated Gal4 gene and enhancer trapping in zebrafish. Proc Natl Acad Sci U S A 105(4):1255–1260. doi:10.1073/pnas.0704963105
2. Plazas PV, Nicol X, Spitzer NC (2013) Activity-dependent competition regulates motor neuron axon pathfinding via PlexinA3. Proc Natl Acad Sci U S A 110(4):1524–1529, doi: 1213048110 [pii] 10.1073/pnas.1213048110
3. Szobota S, Gorostiza P, Del Bene F, Wyart C, Fortin DL, Kolstad KD, Tulyathan O, Volgraf M, Numano R, Aaron HL, Scott EK, Kramer RH, Flannery J, Baier H, Trauner D, Isacoff EY (2007) Remote control of neuronal activity with a light-gated glutamate receptor. Neuron 54(4):535–545, doi: S0896-6273(07)00344-3 [pii] 10.1016/j.neuron.2007.05.010

4. Arrenberg AB, Del Bene F, Baier H (2009) Optical control of zebrafish behavior with halorhodopsin. Proc Natl Acad Sci U S A 106(42):17968–17973, doi: 0906252106 [pii] 10.1073/pnas.0906252106
5. Bergeron SA, Carrier N, Li GH, Ahn S, Burgess HA (2014) Gsx1 expression defines neurons required for prepulse inhibition. Mol Psychiatry. doi: 10.1038/mp.2014.106
6. Wan H, Korzh S, Li Z, Mudumana SP, Korzh V, Jiang YJ, Lin S, Gong Z (2006) Analyses of pancreas development by generation of gfp transgenic zebrafish using an exocrine pancreas-specific elastaseA gene promoter. Exp Cell Res 312(9):1526–1539, doi: S0014-4827(06)00019-X [pii] 10.1016/j.yexcr.2006.01.016
7. Slanchev K, Stebler J, de la Cueva-Mendez G, Raz E (2005) Development without germ cells: the role of the germ line in zebrafish sex differentiation. Proc Natl Acad Sci U S A 102(11):4074–4079, doi: 0407475102 [pii] 10.1073/pnas.0407475102
8. Chu Y, Senghaas N, Koster RW, Wurst W, Kuhn R (2008) Novel caspase-suicide proteins for tamoxifen-inducible apoptosis. Genesis 46(10):530–536. doi:10.1002/dvg.20426
9. Pisharath H, Rhee JM, Swanson MA, Leach SD, Parsons MJ (2007) Targeted ablation of beta cells in the embryonic zebrafish pancreas using E. coli nitroreductase. Mech Dev 124(3):218–229
10. Curado S, Anderson RM, Jungblut B, Mumm J, Schroeter E, Stainier DY (2007) Conditional targeted cell ablation in zebrafish: a new tool for regeneration studies. Dev Dyn 236(4):1025–1035
11. Curado S, Stainier DYR, Anderson RM (2008) Nitroreductase-mediated cell/tissue ablation in zebrafish: a spatially and temporally controlled ablation method with applications in developmental and regeneration studies. Nat Protocols 3(6):948–954
12. Mathias JR, Zhang Z, Saxena MT, Mumm JS (2014) Enhanced cell-specific ablation in zebrafish using a triple mutant of Escherichia coli nitroreductase. Zebrafish. doi:10.1089/zeb.2013.0937
13. Tabor KM, Bergeron SA, Horstick EJ, Jordan DC, Aho V, Porkka-Heiskanen T, Haspel G, Burgess HA (2014) Direct activation of the Mauthner cell by electric field pulses drives ultra-rapid escape responses. J Neurophysiol 112(4):834–844. doi:10.1152/jn.00228.2014
14. Yokogawa T, Hannan MC, Burgess HA (2012) The dorsal raphe modulates sensory responsiveness during arousal in zebrafish. J Neurosci 32:15205–15215
15. Douglass AD, Kraves S, Deisseroth K, Schier AF, Engert F (2008) Escape behavior elicited by single, channelrhodopsin-2-evoked spikes in zebrafish somatosensory neurons. Curr Biol 18(15):1133–1137
16. Zhu P, Narita Y, Bundschuh ST, Fajardo O, Zhang Schärer Y-P, Chattopadhyaya B, Arn Bouldoires E, Stepien AE, Deisseroth K, Arber S, Sprengel R, Rijli FM, Friedrich RW (2009) Optogenetic dissection of neuronal circuits in zebrafish using viral gene transfer and the Tet system. Front Neural Circ 3:21. doi:10.3389/neuro.04.021.2009
17. Kubo F, Hablitzel B, Dal Maschio M, Driever W, Baier H, Arrenberg AB (2014) Functional architecture of an optic flow-responsive area that drives horizontal eye movements in zebrafish. Neuron 81(6):1344–1359, doi:S0896-6273(14)00193-7 [pii] 10.1016/j.neuron.2014.02.043
18. Deschenes I, Trottier E, Chahine M (2001) Implication of the C-terminal region of the alpha-subunit of voltage-gated sodium channels in fast inactivation. J Membr Biol 183(2):103–114
19. Mantegazza M, Yu FH, Catterall WA, Scheuer T (2001) Role of the C-terminal domain in inactivation of brain and cardiac sodium channels. Proc Natl Acad Sci 98(26):15348–15353. doi:10.1073/pnas.211563298
20. Klamer D, Engel JA, Svensson L (2004) Phencyclidine-induced behaviour in mice prevented by methylene blue. Basic Clin Pharmacol Toxicol 94(2):65–72
21. van Bebber F, Paquet D, Hruscha A, Schmid B, Haass C (2010) Methylene blue fails to inhibit Tau and polyglutamine protein dependent toxicity in zebrafish. Neurobiol Dis 39(3):265–271, doi: http://dx.doi.org/10.1016/j.nbd.2010.03.023
22. Vaccaro A, Patten SA, Ciura S, Maios C, Therrien M, Drapeau P, Kabashi E, Parker JA (2012) Methylene blue protects against TDP-43 and FUS neuronal toxicity in C. elegans and D. rerio. PLoS One 7(7):e42117
23. Goll MG, Anderson R, Stainier DYR, Spradling AC, Halpern ME (2009) Transcriptional silencing and reactivation in transgenic zebrafish. Genetics 182(3):747–755. doi:10.1534/genetics.109.102079
24. Burgess HA, Granato M (2008) The neurogenetic frontier—lessons from misbehaving zebrafish. Brief Funct Genomic Proteomic 7(6):474–482
25. Bergeron SA, Hannan MC, Codore H, Fero K, Li G, Moak ZB, Yokogawa T, Burgess HA (2012) Brain selective transgene expression in zebrafish using an NRSE derived motif. Front Neural Circ 6:110. doi:10.3389/fncir.2012.00110

26. Xie X, Mathias JR, Smith MA, Walker SL, Teng Y, Distel M, Koster RW, Sirotkin HI, Saxena MT, Mumm JS (2013) Silencer-delimited transgenesis: NRSE/RE1 sequences promote neural-specific transgene expression in a NRSF/REST-dependent manner. BMC Biol 10:93, doi: 1741-7007-10-93 [pii] 10.1186/1741-7007-10-93
27. O'Malley DM, Kao YH, Fetcho JR (1996) Imaging the functional organization of zebrafish hindbrain segments during escape behaviors. Neuron 17(6):1145–1155
28. Gahtan E, Sankrithi N, Campos JB, O'Malley DM (2002) Evidence for a widespread brain stem escape network in larval zebrafish. J Neurophysiol 87(1):608–614
29. Kettunen P (2012) Calcium imaging in the zebrafish. Adv Exp Med Biol 740:1039–1071. doi:10.1007/978-94-007-2888-2_48

ERRATUM TO

Zebrafish: Methods and Protocols, Methods in Molecular Biology

Koichi Kawakami, E. Elizabeth Patton, and Michael Orger

Koichi Kawakami et al. (eds.), *Zebrafish: Methods and Protocols*, Methods in Molecular Biology, vol. 1451, DOI 10.1007/978-1-4939-3771-4, © Springer Science+Business Media New York 2016

DOI 10.1007/978-1-4939-3771-4_26

The original version of this book was inadvertently published without cover credit line in the copyright page. The credit line should read as "Cover image courtesy of Aristides Arrenberg (Baier lab, UCSF)".

The online version of the updated original book can be found at http://dx.doi.org/10.1007/978-1-4939-3771-4

Koichi Kawakami et al. (eds.), *Zebrafish: Methods and Protocols*, Methods in Molecular Biology, vol. 1451, DOI 10.1007/978-1-4939-3771-4_26, © Springer Science+Business Media New York 2016

Index

A

B

C

Koichi Kawakami et al. (eds.), *Zebrafish: Methods and Protocols*, Methods in Molecular Biology, vol. 1451,
DOI 10.1007/978-1-4939-3771-4, © Springer Science+Business Media New York 2016

FSC
www.fsc.org
MIX
Papier | Fördert gute Waldnutzung
FSC® C083411

Kolibri 360 GmbH
Ferdinand-Jühlke-Straße 7
99095 Erfurt, Deutschland
produktsicherheit@kolibri360.de